T3-BWP-012

STATISTICAL HYPOTHESIS TESTING

THEORY AND METHODS

STATISTICAL HYPOTHESIS TESTING

THEORY AND METHODS

Ning-Zhong Shi & Jian Tao
Northeast Normal University, China

World Scientific

NEW JERSEY • LONDON • SINGAPORE • BEIJING • SHANGHAI • HONG KONG • TAIPEI • CHENNAI

Published by

World Scientific Publishing Co. Pte. Ltd.
5 Toh Tuck Link, Singapore 596224
USA office: 27 Warren Street, Suite 401-402, Hackensack, NJ 07601
UK office: 57 Shelton Street, Covent Garden, London WC2H 9HE

British Library Cataloguing-in-Publication Data
A catalogue record for this book is available from the British Library.

STATISTICAL HYPOTHESIS TESTING
Theory and Methods

Copyright © 2008 by World Scientific Publishing Co. Pte. Ltd.

All rights reserved. This book, or parts thereof, may not be reproduced in any form or by any means, electronic or mechanical, including photocopying, recording or any information storage and retrieval system now known or to be invented, without written permission from the Publisher.

For photocopying of material in this volume, please pay a copying fee through the Copyright Clearance Center, Inc., 222 Rosewood Drive, Danvers, MA 01923, USA. In this case permission to photocopy is not required from the publisher.

ISBN-13 978-981-281-436-4
ISBN-10 981-281-436-1

Printed in Singapore by World Scientific Printers

To my wife Zhuo Guan, and my son Tianji Shi.
— Ning-Zhong Shi

To my wife Wenjie Zhao, and my daughter Xing-Xu Tao.
— Jian Tao

Preface

Statistical hypothesis testing is a basic and important branch of mathematical statistics. The classical theory on testing hypothesis is often based on the large sample, such as Pearson's χ^2 test and Fisher's likelihood ratio test, and its decision criterion depends on the p-value. Modern hypothesis testing theory is based on the thoughts of J. Neyman and E. S. Pearson in the 1930's, and especially credites to the systematization and development of Lehmann (1959, 1986). As distinguished from that of the classical hypothesis testing, modern testing theory depends on the idea of optimization: to minimize the Type II error by controlling the Type I error. The experience gained from decades of teaching and research proves that the latter is theoretically perfect but does not provide more methods for constructing new test statistics. As to testing problem, in fact, a feasible principle is to construct a suitable statistic based on both the background of practical problem and the individual statistical intuition. On the basis of the above principle, a further analysis is made on the properties of the test statistic. Therefore, it is the basic idea of this book to combine the two methods mentioned above reasonably. Moreover, as most of current researches about hypothesis testing are aiming at parametric models, so does this book.

The book consists five chapters, which can be divided into three parts. The first part, Chapter 1, discusses some basic properties of statistical space. Although the term, statistical space, is put forward by this book, its concept has been used in common for a long time, *i.e.*, adding a family of probability measures to a measurable space. Thus there is an essential difference between statistical space and usual probability space: for a given measurable set, the usual probability space concerns measuring 'the probability that a random variable may belong to this set'; the statistical space concerns which measure that is used to measure 'the probability that a random variable may belong to this set' would be better. In this chapter the discussion on the properties of the statistical space is expanded under such basic thought. Considering test depends on estimator in essence, the authors also conclude some methods and properties of estimators in this chapter.

The second part, including Chapters 2 and 3, mainly discusses some basic concepts of test statistics. Chapter 2 focuses on the basic idea and the methods of constructing test statistics, including parametric as well as nonparametric ones. Chapter 3 discusses how to judge whether a test statistic is good or bad, including Neyman-Pearson Lemma viewed as a judgement method, besides comparison the power functions and comparison the robustness etc. Moreover, based on the fact that the interval estimation is very important, the last section of Chapter 3 discusses the relation between the test acceptance region and the interval estimation as well as how to construct an interval estimation.

The third part including Chapters 4 and 5, discusses the test of parametric model. Most of the current testing theories and researches aim at parametric model from which the new testing thought and method, including model selection, originate. This part discusses the test on some models, such as ANOVA and AIC, log linear model and contingency tables, regression analysis together with the test of Logistic model and time series *etc*, as well as model selection. Especially in the section of ANOVA, the reason of introducing the tendency term test is not only that the authors of this book are familiar with the related contents but also that tendency test is becoming more and more important with increasingly development of Bioinformatics research.

The authors of this book hereby would like to express their gratitude to Dr. Jianhua Guo, Dr. Wei Gao, Dr. Baoxue Zhang, Dr. Shurong Zheng, and Dr. Guorong Hu for their help in the preparation of the manuscript.

Ning-Zhong Shi
Jian Tao

Contents

Chapter 1

Statistical Spaces

Mathematical statistics is a science that studies the statistical regularity of random phenomena, essentially by some observation values of random variable (r.v.) X. Sometimes the observation values are also called sample points. The lower case x is used to denote a realization of the r.v. X. Furthermore, the calligraphic letter $\mathcal{X}$ is used to denote the set from which X takes value, and $\mathcal{A}$ is used to denote the σ-algebra generated by some subsets of $\mathcal{X}$. The measurable space $(\mathcal{X}, \mathcal{A})$ is used as a starting point for our further analysis, and will be called the sample space. Considering the need of the research, we define a measure in the measurable space $(\mathcal{X}, \mathcal{A})$ by three methods: $(\mathcal{X}, \mathcal{A}, \nu)$, where ν is a σ-finite measure, is called a **measure space**; $(\mathcal{X}, \mathcal{A}, P)$, where P is a probability measure, is called a **probability space**; and $(\mathcal{X}, \mathcal{A}, \mathcal{P})$, where $\mathcal{P}$ is a family of probability measures or a family of probability distributions, for the sake of convenience, is called a **statistical space**. In many situations, the family of probability distributions depends only on the parameters in probability distribution:

$$\mathcal{P} = \{P_\theta;\ \theta \in \Theta\},$$

where Θ is a parameter space. For example, the family of normal distributions with known variance depends only on the unique mean parameter, the parameter θ hereof is a real number; but when the variance is also unknown, the distribution depends on both the mean and the variance, the parameter θ hereof is a bivariate vector. The primary task of a statistical test is to infer whether the parameter from the distribution is equal to some given constant θ_0, *i.e.* to test the hypothesis $H : \theta = \theta_0$; or to infer whether it belongs to some given subset Θ_0 of Θ, *i.e.*, to test $H : \theta \in \Theta_0$. Before discussing this problem, this chapter will review some basic concepts which will be involved in subsequent chapters. Firstly, analyze some properties of the statistical space; secondly, discuss conditional probability, sufficient statistics and some characteristics of the exponential distribution family; at last, review some basic estimation methods. The readers who are familiar with these contents can skip the chapter.

1.1 Basic Properties of Statistical Spaces

The above-mentioned three types of metric space are consistent in form, but their connotations are quite different in nature: for $A \in \mathcal{A}$, the measure in the measure space is denoted as $\nu(A)$, which represents the size of the set A; the measure in the probability space is denoted as $P(X \in A)$, which represents the possibility of r.v. X taking values in A; the measure in the statistical space is denoted as $P_\theta(X \in A)$, is used to find a parameter θ which is the most "appropriate" to measure the possibility. About the "appropriateness", we can give different criteria according to different problems. First of all, We will study some general properties about statistical spaces.

1.1.1 *Measure in Statistical Spaces*

Let $\mathcal{P}$ and ν be the family of probability distributions and σ-finite measure in the sample space $(\mathcal{X}, \mathcal{A})$, respectively. If

$$A \in \mathcal{A}, \nu(A) = 0 \quad \Rightarrow \quad P(A) = 0, \forall P \in \mathcal{P}, \tag{1.1.1}$$

then $\mathcal{P}$ is said to be absolutely continuous with respect to ν, denoted by $\mathcal{P} \ll \nu$. If the family of probability distributions depends only on the parameter in the probability distributions, then (1.1.1) can be rewritten as

$$A \in \mathcal{A}, \nu(A) = 0 \quad \Rightarrow \quad P_\theta(A) = 0, \forall \theta \in \Theta. \tag{1.1.2}$$

In the later discussion, we will focus on the definition given by (1.1.2). According to the Radon-Nikodym Theorem (cf. Halmos, 1957), if $\mathcal{P} \ll \nu$, then for $\forall \theta \in \Theta$, there exists an $\mathcal{A}$-measurable function f_θ, for $\forall A \in \mathcal{A}$ we have

$$P_\theta(A) = \int_A f_\theta(x) d\nu(x). \tag{1.1.3}$$

Furthermore, f_θ exists uniquely almost everywhere (a.e.) with respect to the measure ν. Then f_θ is called a density function of P_θ and denoted by

$$\frac{dP_\theta}{d\nu} = f_\theta \quad \text{or} \quad dP_\theta = f_\theta d\nu.$$

There are two equivalent formulations about uniqueness a.e. If there also exists f_θ^* s.t. $dP_\theta = f_\theta^* d\nu$, then

$$\nu(x; f_\theta(x) \neq f_\theta^*(x)) = 0;$$

or, for $\forall A \in \mathcal{A}$ we have

$$\int_A f_\theta(x) d\nu(x) = \int_A f_\theta^*(x) d\nu(x).$$

Therefore the uniqueness a.e. is not related to only the measure ν but also the σ-algebra $\mathcal{A}$.

Theorem 1.1.1. *Let $\mathcal{P}$ and ν be the family of probability distributions and σ-finite measure in the sample space $(\mathcal{X}, \mathcal{A})$, respectively. Let g be an $\mathcal{A}$-measurable function. If $\mathcal{P} \ll \nu$, then for $\forall \theta \in \Theta$ we have*

$$\int_{\mathcal{X}} g(x) dP_\theta(x) = \int_{\mathcal{X}} g(x) \frac{dP_\theta}{d\nu} d\nu(x) = \int_{\mathcal{X}} g(x) f_\theta(x) d\nu(x). \tag{1.1.4}$$

Proof. From (1.1.3), (1.1.4) holds obviously when $g(x) = I_A(x)$, for $A \in \mathcal{A}$. When g is a nonnegative simple function, *i.e.*, for $i = 1, \cdots, n$, there exist $A_i \in \mathcal{A}$, satisfying $A_i \cap A_j = \phi$ for $i \neq j$, and $a_i > 0$, such that

$$g(x) = a_1 I_{A_1}(x) + \cdots + a_n I_{A_n}(x),$$

then Eq. (1.1.4) holds, *i.e.*,

$$\sum_{i=1}^n a_i P_\theta(A_i) = \sum_{i=1}^n a_i \int_{\mathcal{X}} I_{A_i}(x) f_\theta(x) d\nu(x).$$

If g a is nonnegative measurable function, then there exists a series of nonnegative simple functions g_n satisfying

$$0 \leq g_1(x) \leq g_2(x) \leq \cdots, \quad \text{and} \quad g_n(x) \to g(x).$$

By the monotone convergence theorem, we have

$$\begin{aligned}\int_{\mathcal{X}} g(x) dP_\theta(x) &= \lim_{n\to\infty} \int_{\mathcal{X}} g_n(x) dP_\theta(x) \\ &= \lim_{n\to\infty} \int_{\mathcal{X}} g_n(x) f_\theta(x) d\nu(x) \\ &= \int_{\mathcal{X}} g(x) f_\theta(x) d\nu(x).\end{aligned}$$

When g is a general measurable function, we can decompose g into a positive part and a negative part. Thus the theorem is proved. □

The technique used in the proof of Theorem 1.1.1 is generally called **I-method**. The method indicates that a proposition still holds for a measurable function if it holds for a measurable set in statistical space. Generally Eq. (1.1.4) is called the **mean** of $g(x)$ when the parameter is θ, and denoted by

$$E_\theta g(X) = \int_{\mathcal{X}} g(x) dP_\theta(x) = \int_{\mathcal{X}} g(x) f_\theta(x) d\nu(x). \tag{1.1.5}$$

When Eq. (1.1.5) is finite,

$$V_\theta g(X) = E_\theta(g(X) - E_\theta g(X))^2 \tag{1.1.6}$$

is called the **variance** of $g(x)$ when the parameter is θ. Especially when $g(x) = x$, $E_\theta X$ and $V_\theta X$ are called the mean and variance of the r.v. X respectively when the parameter is θ. Let $h(x)$ is a measurable function as well. When $E_\theta h(X)$ is finite,

$$CV_\theta(g(X), h(X)) = E_\theta(g(X) - E_\theta g(X))(h(X) - E_\theta h(X)) \tag{1.1.7}$$

is said to be the **covariance** of $g(x)$ and $h(x)$ when the parameter is θ.

It can be seen that the measure ν plays a key role in statistical spaces. Generally we consider only two forms of ν, one is the Counting measure, and the other is the Lebesgue measure.

Let $\boldsymbol{Z}_+$ denote a set of zero and positive integers, and $\mathcal{A}_{\boldsymbol{Z}_+}$ denote a σ-algebra generated by some subsets of $\boldsymbol{Z}_+$, then the Counting measure ν depends on indicator functions defined on $(\boldsymbol{Z}_+, \mathcal{A}_{\boldsymbol{Z}_+})$, *i.e.*, for $A \in \mathcal{A}_{\boldsymbol{Z}_+}$,

$$\nu(A) = \sum_{x \in A} I_A(x),$$

which indicates the number of elements in A. Let $\mathcal{P}$ be the family of probability distributions defined on $(\boldsymbol{Z}_+, \mathcal{A}_{\boldsymbol{Z}_+})$. If $\mathcal{P} \ll \nu$, where ν is a Counting measure, then $\mathcal{P}$ is called a **discrete probability distribution**.

Example 1.1.1. (Binomial distribution $Bi(n, \theta)$, where $\theta \in (0, 1)$) Let $A = \{x;\ x \le n, x \in \boldsymbol{Z}_+\}$, then the probability density distribution (pdf) of the binomial distribution $Bi(n, \theta)$ is

$$P_\theta(X = x) = \binom{n}{x} \theta^x (1-\theta)^{n-x} I_A(x).$$

Its mean and variance are

$$E_\theta X = n\theta, \text{ and } V_\theta X = n\theta(1-\theta),$$

respectively.

Example 1.1.2. (Poisson distribution $P(\theta)$, where $\theta > 0$) The pdf of the Poisson distribution $P(\theta)$ is

$$P_\theta(X = x) = e^{-\theta}\frac{\theta^x}{x!}I_{\boldsymbol{Z}_+}(x).$$

Its mean and variance are

$$E_\theta X = \theta, \text{ and } V_\theta X = \theta,$$

respectively.

Example 1.1.3. (Multinomial distribution $M(n;\boldsymbol{\theta})$, where $\boldsymbol{\theta} = (p_1, \cdots, p_k)$, $p_i > 0$, and $\sum_{i=1}^k p_i = 1$) Let $A = \{\boldsymbol{x} = (x_1, \cdots, x_k);\ x_i \in \boldsymbol{Z}_+, \sum_{i=1}^k x_i = n\}$, then the pdf of the multinomial distribution $M(n;\theta)$ is

$$P_{\boldsymbol{\theta}}(X_1 = x_1, \cdots, X_k = x_k) = \frac{n!}{x_1! \cdots x_k!}p_1^{x_1} \cdots p_k^{x_k} I_A(\boldsymbol{x}).$$

Its mean and variance of X_i are

$$E_{\boldsymbol{\theta}} X_i = np_i, \quad \text{and} \quad V_{\boldsymbol{\theta}} X_i = np_i(1 - p_i),$$

respectively. When $i \neq j$, the covariance of X_i and X_j is

$$CV_{\boldsymbol{\theta}}(X_i, X_j) = -np_ip_j.$$

Let $\boldsymbol{R}$ denote the real space, and $\mathcal{B}$ be a Borel algebra generated by some subsets of $\boldsymbol{R}$. Then the Lebesgue measure ν depends on the length of interval defined on $(\boldsymbol{R}, \mathcal{B})$, *i.e.*, for the half-open interval $(a, b] \in \mathcal{B}$,

$$\nu((a, b]) = b - a.$$

Let $\mathcal{P}$ be a family of probability distributions defined on $(R, \mathcal{B})$. If $\mathcal{P} \ll \nu$, where ν is the Lebesgue measure, then $\mathcal{P}$ is a **continuous probability distribution**.

Example 1.1.4. (Uniform distribution $U(a, b)$, where $a < b$) The pdf of the uniform distribution $U(a, b)$ is

$$f_\theta(x) = \begin{cases} \dfrac{1}{b - a}, & \text{if} \quad x \in [a, b] \\ 0, & \text{otherwise,} \end{cases}$$

where $\boldsymbol{\theta} = (a, b)$. The mean and the variance of X are

$$E_{\boldsymbol{\theta}} X = \frac{1}{2}(a + b), \quad \text{and} \quad V_{\boldsymbol{\theta}} X = \frac{1}{12}(b - a)^2,$$

respectively.

Example 1.1.5. (Normal distribution $N(\mu, \sigma^2)$, where $\sigma^2 > 0$) The pdf of the normal distribution $N(\mu, \sigma^2)$ is

$$f_{\boldsymbol{\theta}}(x) = \frac{1}{\sqrt{2\pi}\sigma} \exp\left\{-\frac{1}{2\sigma^2}(x-\mu)^2\right\},$$

where $\boldsymbol{\theta} = (\mu, \sigma^2)$. The mean and the variance of X are

$$E_{\boldsymbol{\theta}} X = \mu, \quad \text{and} \quad V_{\boldsymbol{\theta}} X = \sigma^2,$$

respectively. Especially when $\mu = 0$ and $\sigma = 1$, it is called a **standard normal distribution** with pdf

$$f_{\boldsymbol{\theta}}(x) = \frac{1}{\sqrt{2\pi}} \exp\left\{-\frac{1}{2}x^2\right\}.$$

Example 1.1.6. (Exponential distribution $E(\mu, \sigma)$, where $\sigma > 0$) The pdf of the exponential distribution $E(\mu, \sigma)$ is

$$f_{\boldsymbol{\theta}}(x) = \begin{cases} \dfrac{1}{\sigma} \exp\left\{-\dfrac{1}{\sigma}(x-\mu)\right\}, & \text{if} \quad x \geq \mu, \\ 0, & \text{otherwise,} \end{cases}$$

where $\boldsymbol{\theta} = (\mu, \sigma)$. The mean and the variance of X are

$$E_{\boldsymbol{\theta}} X = \mu + \sigma, \quad \text{and} \quad V_{\boldsymbol{\theta}} X = \sigma^2,$$

respectively.

Example 1.1.7. (Cauchy distribution $C(\mu, \sigma)$, where $\sigma > 0$) The pdf of the Cauchy distribution $C(\mu, \sigma)$ is

$$f_{\boldsymbol{\theta}}(x) = \frac{1}{\pi} \frac{\sigma}{\sigma^2 + (x-\mu)^2},$$

where $\boldsymbol{\theta} = (\mu, \sigma)$. The mean and the variance of the Cauchy distribution do not exist.

Example 1.1.8. (The χ^2 distribution $\chi^2(n)$ and noncentral χ^2 distribution $\chi^2(n, \alpha^2)$) Let $X_1, \cdots, X_n$ be mutually independent random variables, and $X_i \sim N(\mu_i, 1), i = 1, \cdots, n$. Let $\alpha^2 = \sum_{i=1}^{n} \mu_i^2$, and $X = \sum_{i=1}^{n} X_i^2$. Then the pdf of X is

$$f_{\boldsymbol{\theta}}(x) = \begin{cases} \displaystyle\sum_{s=0}^{\infty} e^{-\alpha^2/2} \frac{(\alpha^2/2)^s}{s!} \frac{1}{2^{n/2+s}\Gamma(n/2+s)} x^{n/2+s-1} e^{-x/2}, & \text{if} \quad x > 0, \\ 0, & \text{otherwise,} \end{cases}$$

where $\boldsymbol{\theta} = (n, \alpha^2)$. Especially when $\alpha = 0$, we call it a χ^2 distribution with n degrees of freedom with pdf

$$f_{\theta}(x) = \begin{cases} \dfrac{1}{2^{n/2}\Gamma(n/2)} x^{n/2-1} e^{-x/2}, & \text{if} \quad x > 0, \\ 0, & \text{otherwise.} \end{cases}$$

Example 1.1.9. (Student's t-distribution $t(n)$ and noncentral Student's t-distribution $t(n, \alpha)$) Let Y and Z be mutually independent random variables, where $Y \sim N(\alpha, 1)$, and $Z \sim \chi^2(n)$. Let $X = Y/\sqrt{Z/n}$. Then X has a noncentral Student's t-distribution with n degrees of freedom, and its pdf is

$$f_{\boldsymbol{\theta}}(x) = \frac{n^{n/2}}{\sqrt{\pi}\Gamma(\frac{n}{2})} \frac{e^{-\alpha^2/2}}{(n + x^2)^{(n+1)/2}} \sum_{s=0}^{\infty} \Gamma\left(\frac{n+s+1}{2}\right) \frac{\alpha^s}{s!} \left(\frac{2x^2}{n+x^2}\right)^{s/2},$$

where $\boldsymbol{\theta} = (n, \alpha)$. Especially when $\alpha = 0$, X has a Student's t-distribution with n degrees of freedom, and its pdf is

$$f_{\theta}(x) = \frac{\Gamma(\frac{n+1}{2})}{\sqrt{n\pi}\Gamma(\frac{n}{2})} \left(1 + \frac{x^2}{n}\right)^{-(n+1)/2}.$$

Example 1.1.10. (F-distribution $F(m, n)$ and noncentral F-distribution $F(m, n, \alpha^2)$) Let Y and Z be mutually independent random variable, where $Y \sim \chi^2(m, \alpha^2), Z \sim \chi^2(n)$. Let $X = m^{-1}Y/(n^{-1}Z)$. Then X has a noncentral F-distribution with m and n degrees of freedom, its pdf is

$$f_{\boldsymbol{\theta}}(x) = \begin{cases} e^{-\frac{\alpha^2}{2}} \displaystyle\sum_{s=0}^{\infty} \frac{(\frac{\alpha^2}{2})^s}{s!} \frac{n^{\frac{n}{2}} m^{\frac{m}{2}+s}}{\mathrm{B}(\frac{m}{2}+s, \frac{n}{2})} \frac{x^{\frac{m}{2}+s-1}}{(n+mx)^{\frac{m+n}{2}+s}}, & \text{if } x > 0 \\ 0, & \text{otherwise,} \end{cases}$$

where $\boldsymbol{\theta} = (m, n, \alpha^2)$ and $\mathrm{B}(\frac{m}{2}+s, \frac{n}{2}) = \dfrac{\Gamma(\frac{m}{2}+s)\Gamma(\frac{n}{2})}{\Gamma(\frac{m+n}{2}+s)}$. Especially when $\alpha = 0$, X has an F-distribution with m and n degrees of freedom, its pdf

$$f_{\boldsymbol{\theta}}(x) = \begin{cases} \dfrac{n^{\frac{n}{2}} m^{\frac{m}{2}}}{\mathrm{B}(\frac{m}{2}, \frac{n}{2})} \dfrac{x^{m/2-1}}{(n+mx)^{(m+n)/2}}, & \text{if} \quad x > 0, \\ 0, & \text{otherwise,} \end{cases}$$

where $\mathrm{B}(\frac{m}{2}, \frac{n}{2}) = \dfrac{\Gamma(\frac{m}{2})\Gamma(\frac{n}{2})}{\Gamma(\frac{m+n}{2})}$.

Example 1.1.11. (Multidimensional normal distribution $N(\mu, \Sigma)$, where μ is an n-dimensional vector, and Σ is an $n \times n$ positive definite matrix) The pdf of the multidimensional normal distribution $N(\boldsymbol{\mu}, \boldsymbol{\Sigma})$ is

$$f_{\boldsymbol{\theta}}(\boldsymbol{x}) = (2\pi)^{-n/2} \frac{1}{\sqrt{\det \boldsymbol{\Sigma}}} \exp\left\{ -\frac{1}{2}(\boldsymbol{x} - \boldsymbol{\mu})' \boldsymbol{\Sigma}^{-1} (\boldsymbol{x} - \boldsymbol{\mu}) \right\},$$

where $\boldsymbol{\theta} = (\boldsymbol{\mu}, \boldsymbol{\Sigma})$, $\boldsymbol{x} = (x_1, \cdots, x_n)'$, $E_{\boldsymbol{\theta}}\boldsymbol{X} = \boldsymbol{\mu}$, $\boldsymbol{\Sigma} = (\sigma_{ij})$, and $\sigma_{ij} = E_{\boldsymbol{\theta}}(X_i - \mu_i)(X_j - \mu_j)$.

Example 1.1.12. (Beta distribution $Be(a, b)$, where $a > 0, b > 0$) The pdf of the beta distribution $Be(a, b)$ is

$$f_{\boldsymbol{\theta}}(x) = \begin{cases} \dfrac{\Gamma(a+b)}{\Gamma(a)\Gamma(b)} x^{a-1}(1-x)^{b-1}, & \text{if} \quad 0 \le x \le 1 \\ 0, & \text{otherwise,} \end{cases}$$

where $\boldsymbol{\theta} = (a, b)$. The mean and variance of X are

$$E_{\boldsymbol{\theta}}X = \frac{a}{a+b}, \quad \text{and} \quad V_{\boldsymbol{\theta}}X = \frac{ab}{(a+b+1)(a+b)^2},$$

respectively.

It can be seen from the above discussion that it is important that the family of probability distributions is absolutely continuous with respect to a σ-finite measure in statistical spaces, since it forms the foundation of probability calculation which depends on parameters. In that way, how to judge whether there exists a σ-finite measure or not for a given family of probability distributions, such that it is absolutely continuous about the σ-finite measure? This will involve some notions of the family of probability distributions, such as divisibility.

1.1.2 *Equivalence and Divisibility of Family of Probability Distributions*

Now we discuss the relation between two measures. Let ν and λ be two σ-finite measures defined in the sample space $(\mathcal{X}, \mathcal{A})$. If $\nu \ll \lambda$ and $\lambda \ll \nu$, then ν and λ are said to be **mutually absolutely continuous**. The necessity of the assumption of σ-finiteness of measure can be seen in the following theorems.

Theorem 1.1.2. *Let $(\mathcal{X}, \mathcal{A}, \nu)$ be a measure space. If ν is σ-finite, then there exists a probability measure which is mutually absolutely continuous with respect to ν.*

Proof. Since $\mathcal{A}$ is a σ-algebra, and ν is σ-finite, then there exists a series of sets $\{A_i\}$, such that

$$A_i \in \mathcal{A}, \quad \nu(A_i) < \infty, \quad A_1 \subset A_2 \subset \cdots, \quad \text{and} \quad \bigcup_{i=1}^{\infty} A_i = \mathcal{X}.$$

Without loss of generality, let $\nu(A_1) > 0$. For $\forall A \in \mathcal{A}$, let

$$\lambda(A) = \sum_{i=1}^{\infty} \frac{1}{2^i} \frac{\nu(A \cap A_i)}{\nu(A_i)}.$$

Obviously λ is a probability measure, and we have $\nu \ll \lambda$ and $\lambda \ll \nu$.

Let $\mathcal{P}$ and ν be a family of probability distributions and a σ-finite measure respectively defined in the sample space $(\mathcal{X}, \mathcal{A})$. If 1) $\mathcal{P} \ll \nu$; and 2) $\nu(A) = 0$ if $P_\theta(A) = 0$ for $\forall A \in \mathcal{A}$ and $\forall \theta \in \Theta$, then $\mathcal{P}$ and ν are **equivalent**. □

Theorem 1.1.3. *Let $\mathcal{P}$ be a family of probability distributions defined in the probability space $(\mathcal{X}, \mathcal{A})$. $\mathcal{P}$ is absolutely continuous with respect to some σ-infinite measure, and its necessary and sufficient condition is that there exists a probability measure that is equivalent to $\mathcal{P}$ with the form*

$$\lambda = \sum_{i=1}^{\infty} c_i P_{\theta_i} \tag{1.1.8}$$

where $\theta_i \in \Theta$, $c_i > 0$, and $\sum\limits_{i=1}^{\infty} c_i = 1$.

Proof. The sufficiency is obviously given by the condition of equivalence, so it suffices to prove the necessity as below.

Let ν be a σ-finite measure, satisfying $\mathcal{P} \ll \nu$. From Theorem 1.1.2, we can consider ν as a probability measure. Let

$$\mathcal{Q} = \left\{ Q;\ Q = \sum_{i=1}^{\infty} c_i P_{\theta_i}, \theta_i \in \Theta, c_i > 0, \sum_{i=1}^{\infty} c_i = 1 \right\}.$$

Then $\mathcal{Q}$ is also a family of probability distributions defined in $(\mathcal{X}, \mathcal{A})$, and we have $\mathcal{Q} \ll \nu$. Let

$$\sup_{Q \in \mathcal{Q}} \nu \left\{ x;\ \frac{dQ(x)}{d\nu(x)} > 0 \right\} = \alpha.$$

From the definition of supremum, there exists a series of probability distributions $\{Q_n\}$ in $\mathcal{Q}$, such that

$$\lim_{n \to \infty} \nu \left\{ x;\ \frac{dQ_n(x)}{d\nu(x)} > 0 \right\} = \alpha. \tag{1.1.9}$$

Furthermore, let $\lambda = \sum_{n=1}^{\infty} \frac{1}{2^n} Q_n$, then $\lambda \in \mathcal{Q}$. And from Eq. (1.1.9), we have

$$\nu\left\{x;\ \frac{d\lambda(x)}{d\nu(x)} > 0\right\} = \alpha.$$

Now we prove that λ and $\mathcal{P}$ are equivalent. Since $\lambda \in \mathcal{Q}$, for $\forall A \in \mathcal{A}$, then we have $\lambda(A) = 0$ if $P_\theta(A) = 0$ for $\forall\theta \in \Theta$. We use the proof by contradiction method to show that $\mathcal{P} \ll \lambda$. Suppose that there exists $A \in \mathcal{A}$, and $\theta \in \Theta$, such that $\lambda(A) = 0$ and $P_\theta(A) > 0$. Let $Q_0 = \frac{1}{2}(\lambda + P_\theta)$. Obviously $Q_0 \in \mathcal{Q}$. From $\lambda(A) = 0$, $d\lambda(x)/d\nu(x) = 0$ holds almost everywhere in A. Therefore

$$\begin{aligned}
\nu\left\{x;\ \frac{dQ_0(x)}{d\nu(x)} > 0\right\} &= \nu\left\{x;\ \frac{d\lambda(x)}{d\nu(x)} > 0 \quad \text{or} \quad \frac{dP_\theta(x)}{d\nu(x)} > 0\right\} \\
&\geq \nu\left\{x;\ \frac{d\lambda(x)}{d\nu(x)} > 0\right\} + \nu\left\{x \in A;\ \frac{dP_\theta(x)}{d\nu(x)} > 0\right\} \\
&> \nu\left\{x;\ \frac{d\lambda(x)}{d\nu(x)} > 0\right\} \\
&= \alpha.
\end{aligned}$$

This contradicts with the definition of α. □

Though the above theorem provides a judging criterion, the conditions themselves are very difficult to be verified. So it is necessary to make a further analysis of the family of probability distributions. At first we review the definition of distance.

Let d be a bivariate nonnegative function defined on $\mathcal{X}$. If for $\forall x, y, z \in \mathcal{X}$, d satisfies the following conditions

(1) $d(x, x) = 0$,
(2) $d(x, y) = d(y, x)$,
(3) $d(x, z) \leq d(x, y) + d(y, z)$,

then d is called a **quasi-distance**, and $\mathcal{X}$ a **quasi-distance space**. If an additional condition is attached to Condition (1), *i.e.*, $d(x, y) = 0$ iff $x = y$, then d is called a **distance**, and $\mathcal{X}$ a **distance space**.

Example 1.1.13. Let$(\mathcal{X}, \mathcal{A}, P)$ be a probability space. For $A_1, A_2 \in \mathcal{A}$, let

$$A_1 \triangle A_2 = (A_1 - A_2) \cup (A_2 - A_1),$$

$$d(A_1, A_2) = P(A_1 \triangle A_2).$$

Then the σ-algebra $\mathcal{A}$ forms a quasi-distance space about d.

Example 1.1.14. Let $(\mathcal{X}, \mathcal{A}, \mathcal{P})$ be a statistical space. For $\theta_1, \theta_2 \in \Theta$, let

$$d(\theta_1, \theta_2) = \sup_{A \in \mathcal{A}} |P_{\theta_1}(A) - P_{\theta_2}(A)|.$$

Then the family of probability distributions $\mathcal{P}$ forms a quasi-distance space about d. If θ and P_θ are one-to-one, then $\mathcal{P}$ forms a quasi-distance space about d.

Example 1.1.15. Let $\mathcal{P}$ and ν be the family of probability distributions and σ-finite measure in the sample space $(\mathcal{X}, \mathcal{A})$ respectively, and $\mathcal{P} \ll \nu$. Let $\mathcal{F}$ be a set of all the pdfs, *i.e.*

$$\mathcal{F} = \left\{ f_\theta;\ f_\theta = \frac{dP_\theta}{d\nu}, \theta \in \Theta \right\}.$$

For $f_{\theta_1}, f_{\theta_2} \in \mathcal{F}$, let

$$d^*(f_{\theta_1}, f_{\theta_2}) = \int_{\mathcal{X}} |f_{\theta_1}(x) - f_{\theta_2}(x)| d\nu(x).$$

Then $\mathcal{F}$ forms a quasi-distance space about d^*, and forms a distance space almost everywhere, *i.e.*,

$$d^*(f_{\theta_1}, f_{\theta_2}) = 0 \quad \Longleftrightarrow \quad f_{\theta_1}(x) = f_{\theta_2}(x) \quad \text{a.e. } \nu.$$

It can be verified that, if θ and P_θ are one-to-one, then there exists an equivalent relation between the two distances defined in Examples 1.1.4 and 1.1.5, *i.e.*, for $\theta_1, \theta_2 \in \Theta$ we have

$$\begin{aligned} d(\theta_1, \theta_2) &= \sup_{A \in \mathcal{A}} |P_{\theta_1}(A) - P_{\theta_2}(A)| \\ &= \frac{1}{2} \int_{\mathcal{X}} |f_{\theta_1}(x) - f_{\theta_2}(x)| d\nu(x) \\ &= \frac{1}{2} d^*(f_{\theta_1}, f_{\theta_2}), \end{aligned}$$

where $f_{\theta_i} = dP_{\theta_i}/d\nu$, $i = 1, 2$.

Let $\mathcal{X}$ form a quasi-distance space about d. If there exists a countable subset $\mathcal{X}_0$ of $\mathcal{X}$, we can find an $x_0 \in \mathcal{X}_0$, such that $d(x, x_0) < \varepsilon$, the $\mathcal{X}$ is called **divisible** about d, and $\mathcal{X}_0$ a **countable dense subset** of $\mathcal{X}$.

Theorem 1.1.4. *Let $(\mathcal{X}, \mathcal{A}, \mathcal{P})$ be a statistical space. Suppose that $\mathcal{P}$ is divisible about the distance d defined in Example 1.1.14, and the corresponding countable dense subset is $\mathcal{P}_0 = \{P_{\theta_n}\}$. For $A \in \mathcal{A}$, let*

$$\lambda(A) = \sum_{n=1}^{\infty} \frac{1}{2^n} P_{\theta_n}(A). \tag{1.1.10}$$

then $\mathcal{P}$ is equivalent to λ.

Proof. From Eq. (1.1.10), λ is also a probability measure defined on $(\mathcal{X}, \mathcal{A})$, and for $A \in \mathcal{A}$, we must have $\lambda(A) = 0$ if $P_\theta(A) = 0$ for $\forall \theta \in \Theta$. Now we need to prove that $\mathcal{P} \ll \lambda$.

For $A \in \mathcal{A}$, if $\lambda(A) = 0$, from Eq. (1.1.10), $P_{\theta_n}(A) = 0$ holds for all $P_{\theta_n} \in \mathcal{P}_0$. We will Prove that $P_\theta(A) = 0$, for $\forall P_\theta \in \mathcal{P}$. Since $\mathcal{P}_0$ is a countable dense subset of $\mathcal{P}$, for $\forall \varepsilon > 0$, there exists $P_{\theta_n} \in \mathcal{P}_0$, such that $d(\theta, \theta_n) < \varepsilon$. Since $P_{\theta_n}(A) = 0$,

$$P_\theta(A) = |P_\theta(A) - P_{\theta_n}(A)| \leq d(\theta, \theta_n) < \varepsilon.$$

By the arbitrariness of $\varepsilon > 0$, $P_\theta(A) = 0$. □

Theorem 1.1.5. *Let $(\mathcal{X}, \mathcal{A}, \mathcal{P})$ be a statistical space. If $\mathcal{P}$ is divisible about the distance d defined in Example 1.1.14, then there exists a σ-finite measure ν defined on $(\mathcal{X}, \mathcal{A})$, such that $\mathcal{P} \ll \nu$.*

The above theorem follows the results of Theorems 1.1.3 and 1.1.4 directly, and gives a condition that is easier to be verified. Let $(\mathcal{X}, \mathcal{A}, \mathcal{P})$ be a statistical space. If the parameter Θ is an at most countable set, then it obviously satisfies the conditions of Theorem 1.1.5; otherwise, it suffices to assume that Θ_0 be a subset formed by the rational numbers in the parameter space Θ, let $\mathcal{P}_0 = \{P_\theta; \theta \in \Theta_0\}$, and then verify whether $\mathcal{P}_0$ can form a countable dense subset of $\mathcal{P}$ about the distance defined in Example 1.1.14 or not. As a result, we have a very clear insight into the measure in statistical spaces and the relation between the measure and the parameter.

1.2 Conditional Probability and Sufficient Statistics

First we will review the conditional probability in the classical probability model by a simple example. When tossing a coin, there are two possible outcomes: Head (H) or Tail (T). And then there are four possible cases when tossing two times: {HH,HT,TH,TT}. Let A denote the event {H occurs at least one time}, and B the event {H occurs exactly once}. Suppose that the coin is even, then $P(A) = 3/4, P(AB) = 1/2$; the conditional probability of B given A is

$$P(B|A) = \frac{2}{3} = \left(\frac{1}{2}\right) \Big/ \left(\frac{3}{4}\right) = P(AB)/P(A),$$

which is equivalent to the following product-form

$$P(AB) = P(B|A)P(A). \tag{1.2.1}$$

This section will discuss a general conditional probability which involves the concept of statistics.

1.2.1 *Statistics and Random Variables*

In the former discussion, we have used the terms of "Random Variable" and "Statistics" in a non-formal way, now we will define them more exactly. Let $(\mathcal{X}, \mathcal{A}, P)$ be a probability space, and $(\mathcal{Y}, \mathcal{B})$ a measurable space. Let t be a mapping from $(\mathcal{X}, \mathcal{A})$ to $(\mathcal{Y}, \mathcal{B})$. If the mapping is measurable, *i.e.*

$$t^{-1}(B) \in \mathcal{A}, \quad \forall B \in \mathcal{B},$$

and then t is called a **statistic**. Especially, when $\mathcal{Y}$ is a subset of the real space, t is called a **random variable**. If t is a statistic, let

$$\mathcal{A}_t = \{t^{-1}(B);\ B \in \mathcal{B}\} \quad \text{and} \quad Q(B) = P(t^{-1}(B)),\ \forall B \in \mathcal{B}. \tag{1.2.2}$$

Obviously $\mathcal{A}_t$ is also a σ-algebra, and we have $\mathcal{A}_t \subset \mathcal{A}$; $Q(B)$ is a probability measure defined in $(\mathcal{Y}, \mathcal{B})$, which is called an **induced probability measure** by t . $(\mathcal{Y}, \mathcal{B}, Q)$ is called an **induced probability space** by t. Sometimes $(\mathcal{X}, \mathcal{A}_t, P)$ is also called an **induced probability space** by t.

Theorem 1.2.1. *For a given probability space $(\mathcal{X}, \mathcal{A}, P)$, let $(\mathcal{Y}, \mathcal{B}, Q)$ be the induced probability space by t. Then for any $\mathcal{B}$-measurable function h we have*

$$\int_{\mathcal{X}} h(t(x))dP(x) = \int_{\mathcal{Y}} h(y)dQ(y). \tag{1.2.3}$$

Proof. Obviously $h(t(x))$ is an $\mathcal{A}_t$-measurable function, thus the left-hand side of Eq. (1.2.3) is integrable. When $h(y) = I_B(y), B \in \mathcal{B}$, Eq. (1.2.3) is equivalent to $P(t^{-1}(B)) = Q(B)$, and by Eq. (1.2.2), it holds obviously. And then by the I-method, we can prove the conclusion. □

From the above theorem, we sometimes write $Q(B) = P(t(X) \in B)$ for $B \in \mathcal{B}$. The situation with two statistics will be considered as below. Let t and u be statistics from $(\mathcal{X}, \mathcal{A}, P)$ to measurable space $(\mathcal{Y}, \mathcal{B})$ and $(\mathcal{Z}, \mathcal{C})$, respectively. Let $\mathcal{A}_{tu}$ be a σ-algebra generated by $\{t^{-1}(B) \cap u^{-1}(C); B \in \mathcal{B}, C \in \mathcal{C}\}$. Obviously $\mathcal{A}_{tu} \subset \mathcal{A}, (\mathcal{X}, \mathcal{A}_{tu}, P)$ also forms a probability space. If for $\forall B \in \mathcal{B}, \forall C \in \mathcal{C}$ we have

$$P(t^{-1}(B) \cap u^{-1}(C)) = P(t^{-1}(B))P(u^{-1}(C)),$$

then t and u are said to be **mutually independent**. According to Theorem 1.2.1, the above equation can be also written as

$$P(t(X) \in B, u(X) \in C) = P(t(X) \in B)P(u(X) \in C). \tag{1.2.4}$$

Especially when $\mathcal{Y}$ and $\mathcal{Z}$ are both subsets of the real space, the r.v.'s $t(X)$ and $u(X)$ are said to be mutually independent. The definition can easily be extended to the general situation.

Let t_i be a statistic from $(\mathcal{X}, \mathcal{A}, P)$ to measurable space $(\mathcal{Y}_i, \mathcal{B}_i)$, $i = 1, \cdots, n$. If for $\forall B_i \in \mathcal{B}_i, i = 1, \cdots, n$, we have

$$P(\bigcap_{i=1}^{n} t_i^{-1}(B_i)) = \prod_{i=1}^{n} P(t_i^{-1}(B_i)),$$

then $t_1, \cdots, t_n$ are said to be **mutually independent**. Especially when $\mathcal{Y}_i, i = 1, \cdots, n$, all are the subsets of the real space, the r.v.'s $t_1(X), \cdots, t_n(X)$ are said to be mutually independent.

Now the statistical space $(\mathcal{X}, \mathcal{A}, \mathcal{P})$ is considered, where $\mathcal{P} = \{P_\theta;\ \theta \in \Theta\}$. Similar to Eq. (1.2.2), for $\theta \in \Theta$, let $Q_\theta(B) = P_\theta(t^{-1}(B))$ for $B \in \mathcal{B}$, then $\mathcal{Q} = \{Q_\theta;\ \theta \in \Theta\}$ also forms a family of probability distributions, which will be called an **induced family of probability distributions** by t and $(\mathcal{Y}, \mathcal{B}, \mathcal{Q})$ an **induced statistical space** by t. If P_θ and θ are one-to-one, then Q_θ and θ are also one-to-one. Therefore, when we are making a statistical inference to the parameter θ, to simplify the problem, we usually find an appropriate statistic t, and then study the problem after transformed it into the induced statistical space by t.

Example 1.2.1. Let $X_1, \cdots, X_n$ be the mutually independent r.v.'s from a normal distribution with mean θ and variance 1, which is sometimes denoted as

$$X_1, \cdots, X_n \overset{\text{i.i.d.}}{\sim} N(\theta, 1), \tag{1.2.5}$$

and then $\boldsymbol{X} = (X_1, \cdots, X_n)'$ has an n-dimensional normal distribution with n-dimensional mean vector $(\theta, \cdots, \theta)'$ and covariance matrix $\boldsymbol{\Sigma} = \boldsymbol{I}$. Since we only make a statistical inference to the parameter θ, the sample mean can be considered as a statistic, *i.e.*,

$$t(\boldsymbol{X}) = \bar{X} = \frac{1}{n}\sum_{i=1}^{n} X_i.$$

Then $t(\boldsymbol{X}) \sim N(\theta, 1/n)$, which is a univariate normal distribution. The parameter θ does not change, but the variance becomes smaller than that in Eq. (1.2.5). Considering intuitively, when we make statistical inference to θ we can obtain a higher precision by using $t(\boldsymbol{X})$ than by using X_i; the same precision as using $\boldsymbol{X} = (X_1, \cdots, X_n)'$, but the corresponding computation is more convenient. We will further discuss the problem later.

Let μ be a σ-finite measure defined on $(\mathcal{X}, \mathcal{A})$, satisfying $\mathcal{P} \ll \mu$. For $\theta \in \Theta$, the pdf is $dP_\theta/d\mu = f_\theta$. Similar to Eq. (1.2.2), for $B \in \mathcal{B}$, let

$$\nu(B) = \mu(t^{-1}(B)).$$

And then ν is a σ-finite measure defined on $(\mathcal{Y}, \mathcal{B})$. It is easy to verify that $\mathcal{Q} \ll \nu$. For $\theta \in \Theta$, let $dQ_\theta/d\nu = g_\theta$. And then Eq. (1.2.3) in Theorem 1.2.1 can be written as

$$\int_{\mathcal{X}} h(t(x)) f_\theta(x) d\mu(x) = \int_{\mathcal{Y}} h(y) g_\theta(y) d\nu(y). \tag{1.2.6}$$

1.2.2 *Conditional Probability*

For given probability space $(\mathcal{X}, \mathcal{A}, P)$, let $(\mathcal{Y}, \mathcal{B}, Q)$ be the induced probability space by t. Let h be a measurable function defined in $(\mathcal{X}, \mathcal{A}, P)$, satisfying $Eh(X) < \infty$. For $\forall B \in \mathcal{B}$, let

$$R(B) = \int_{t^{-1}(B)} h(x) dP(x). \tag{1.2.7}$$

And then R is a measure defined on $(\mathcal{Y}, \mathcal{B})$. From Eq. (1.2.2), $R \ll Q$. Applying the Radon-Nikodym Theorem, there exists a $\mathcal{B}$-measurable function $g = dR/dQ$, s.t. for $B \in \mathcal{B}$,

$$R(B) = \int_B g(y) dQ(y). \tag{1.2.8}$$

Comparing Eq. (1.2.7) with Eq. (1.2.8), we can get that, for $\forall B \in \mathcal{B}$,

$$\int_{t^{-1}(B)} h(x) dP(x) = \int_B g(y) dQ(y). \tag{1.2.9}$$

g in Eq. (1.2.9) is unique almost everywhere with respect to the measure Q. The function $g(y)$ in Eq. (1.2.9) is called a **conditional mean** of $h(x)$ when $t(x) = y$ is given, which is denoted as

$$E(h(X)|y). \tag{1.2.10}$$

Obviously this is a function of y. Equation (1.2.9) can be rewritten as, for $\forall B \in \mathcal{B}$,

$$\int_{t^{-1}(B)} h(x) dP(x) = \int_B E(h(X)|y) dQ(y). \tag{1.2.11}$$

Especially, when $h(x) = I_A(x)$, for $A \in \mathcal{A}$, let $P(A|y) = E(I_A(X)|y)$, Eq. (1.2.11) can be rewritten as, for $\forall B \in \mathcal{B}$

$$P\{A \cap t^{-1}(B)\} = \int_B P(A|y) dQ(y). \tag{1.2.12}$$

$P(A|y)$ in Eq. (1.2.12) is called a **conditional probability** of A given $t(x) = y$.

The above result is applicable to the statistical space. Let $(\mathcal{X}, \mathcal{A}, \mathcal{P})$ be a given statistical space, where $\mathcal{P} = \{P_\theta; \theta \in \Theta\}$; let $(\mathcal{Y}, \mathcal{B}, \mathcal{Q})$ be the induced statistical space by t, then $\mathcal{Q}$ can be denoted as $\mathcal{Q} = \{Q_\theta; \theta \in \Theta\}$. Equations (1.2.11) and (1.2.12) can be written as

$$\int_{t^{-1}(B)} h(x)dP_\theta(x) = \int_B E_\theta(h(X)|y)dQ_\theta(y), \tag{1.2.13}$$

and

$$P_\theta(A \cap t^{-1}(B)) = \int_B P_\theta(A|y)dQ_\theta(y), \tag{1.2.14}$$

respectively. Therefore, both the conditional mean and the conditional probability depend on θ in the statistical space.

Example 1.2.2. Let $(\mathcal{X} \times \mathcal{Y}, \mathcal{A} \times \mathcal{B}, \mu \times \nu)$ denote the direct product space of two σ-infinite measure spaces $(\mathcal{X}, \mathcal{A}, \mu)$ and $(\mathcal{Y}, \mathcal{B}, \nu)$, and $(\mathcal{X} \times \mathcal{Y}, \mathcal{A} \times \mathcal{B}, \mathcal{P})$ the statistical space corresponding to the direct product space, where $\mathcal{P} = \{P_\theta; \theta \in \Theta\}$, and $\mathcal{P} \ll \mu \times \nu$. Let $dP_\theta(x,y) = f_\theta(x,y)d\mu(x)d\nu(y)$. Let h be an $\mathcal{A} \times \mathcal{B}$ measurable function, $E_\theta h(X,Y) < \infty$.

Let $t(x,y) = y$, then t is a statistic from $\mathcal{A} \times \mathcal{B}$ to $\mathcal{B}$, and the induced statistical space by t can be denoted as $(\mathcal{Y}, \mathcal{B}, \mathcal{Q})$. For $\forall B \in \mathcal{B}$, since $Q_\theta(B) = P_\theta(t^{-1}(B))$, we have

$$Q_\theta(B) = \int_{\mathcal{X} \times B} f_\theta(x,y)d\mu(x)d\nu(y) = \int_B \left[\int_{\mathcal{X}} f_\theta(x,y)d\mu(x)\right] d\nu(y). \tag{1.2.15}$$

Obviously $\mathcal{Q} \ll \nu$. For $\theta \in \Theta$, let $q_\theta = dQ_\theta/d\nu$, from the above equation we have

$$q_\theta(y) = \int_{\mathcal{X}} f_\theta(x,y)d\mu(x). \tag{1.2.16}$$

q_θ is called a **marginal density** of y. From Eqs. (1.2.13) and (1.2.15), we have

$$\int_B \left[\int_{\mathcal{X}} h(x,y)f_\theta(x,y)d\mu(x)\right] d\nu(y) = \int_B q_\theta(y)E_\theta(h(X,Y)|y)d\nu(y).$$

From Theorem 1.1.3, we can set that $\nu\{y; q_\theta(y) = 0\} = 0$. Since the above equation holds for any $B \in \mathcal{B}$, we have

$$E_\theta(h(X,Y)|y) = \begin{cases} \dfrac{1}{q_\theta(y)} \displaystyle\int_{\mathcal{X}} h(x,y)f_\theta(x,y)d\mu(x), & \text{if } q_\theta(y) > 0, \\ E_\theta h(X,Y), & \text{if } q_\theta(y) = 0. \end{cases} \tag{1.2.17}$$

a.e. with respect to the measure ν. Especially when $h(x,y) = I_C(x,y)$ for $C \in \mathcal{A} \times \mathcal{B}$, we have

$$P_\theta(C|y) = \begin{cases} \dfrac{1}{q_\theta(y)} \displaystyle\int_{C_y} f_\theta(x,y)d\mu(x), & \text{if} \quad q_\theta(y) > 0, \\ P_\theta(C), & \text{if} \quad q_\theta(y) = 0, \end{cases} \tag{1.2.18}$$

where $C_y = \{x; (x,y) \in C\}$, and $y \in \mathcal{Y}$. When $q_\theta(y) > 0$, obviously $P_\theta(\cdot|y) \ll \mu$. Let $f_\theta(x|y)$ denote its pdf, then Eq. (1.2.18) can be written as

$$\int_{C_y} f_\theta(x|y)d\mu(x) = \frac{1}{q_\theta(y)} \int_{C_y} f_\theta(x,y)d\mu(x).$$

Therefore, we have

$$f_\theta(x,y) = f_\theta(x|y)q_\theta(y) \tag{1.2.19}$$

a.e. with respect to the measure μ. That is corresponding to Eq. (1.2.1) under the classical probability model. Therefore, generally, the conditional mean of $h(x)$ can be rewritten as

$$E_\theta(h(X)|y) = \int_{\mathcal{X}} h(x) f_\theta(x|y) d\mu(x).$$

Taking the mean value with respect to y on both sides of the above equation, we have

$$\begin{aligned} E_\theta E_\theta(h(X)|Y) &= \int_{\mathcal{Y}} \left[\int_{\mathcal{X}} h(x) f_\theta(x|y) d\mu(x) \right] q_\theta(y) d\nu(y) \\ &= \int_{\mathcal{X}} h(x) \int_{\mathcal{Y}} f_\theta(x,y) d\nu(y) d\mu(x) \\ &= \int_{\mathcal{X}} h(x) s_\theta(x) d\mu(x) \\ &= E_\theta h(X), \end{aligned}$$

where $s_\theta(x) = \int f_\theta(x,y)d\nu(y)$ is the marginal density of x.

Sometimes, we need to consider the conditional probability in the same space, just as the conditional density defined in Eq. (1.2.1) under the classical probability model at the beginning of this section. For the given statistical space $(\mathcal{X}, \mathcal{A}, \mathcal{P})$, let $\mathcal{A}_0$ be a sub σ-algebra of $\mathcal{A}$, *i.e.*, $A_0 \in \mathcal{A}_0$, then we must have $A_0 \in \mathcal{A}$. For $\theta \in \Theta, \forall A_0 \in \mathcal{A}_0$, let

$$S_\theta(A_0) = \int_{A_0} h(x) dP_\theta(x), \tag{1.2.20}$$

then $S_\theta \ll P_\theta$. Let the pdf be $dS_\theta/dP_\theta = s_\theta$, which is an $\mathcal{A}_0$ measurable function. For $\forall A_0 \in \mathcal{A}_0$, we also have

$$S_\theta(A_0) = \int_{A_0} s_\theta(x)dP_\theta(x). \tag{1.2.21}$$

Combining Eq. (1.2.20) with Eq. (1.2.21), we have

$$\int_{A_0} h(x)dP_\theta(x) = \int_{A_0} s_\theta(x)dP_\theta(x) \tag{1.2.22}$$

for $\forall A_0 \in \mathcal{A}_0$. Note that h in the above equation is $\mathcal{A}$-measurable, while s_θ is $\mathcal{A}_0$-measurable, which generally depends on the parameter θ. Generally, s_θ satisfying Eq. (1.2.22) is called the **conditional mean** of h given $\mathcal{A}_0$, and denoted as

$$E_\theta(h(X)|\mathcal{A}_0, x).$$

Therefore Eq. (1.2.22) can be written as

$$\int_{A_0} h(x)dP_\theta(x) = \int_{A_0} E_\theta(h(X)|\mathcal{A}_0, x)dP_\theta(x) \tag{1.2.23}$$

for $\forall A_0 \in \mathcal{A}_0$. Especially when $h(x) = I_A(x), A \in \mathcal{A}$,

$$P_\theta(A \cap A_0) = \int_{A_0} E_\theta(I_A(X)|\mathcal{A}_0, x)dP_\theta(x) = \int_{A_0} P_\theta(A|\mathcal{A}_0, x)dP_\theta(x), \tag{1.2.24}$$

of which $P_\theta(A|\mathcal{A}_0, x)$ is called the **conditional probability** of A given $\mathcal{A}_0$.

In Eq. (1.2.23), let $A_0 = \mathcal{X}$, we can get the following theorem similar to Example 1.2.2.

Theorem 1.2.2. *For the given statistical space $(\mathcal{X}, \mathcal{A}, \mathcal{P})$, let $\mathcal{A}_0$ be a sub σ-algebra of $\mathcal{A}$, then for $\mathcal{A}$-integrable function h and $\forall \theta \in \Theta$, we have*

$$E_\theta E_\theta(h(X)|\mathcal{A}_0, X) = E_\theta h(X).$$

Note that when $\mathcal{A}_0 = \mathcal{A}_t$, applying Theorem 1.2.1, from Eqs. (1.2.13) and (1.2.23) we have

$$\int_{A_0} E_\theta(h(X)|t(x))dP_\theta(x) = \int_{A_0} E_\theta(h(X)|\mathcal{A}_0, x)dP_\theta(x)$$

for $\forall A_0 \in \mathcal{A}_t$. The above equation implies that $P_\theta(D) = 0$ for $D \in \mathcal{A}_t$ with $D = \{x;\ E_\theta(h(X)|t(x)) \neq E_\theta(h(X)|\mathcal{A}_0, x)\}$.

Example 1.2.3. For the given statistical space $(\mathcal{X}, \mathcal{A}, \mathcal{P})$ with $\mathcal{P} = \{P_\theta; \theta \in \Theta\}$, let h be an $\mathcal{A}$-measurable function satisfying $E_\theta h(X) < \infty$ for $\forall \theta \in \Theta$.

(i) If $\mathcal{A}_0 = \{\phi, \mathcal{X}\}$, then obviously

$$E_\theta(h(X)|\mathcal{A}_0, x) = E_\theta h(X).$$

That is the most common conditional mean.

(ii) For $A_0 \in \mathcal{A}$, let $\mathcal{A}_0 = \{\phi, A_0, A_0^c, \mathcal{X}\}$, where A_0^c denotes the complement of A_0 in $\mathcal{X}$. Since the conditional mean must be an $\mathcal{A}_0$-measurable function, the necessary and sufficient condition is that there exist two constants c_1 and c_2, s.t.

$$E_\theta(h(X)|\mathcal{A}_0, x) = \begin{cases} c_1, & \text{if} \quad x \in A_0, \\ c_2, & \text{if} \quad x \in A_0^c. \end{cases}$$

Substituting into Eq. (1.2.23) yields

$$E_\theta(h(X)|\mathcal{A}_0, x) = \begin{cases} \dfrac{1}{P_\theta(A_0)} \displaystyle\int_{A_0} h(x) dP_\theta(x), & \text{if} \quad x \in A_0 \\ \dfrac{1}{P_\theta(A_0^c)} \displaystyle\int_{A_0^c} h(x) dP_\theta(x), & \text{if} \quad x \in A_0^c. \end{cases}$$

Especially when $h(x) = I_A(x)$, for $A \in \mathcal{A}$, we have

$$P_\theta(A|\mathcal{A}_0, x) = \begin{cases} \dfrac{1}{P_\theta(A_0)} P_\theta(A_0 \cap A), & \text{if} \quad x \in A_0 \\ \dfrac{1}{P_\theta(A_0^c)} P_\theta(A_0^c \cap A), & \text{if} \quad x \in A_0^c. \end{cases}$$

When $x \in A_0$, we will use $P_\theta(A|A_0)$ to denote $P_\theta(A|\mathcal{A}_0, x)$, we have

$$P_\theta(A \cap A_0) = P_\theta(A|A_0) P_\theta(A_0),$$

which coincides with Eq. (1.2.1) defined in the classical probability model.

From the definition of conditional mean, the conditional mean is an integral under the probability measure, so it is easy to prove the following properties.

(1) For constants a and b,

$$E_\theta(ah(X) + bg(X)|\mathcal{A}_0, x) = aE_\theta(h(X)|\mathcal{A}_0, x) + bE_\theta(g(X)|\mathcal{A}_0, x).$$

(2) If $h(x) \geq 0$, then $E_\theta(h(X)|\mathcal{A}_0, x) \geq 0$.

(3) **Monotone convergence theorem**. For an integrable function h and a series of measurable functions $\{h_n\}$, if $h_n \uparrow h$, then $E_\theta(h_n(X)|\mathcal{A}_0, x) \uparrow E_\theta(h(X)|\mathcal{A}_0, x)$.

(4) **Dominated convergence theorem.** For a function h and a series of integrable functions $\{h_n\}$, if there exists a nonnegative integrable function g satisfying $|h_n(x)| \leq g(x)$, then $E_\theta(h_n(X)|\mathcal{A}_0, x) \to E_\theta(h(X)|\mathcal{A}_0, x)$ when $h_n \to h$.

Theorem 1.2.3. *Let $(\mathcal{X}, \mathcal{A}, \mathcal{P})$ be a statistical space, $\mathcal{A}_0 \subset \mathcal{A}$ a sub σ-algebra, f an $\mathcal{A}$-measurable function, and g an $\mathcal{A}_0$-measurable function. If $E_\theta g(X)h(X) < \infty$ for $\theta \in \Theta$, then we have*

$$E_\theta(g(X)h(X)|\mathcal{A}_0, x) = g(x)E_\theta(f(X)|\mathcal{A}_0, x)$$

a.e. with respect to $\mathcal{A}_0$ and P_θ.

Proof. At first, we consider the case where g is a simple function, *i.e.* $g(x) = I_{A_0}(x)$ for $A_0 \in \mathcal{A}_0$. For $\forall A \in \mathcal{A}_0$, Note that $A \cap A_0 \in \mathcal{A}_0$, we have

$$\begin{aligned}\int_A I_{A_0}(x)E_\theta(h(X)|\mathcal{A}_0, x)dP_\theta(x) &= \int_{A\cap A_0} E_\theta(h(X)|\mathcal{A}_0, x)dP_\theta(x) \\ &= \int_{A\cap A_0} h(x)dP_\theta(x) \\ &= \int_A E_\theta(I_{A_0}(X)h(X)|\mathcal{A}_0, x)dP_\theta(x).\end{aligned}$$

From the above properties (1) and (3), based on the equivalent formulation about uniqueness a.e. discussed in the first section, we can prove the theorem by the I-method. □

1.2.3 *Sufficient Statistics*

Let $\boldsymbol{x}$ be a sample from the statistical space $(\mathcal{X}, \mathcal{A}, \mathcal{P})$, where $\mathcal{P} = \{P_\theta; \theta \in \Theta\}$. Our aim is to make statistical inference to the parameter by using the sample $\boldsymbol{x}$. In practical inference, we need to find an appropriate statistic t, and then make inference to θ based on $t(\boldsymbol{x})$ and the induced statistical space by t. It is important to find "appropriate statistics", since we can find many seemingly reasonable statistics in principle. Recall Example 1.2.1, let $x_1, \cdots, x_n$ be an independent sample from $N(\theta, 1)$, and $\boldsymbol{x} = (x_1, \cdots, x_n)$. We have defined the sample mean, *i.e.* $t(\boldsymbol{x}) = \bar{x}$, to estimate θ. We can also define $t_1(\boldsymbol{x}) = x_1, t_{\max}(x) = \max\{x_i\}$ and so on. And then which statistics are "appropriate"? What are the criteria for the appropriateness? Fisher proposed the concept of sufficient statistics in 1922. Halmos and Savage (1949), Bahadur (1954) gave an exact formulation about it and the proof of some equivalent propositions. The sufficiency of statistics has been one of the most important concepts in statistics.

Basically speaking, a sufficient statistic is a statistic without losing any information about the parameter θ, which can be characterized by a conditional probability as follows: for $\forall A \in \mathcal{A}$, if the conditional probability $P_\theta(A|t)$ is independent of θ, then t is called a sufficient statistic of the parameter θ or t is called a **sufficient statistic** about the family of distributions $\mathcal{P}$. This implies that the measurement of the statistical space is independent of the parameter θ given t, *i.e.*, the statistic t carries all the information about the parameter θ.

Example 1.2.4. (Continuation of Example 1.2.1) From Eq. (1.2.19), when $\bar{x} = t$, the conditional pdf of $\boldsymbol{x}$ is

$$f_\theta(\boldsymbol{x}|t) = \begin{cases} \dfrac{\left(\dfrac{1}{2\pi}\right)^{n/2} \exp\left\{-\dfrac{1}{2}\displaystyle\sum_{i=1}^{n}(x_i-\theta)^2\right\}}{\left(\dfrac{n}{2\pi}\right)^{1/2} \exp\left\{-\dfrac{n}{2}(t-\theta)^2\right\}}, & \text{if} \quad \bar{x} = t, \\ 0, & \text{otherwise}, \end{cases}$$

$$= \begin{cases} \dfrac{1}{\sqrt{n}}\left(\dfrac{1}{2\pi}\right)^{(n-1)/2} \exp\left\{-\dfrac{1}{2}\displaystyle\sum_{i=1}^{n}(x_i-t)^2\right\}, & \text{if} \quad \bar{x} = t \\ 0, & \text{otherwise}, \end{cases}$$

which is independent of the mean θ, so $t(\boldsymbol{x}) = \bar{x}$ is a sufficient statistic for the mean independent of parameter θ.

Example 1.2.5. Let the statistical space be $(\mathcal{X}, \mathcal{A}, \mathcal{P}) = (\mathcal{R}^n, \mathcal{B}^n, \mathcal{P})$, where $\mathcal{B}^n$ is a σ-algebra generated by the Borel sets in the n-dimensional Euclidean space $\mathcal{R}^n$. For the sample $\boldsymbol{x} = (x_1, \cdots, x_n)$, let $y_1 \leq y_2 \leq \cdots \leq y_n$ be the result of arranging $x_1, \cdots, x_n$ from the smallest to the largest. Let $\boldsymbol{y} = (y_1, \cdots, y_n)$, $\boldsymbol{t}(\boldsymbol{x}) = \boldsymbol{y}$ is called the **order statistics** of $\boldsymbol{x}$. Then the range of $\boldsymbol{y}$ is

$$\mathcal{Y} = \{\boldsymbol{y};\ y_1 \leq y_2 \leq \cdots \leq y_n\}.$$

Obviously $\mathcal{Y} \subset \mathcal{B}^n$. Let $\mathcal{B}$ be a σ-algebra generated by all the Borel sets in $\mathcal{Y}$, and $\mathcal{A}_t$ be the induced σ-algebra by $\boldsymbol{t}$ in $\mathcal{A}$. For $A \in \mathcal{A}_t$, A is a Borel set symmetrical relative to the n coordinates, *i.e.*, $\forall \boldsymbol{x} = (x_1, \cdots, x_n) \in A$ implying $(x_{i_1}, \cdots, x_{i_n}) \in A$, where $(i_1, \cdots, i_n)$ is a permutation of $(1, \cdots, n)$.

For $P_\theta \in \mathcal{P}$, if the probability keeps unchanged for different permutation, *i.e.*, for $\forall A \in \mathcal{A}$, we have

$$P_\theta((X_1, \cdots, X_n) \in A) = P_\theta((X_{i_1}, \cdots, X_{i_n}) \in A),$$

then P_θ is called a **symmetrical distribution.** Obviously, when $x_1, \cdots, x_n$ are i.i.d. r.v.s, the joint distribution of $\boldsymbol{x} = (x_1, \cdots, x_n)$ is a symmetrical distribution. We will explain that when all the distributions in $\mathcal{P}$ are symmetrical, the order statistic t is sufficient for $\mathcal{P}$.

For $\boldsymbol{x} \in \mathcal{X}$ and $A \in \mathcal{A}$, let $\#(A, \boldsymbol{x})$ denote the number of permutations satisfying $(x_{i_1}, \cdots, x_{i_n}) \in A$, which is a symmetrical function about $\boldsymbol{x}$-coordinate. Let $\boldsymbol{y} = \boldsymbol{t}(\boldsymbol{x})$, and then $\#(A, \boldsymbol{x}) = \#(A, \boldsymbol{y})$. For $\forall A_0 \in \mathcal{A}_{\boldsymbol{t}}$ and $\forall P_\theta \in \mathcal{P}$, from the symmetry we have

$$P_\theta(A \cap A_0) = \int_{A_0} I_A(x_1, \cdots, x_n) dP_\theta(\boldsymbol{x}) = \int_{A_0} I_A(x_{i_1}, \cdots, x_{i_n}) dP_\theta(\boldsymbol{x}). \tag{1.2.25}$$

Note that $\#(A, \boldsymbol{y}) = \sum I_A(x_{i_1}, \cdots, x_{i_n})$, where the summation runs over the $n!$ permutations. After taking summation on both sides of Eq. (1.2.25), we have

$$P_\theta(A \cap A_0) = \int_{A_0} \frac{\#(A, \boldsymbol{y})}{n!} dP_\theta.$$

Then $P_\theta(A|\boldsymbol{y}) = \#(A, \boldsymbol{y})/n!$ is independent of θ.

From the above two examples we can see that, though we can calculate a sufficient statistic directly according to its definition, its calculation is quite complicated. Now we offer a decision theorem, which is convenient to use and usually called as the **Neyman's Factorization Theorem.**

Theorem 1.2.4. *Let $(\mathcal{X}, \mathcal{A}, \mathcal{P})$ be a statistical space, where exists a σ-infinite measure λ, s.t. $\mathcal{P} \ll \lambda$. Let $\mathcal{A}_t$ be a sub σ-algebra of $\mathcal{A}$ induced by t. The t is a sufficient statistic for $\mathcal{P}$ if and only if for $\forall \theta \in \Theta$, there exists an $\mathcal{A}_t$-measurable function g_θ, s.t.*

$$\frac{dP_\theta}{d\lambda} = g_\theta, \tag{1.2.26}$$

a.e. with respect to $\mathcal{A}$ and λ, where λ is given by Eq. (1.1.7) in Theorem 1.1.3.

Proof. From Theorem 1.1.2, we can consider both P_θ and λ as probability measures on $(\mathcal{X}, \mathcal{A}_t)$. From Theorem 1.1.3, $P_\theta \ll \lambda$. Then there exists a Radon-Nikodym derivative g_θ on $(\mathcal{X}, \mathcal{A}_t)$. Obviously g_θ is $\mathcal{A}_t$-measurable, and for any integrable function h on $(\mathcal{X}, \mathcal{A}_t, P_\theta)$ we have

$$\int_{\mathcal{X}} h(x) dP_\theta(x) = \int_{\mathcal{X}} h(x) g_\theta(x) d\lambda(x). \tag{1.2.27}$$

Necessity. Let t be a sufficient statistic for $\mathcal{P}$, *i.e.* for $\forall A \in \mathcal{A}$ there exists a conditional probability $P(A|\mathcal{A}_t, x)$ independent of θ. Then we will prove Eq. (1.2.26). From the definition of λ and Theorem 1.1.3, for $\forall A_0 \in \mathcal{A}_t$ we have

$$\begin{aligned}\lambda(A \cap A_0) &= \sum c_i P_{\theta_i}(A \cap A_0) \\ &= \sum c_i \int_{A_0} P(A|\mathcal{A}_t, x) dP_{\theta_i}(x) \\ &= \int_{A_0} P(A|\mathcal{A}_t, x) d\lambda(x).\end{aligned}$$

This implies

$$\lambda(A|\mathcal{A}_t, x) = P(A|\mathcal{A}_t, x) \tag{1.2.28}$$

a.e. with respect to $\mathcal{A}_t$, and λ. We use E_λ to denote the mean when the probability measure is λ, and then for $\forall \theta \in \Theta$, and $\forall A \in \mathcal{A}$ we have

$$\begin{aligned}P_\theta(A) &= \int_{\mathcal{X}} P(A|\mathcal{A}_t, x) dP_\theta(x) \\ &= \int_{\mathcal{X}} \lambda(A|\mathcal{A}_t, x) g_\theta(x) d\lambda(x) \\ &= \int_{\mathcal{X}} E_\lambda(I_A(X)|\mathcal{A}_t, x) g_\theta(x) d\lambda(x) \\ &= \int_{\mathcal{X}} E_\lambda(I_A(X) g_\theta(X)|\mathcal{A}_t, x) d\lambda(x) \\ &= \int_{\mathcal{X}} I_A(x) g_\theta(x) d\lambda(x) \\ &= \int_A g_\theta(x) d\lambda(x),\end{aligned}$$

where the second equality follows from Eqs. (1.2.27) and (1.2.28); the fourth follows from Theorem 1.2.2; and the fifth follows from Theorem 1.2.1. Since $A \in \mathcal{A}$ is arbitrary, Eq. (1.2.26) holds.

Sufficiency. Suppose Eq. (1.2.26) holds, we will prove that for $\forall \theta \in \Theta$ and $A \in \mathcal{A}$, we have

$$P_\theta(A|\mathcal{A}_t, x) = \lambda(A|\mathcal{A}_t, x) \tag{1.2.29}$$

a.e. with respect to $\mathcal{A}_t$, and λ. For $\forall A_0 \in \mathcal{A}_t$, we have

$$\begin{aligned}\int_{A_0} \lambda(A|\mathcal{A}_t, x)dP_\theta(x) &= \int_{A_0} \lambda(A|\mathcal{A}_t, x)g_\theta(x)d\lambda(x)\\ &= \int_{A_0} E_\lambda(I_A(X)|\mathcal{A}_t, x)g_\theta(x)d\lambda(x)\\ &= \int_{A_0} E_\lambda(I_A(X)g_\theta(X)|\mathcal{A}_t, x)d\lambda(x)\\ &= \int_{A_0} I_A(x)g_\theta(x)d\lambda(x)\\ &= \int_{A\cap A_0} g_\theta(x)d\lambda(x)\\ &= P_\theta(A\cap A_0).\end{aligned}$$

Since $A_0 \in \mathcal{A}_t$ is arbitrary, from the definition of conditional probability, Eq. (1.2.29) holds. □

Theorem 1.2.5. (Neyman's Factorization Theorem) *Let $(\mathcal{X}, \mathcal{A}, \mathcal{P})$ be a statistical space, and ν be a σ-finite measure satisfying $\mathcal{P} \ll \nu$. Let $\mathcal{A}_t$ be a sub σ-algebra of $\mathcal{A}$ induced by t. The t is a sufficient statistic for $\mathcal{P}$ if and only if for $\forall\theta \in \Theta$ there exists an $\mathcal{A}_t$ measurable function g_θ and $\mathcal{A}$-measurable function h independent of θ, s.t.*

$$\frac{dP_\theta(x)}{d\nu(x)} = g_\theta(x)h(x) \tag{1.2.30}$$

a.e. with respect to both $\mathcal{A}$ and ν.

Proof. Let λ be the probability measure in Theorem 1.2.3, we will prove that Eq. (1.2.30) is equivalent to Eq. (1.2.26). From Theorem 1.1.3, $\mathcal{P}$ and λ are equivalent, so $\lambda \ll \nu$. Let

$$\frac{d\lambda(x)}{d\nu(x)} = h(x),$$

then h is an $\mathcal{A}$-measurable function independent of θ. Hence Eq. (1.2.26) implies Eq. (1.2.30). On the contrary, from Eq. (1.2.30)

$$\frac{d\lambda(x)}{d\nu(x)} = \sum_{i=1}^{\infty} c_i \frac{dP_{\theta_i}(x)}{d\nu(x)} = \sum_{i=1}^{\infty} c_i g_{\theta_i}(x)h(x).$$

Let $s(x) = \sum\limits_{i=1}^{\infty} c_i g_{\theta_i}(x)$, for $\forall\theta \in \Theta$ and $x \in \mathcal{X}$, let

$$g_\theta^*(x) = \begin{cases} \dfrac{g_\theta(x)}{s(x)}, & \text{when } 0 < s(x) < \infty \\ 0, & \text{otherwise.} \end{cases}$$

And then g_θ^* is an $\mathcal{A}_t$-measurable function. Since λ is a probability measure, we have $d\lambda(x)/d\nu(x) = s(x)h(x) > 0$ a.e. with respect to both $\mathcal{A}$ and λ, thus

$$\begin{aligned}\frac{dP_\theta(x)}{d\lambda(x)} &= \frac{dP_\theta(x)}{d\nu(x)} \cdot \frac{d\nu(x)}{d\lambda(x)} \\ &= g_\theta(x)h(x) \cdot \frac{1}{s(x)h(x)} \\ &= g_\theta^*(x).\end{aligned}$$

The conclusion follows from Theorem 1.2.3. □

The above theorem shows that a statistic $t(x)$ from the statistical space $(\mathcal{X}, \mathcal{A}, \mathcal{P})$ to the statistical space $(\mathcal{Y}, \mathcal{B}, \mathcal{Q})$ is sufficient for θ if there exists a $\mathcal{B}$-measurable function g_θ and an $\mathcal{A}$-measurable function h which is independent of θ, s.t.

$$f_\theta(x) = g_\theta(t(x))h(x), \tag{1.2.31}$$

where $f_\theta = dP_\theta/d\nu$ for $\forall\theta \in \Theta$. The example below shows that this theorem provides a convenient decision criterion.

Example 1.2.6. (Continuation of Examples 1.2.1 and 1.2.4) Let $x_1, \cdots, x_n$ be i.i.d. r.v.s from $N(\theta, 1)$, and $\boldsymbol{x} = (x_1, \cdots, x_n)$. Let $t(\boldsymbol{x}) = \bar{x}$. And then the joint distribution can be factorized as

$$\begin{aligned}f_\theta(\boldsymbol{x}) &= \left(\frac{1}{2\pi}\right)^{\frac{n}{2}} \exp\left\{-\frac{1}{2}\sum_{i=1}^n (x_i - \theta)^2\right\} \\ &= \exp\left\{-\frac{n}{2}(t-\theta)^2\right\} \cdot \left(\frac{1}{2\pi}\right)^{\frac{n}{2}} \exp\left\{-\frac{1}{2}\sum_{i=1}^n (x_i - t)^2\right\}.\end{aligned}$$

From (1.2.31), t is a sufficient statistic for θ. However, both t_1 and $t_{\max}$ mentioned at the beginning of this section are not sufficient for θ.

If $x_1, \cdots, x_n$ are i.i.d. r.v.s from $N(\mu, \sigma^2)$ with the mean μ known, then from the joint distribution, $S_\mu^2 = \sum_{i=1}^n (x_i - \mu)^2$ is a sufficient statistic for σ^2. When μ is unknown, $\boldsymbol{\theta} = (\mu, \sigma^2)$, the joint distribution can be factorized as

$$\begin{aligned}f_{\boldsymbol{\theta}}(\boldsymbol{x}) &= \left(\frac{1}{2\pi\sigma^2}\right)^{n/2} \exp\left\{-\frac{1}{2\sigma^2}\sum_{i=1}^n (x_i - \mu)^2\right\} \\ &= \left(\frac{1}{2\pi\sigma^2}\right)^{n/2} \exp\left\{-\frac{1}{2\sigma^2}\left[\sum_{i=1}^n x_i^2 - 2\mu\sum_{i=1}^n x_i + n\mu^2\right]\right\}.\end{aligned}$$

From Eq. (1.2.31), $(\sum_{i=1}^n x_i, \sum_{i=1}^n x_i^2)$ is a sufficient statistic for $\boldsymbol{\theta}$. Let $\boldsymbol{t}(\boldsymbol{x}) = (\bar{x}, S_{\bar{x}}^2)$. Since the one-to-one mapping of a sufficient statistic is still sufficient, $\boldsymbol{t}$ is also a sufficient statistic for $\boldsymbol{\theta}$.

Example 1.2.7. (Continuation of Example 1.2.5) Let $x_1, \cdots, x_n$ be a set of samples from $(\mathcal{X}, \mathcal{A}, \mathcal{P})$, where for $\forall\theta \in \Theta$, P_θ is a symmetric distribution, and f_θ denotes the pdf. Let $\boldsymbol{t}(\boldsymbol{x}) = \boldsymbol{y}$ be the order statistics of $\boldsymbol{x}$, then

$$f_\theta(\boldsymbol{x}) = g_\theta(\boldsymbol{y}) \cdot \frac{1}{n!},$$

where $g_\theta(\boldsymbol{y})$ is the pdf of $\boldsymbol{t}$, *i.e.*

$$g_\theta(\boldsymbol{y}) = \begin{cases} n! f_\theta(y_1, \cdots, y_n), & \text{if} \quad y_1 \le \cdots \le y_n, \\ 0, & \text{otherwise.} \end{cases}$$

From Eq. (1.2.31) we know that $\boldsymbol{t}$ is a sufficient statistic for θ. It can be seen that the decision method is easier to use than that of Example 1.2.5.

Example 1.2.8. Let $x_1, \cdots, x_n$ be a set of samples from uniform distribution $U(0, \theta)$, then the joint distribution is

$$f_\theta(\boldsymbol{x}) = \begin{cases} \dfrac{1}{\theta^n}, & \text{if} \quad \max\{x_i\} \le \theta, \\ 0, & \text{if} \quad \text{otherwise.} \end{cases}$$

Let $x_{(n)} = \max\{x_i\}$, then the above equation can be written as

$$f_\theta(\boldsymbol{x}) = \frac{1}{\theta^n} I_{[0,\theta]}(x_{(n)}).$$

From Eq. (1.2.31), $x_{(n)}$ is a sufficient statistic for θ.

Minimal sufficient statistics. From Eq. (1.2.31), for a given statistical space, we can construct many sufficient statistics. Consider the problem discussed in the Example 1.2.1, Example 1.2.4 and Example 1.2.6. Let $x_1, \cdots, x_n$ be i.i.d. from $N(\theta, 1)$, and $\boldsymbol{x} = (x_1, \cdots, x_n)$. Then the sample space is $(\mathcal{R}^n, \mathcal{B}^n)$, where $\mathcal{B}^n$ denotes σ-algebra generated by the Borel sets of the n-dimensional Euclidean space $\mathcal{R}^n$. By Example 1.2.4 and Example 1.2.6, we know that $t(\boldsymbol{x}) = \bar{x}$ is a sufficient statistic. The complete data set itself, $t_1(\boldsymbol{x}) = \boldsymbol{x}$ is obviously a sufficient statistic, which is called a **trivial sufficient statistic** since it does not compress the data set. In fact, let

$$\boldsymbol{t}_i(\boldsymbol{x}) = \left(\frac{1}{i}(x_1 + \cdots + x_i), x_{i+1}, \cdots, x_n\right),$$

where $i = 1, \cdots, n$. Then from Eq. (1.2.31), $\boldsymbol{t}_i(\boldsymbol{x})$ for $i = 1, \cdots, n$ are all sufficient statistics for θ. Which one is the best? What is the decision criterion?

Generally speaking, in all the sufficient statistics, the stronger the function of data compression a sufficient statistic has, the better it is, since its calculation is more convenient. Let $\mathcal{A}_i$ be a σ-algebra generated by $\boldsymbol{t}_i$ in $\mathcal{B}^n$, then essentially the $\mathcal{A}_i$ is a σ-algebra generated by the Borel sets in $(n-i+1)$-dimensional Euclidean space, obviously we have

$$\mathcal{A}_1 \supset \mathcal{A}_2 \supset \cdots \supset \mathcal{A}_n.$$

As far as the function of data compression is concerned, $\boldsymbol{t}_{i+1}$ is superior to $\boldsymbol{t}_i$. Especially $\boldsymbol{t}_{i+1}$ can be obtained by a function of $\boldsymbol{t}_i$, say let $\boldsymbol{t}_i(\boldsymbol{x}) = (y_1, \cdots, y_{n-i+1})$, and then

$$\boldsymbol{t}_{i+1}(\boldsymbol{x}) = \left(\frac{1}{i+1}(iy_1 + y_2), y_3, \cdots, y_{n-i+1}\right),$$

has $n-i$ components. The property can be abstracted as follows: let t be a sufficient statistic on $(\mathcal{A}, \mathcal{X}, \mathcal{P})$, if for any other sufficient statistic s, there exists a measurable mapping h, s.t.

$$t(x) = h(s(x)), \tag{1.2.32}$$

then t is called a **minimal sufficient statistic**. Roughly speaking, the minimal sufficient statistic is a sufficient statistic that cannot be compressed any more without losing information about the unknown parameter.

Bahadur (1954) proved that, for a given statistical space $(\mathcal{A}, \mathcal{X}, \mathcal{P})$, there exists a minimal sufficient statistic in $\mathcal{P}$ if the parameter space Θ is divisible for the quasi-distance defined in Example 1.1.13. However, the problem how to decide whether a statistic is sufficient or not has not been solved. This problem will be discussed in the next section.

1.3 Exponential Family and Completeness

Many probability density functions can be written in a uniform form, which have the same properties. A collection of the functions forms a so-called exponential family. Before discussing it in detail, we will review the Laplace transform first. Let $(\mathcal{X}, \mathcal{A}, \nu)$ be a measure space, t_i and h be $\mathcal{A}$-measurable functions, where $i = 1, \cdots, k$, if

$$f(u) = \int_{\mathcal{X}} h(x) \exp\left\{\sum_{i=1}^{k} u_i t_i(x)\right\} d\nu(x) \tag{1.3.1}$$

is finite, (1.3.1) is called a **generalized Laplace transform** (cf. Widder, 1946).

Let $\mathcal{U} = \{u; f(u) < \infty\}$, from the property that the exponential function is convex we know $\mathcal{U}$ is a convex set, *i.e.* for $\forall \alpha \in [0,1]$ and $u, v \in \mathcal{U}$, we have

$$\left| h(x) \exp\left\{ \sum_{i=1}^{k} (\alpha u_i + (1-\alpha) v_i) t_i(x) \right\} \right|$$
$$\leq \alpha \left| h(x) \exp\left\{ \sum_{i=1}^{k} u_i t_i(x) \right\} \right| + (1-\alpha) \left| h(x) \exp\left\{ \sum_{i=1}^{k} v_i t_i(x) \right\} \right|.$$

The integrability of the right side determines the integrability of the left. By the properties of the Laplace transform, we can get the following theorem.

Theorem 1.3.1. *The f is continuous at interior points in $\mathcal{U}$, and all its any-order partial derivatives with respect to u_i exist. Furthermore, the order of differentiation and integration can be interchanged, i.e.*

$$\frac{\partial f(u)}{\partial u_i} = \int_{\mathcal{X}} h(x) t_i(x) \exp\left\{ \sum_{i=1}^{k} u_i t_i(x) \right\} d\nu(x).$$

1.3.1 *Exponential Family*

Let $(\mathcal{X}, \mathcal{A}, \mathcal{P})$ be a statistical space, for a σ-finite measure ν on $(\mathcal{X}, \mathcal{A})$ we have $\mathcal{P} \ll \nu$. Let u, u_i be two functions in the parameter space Θ, $i = 1, \cdots, k$, where $u(\theta) > 0$ holds for $\forall \theta \in \Theta$. If for $\forall \theta \in \Theta$,

$$\frac{dP_\theta}{d\nu} = u(\theta) \exp\left\{ \sum_{i=1}^{k} u_i(\theta) t_i(x) \right\} h(x), \tag{1.3.2}$$

and the support set $\{x; dP_\theta/d\nu > 0\}$ is independent of θ, then $\mathcal{P}$ is called an **exponential family**. If $\Theta \subset R^k$, and for $\forall \theta \in \Theta$, we have $u_i(\theta) = \theta_i, i = 1, \cdots, k$, then Eq. (1.3.2) can be written as

$$\frac{dP_\theta}{d\nu} = u(\theta) \exp\left\{ \sum_{i=1}^{k} \theta_i t_i(x) \right\} h(x), \tag{1.3.3}$$

$\mathcal{P}$ is called a **natural exponential family**.

Example 1.3.1. Consider the normal distribution $N(\mu, \sigma^2)$, where $\boldsymbol{\theta} = (\mu, \sigma^2)$. The probability density is

$$\begin{aligned} f_{\boldsymbol{\theta}}(x) &= \frac{1}{\sqrt{2\pi}\sigma} \exp\left\{ -\frac{(x-\mu)^2}{2\sigma^2} \right\} \\ &= \frac{1}{\sqrt{2\pi}\sigma} \exp\left\{ -\frac{\mu^2}{2\sigma^2} \right\} \exp\left\{ -\frac{1}{2\sigma^2} x^2 + \frac{\mu}{\sigma^2} x \right\}. \end{aligned} \tag{1.3.4}$$

$\mathcal{P} = \{P_{\boldsymbol{\theta}};\ \boldsymbol{\theta} \in \boldsymbol{R} \times \boldsymbol{R}_+\}$ belongs to the exponential family with $t_1(x) = x, t_2(x) = x^2$. If let $\theta_1 = \mu/\sigma^2, \theta_2 = -1/(2\sigma^2)$, then (1.3.4) has the form (1.3.3), *i.e.*, the form of the natural exponential family. Solving inversely we have $\mu = -\theta_2/(2\theta_1), \sigma^2 = -1/(2\theta_1)$.

Example 1.3.2. Uniform distribution $U(0, \theta)$ does not belong to the exponential family, since its support set depends on θ. The density function of two-parameter exponential distribution is

$$f_{\theta}(x) = \begin{cases} \dfrac{1}{\sigma} \exp\left\{\dfrac{1}{\sigma}(x - \mu)\right\}, & \text{if} \quad x \geq \mu. \\ 0, & \text{otherwise.} \end{cases}$$

Obviously, it does not belong to the exponential family either, since its support set depends on θ, where $\theta = (\mu, \sigma)$. But when the parameter μ in the exponential distribution is known, then the above distribution becomes a one-parameter exponential distribution and belongs to the exponential family, with $t(x) = x - \mu$.

It is easy to verify that some well-known probability distributions (such as binomial, Poisson, multinomial, and multidimensional normal) all belong to the exponential family.

Since the natural exponential family is a probability measure, from Eq. (1.3.3) we have

$$\frac{1}{u(\theta)} = \int_{\mathcal{X}} h(x) \exp\left\{\sum_{i=1}^{k} \theta_i t_i(x)\right\} d\nu(x). \tag{1.3.5}$$

The set $S = \{\theta;\ 1/u(\theta) < \infty\}$ is called a **natural parameter space**.

Theorem 1.3.2. *Let $(\mathcal{X}, \mathcal{A}, \mathcal{P})$ be a statistical space. If $\mathcal{P}$ belongs to the natural exponential family, then*

(i) The natural parameter space is a convex set.

(ii) $u(\theta)$ is continuous at the interior points in S and its all partial derivatives with respect to θ_i exist.

(iii) For the interior points in S we have

$$E_{\theta} t_i(X) = -\frac{\partial}{\partial \theta_i} \ln u(\theta), \quad i = 1, \cdots, k;$$

$$V_{\theta} t_i(X) = -\frac{\partial^2}{\partial \theta_i^2} \ln u(\theta), \quad i = 1, \cdots, k;$$

$$CV_{\theta}(t_i(X), t_j(X)) = -\frac{\partial^2}{\partial \theta_i \partial \theta_j} \ln u(\theta), \quad i, j = 1, \cdots, k;\ i \neq j.$$

Proof. From Eq. (1.3.1), we know that Eq. (1.3.5) is a generalized Laplace transform. From Theorem 1.3.1, we know both (i) and (ii) hold. Now we prove (iii). Taking first partial derivative on both sides of Eq. (1.3.5), by Theorem 1.3.1, we have

$$-\frac{1}{u^2(\theta)} \cdot \frac{\partial u(\theta)}{\partial \theta_i} = \int_{\mathcal{X}} h(x) t_i(x) \exp\left\{\sum_{i=1}^{k} \theta_i t_i(x)\right\} d\nu(x).$$

Thus we can get $E_\theta t_i(X)$. Similarly, taking second partial derivative yields the other two equalities. □

Example 1.3.3. Consider a binomial distribution $X \sim Bi(n, \theta)$. The density function with respect to the Counting measure is

$$\begin{aligned} f_\theta(x) = P_\theta(X = x) &= \binom{n}{x} \theta^x (1-\theta)^{n-x} \\ &= (1-\theta)^n \exp\left[x \ln \frac{\theta}{1-\theta}\right] \binom{n}{x}. \end{aligned} \tag{1.3.6}$$

This is a form of the exponential family, furthermore, let

$$\omega = \ln \frac{\theta}{1-\theta}, \quad \text{and} \quad \theta = \frac{e^w}{1+e^w}, \tag{1.3.7}$$

Eq. (1.3.6) becomes

$$f_\omega(x) = (1+e^w)^{-n} e^{wx} \binom{n}{x}.$$

This is a form of natural exponential family with $t(x) = x$ and $u(w) = (1+e^w)^{-n}$. Applying Theorem 1.3.2, we have

$$E_w X = -\frac{\partial}{\partial w} u(w) = n \frac{e^w}{1+e^w}.$$

By Eq. (1.3.7) and Theorem 1.2.1, it can be transformed into $E_\theta X = n\theta$. This coincides with Example 1.1.1.

1.3.2 *Completeness*

For a given statistical space $(\mathcal{X}, \mathcal{A}, \mathcal{P})$, let $(\mathcal{Y}, \mathcal{B}, \mathcal{Q})$ be the induced statistical space by the statistic t. If t is a sufficient statistic for θ, from the discussion above, it suffices to consider measurable mappings and measurable functions based on t when we make inferences about θ. But now there are two problems that need to be considered further. The first one, if there are two integrable functions h_1 and h_2, s.t. $E_\theta h_1(t(X)) = E_\theta h_2(t(X))$ for

$\forall\theta \in \Theta$, whether does $h_1(t) = h_2(t)$ holds almost everywhere? The second one, how to judge whether t is the minimal sufficient statistic or not? Both problems involve the concept of completeness.

For a $\mathcal{B}$-measurable function g, if

$$E_\theta g(t(X)) = 0 \text{ for } \forall\theta \in \Theta,$$

then we must have

$$g(t(x)) = 0$$

a.e., then t is said to be a **complete statistic**. Let $\mathcal{A}_t$ be the induced sub σ-algebra of $\mathcal{A}$ by the statistic t, and then $\mathcal{A}_t$ is said to be **complete**.

Example 1.3.4. A Bernoulli experiment means that there are only two outcomes, success and failure. Let the probability of success be $\theta, 0 < \theta < 1$, and then the probability of failure is $1 - \theta$. If 1 denotes success, and 0 denotes failure, *i.e.* $P_\theta(X = 1) = \theta, P_\theta(X = 0) = 1 - \theta$. Repeat Bernoulli experiment for n times, then $t = x_1 + \cdots + x_n$ is a sufficient statistic for θ, and from the binomial distribution $Bi(n, \theta)$ (cf. Examples 1.1.1 and 1.3.3). If $g(t)$ satisfies $E_\theta g(t(X)) = 0$ for $\forall\theta \in (0, 1)$, or equivalently,

$$\sum_{t=0}^{n} g(t) \binom{n}{t} w^t = 0, \quad \forall w \in (0, \infty)$$

where $w = (1 - \theta)^{-1}\theta$. Since the left side of the equation is a polynomial of w, then the coefficients of all the polynomials must be zero, so $g(t) = 0, t = 0, 1, \cdots, n$. Since the statistic t is complete, t is then called a **complete sufficient statistic**.

Example 1.3.5. We have discussed the sufficiency of order statistics in Example 1.2.5, now we will discuss their completeness. Let $x_1, \cdots, x_n$ be an i.i.d. sample from $(\mathcal{R}, \mathcal{B}, \mathcal{P})$, where $\mathcal{B}$ is a σ-algebra generated by the Borel sets in Euclidean space, and $\mathcal{P}$ is a family of all continuous distributions, *i.e.*, $\mathcal{P} \ll \nu$, where ν is the Lebesgue measure. For $P \in \mathcal{P}$, let $dP/d\nu = f$, and then the statistical space is $(\mathcal{R}^n, \mathcal{B}^n, \mathcal{P}^n)$, and we have $P^n \in \mathcal{P}^n$,

$$\frac{dP^n}{d\nu^n} = \prod_{i=1}^{n} f(x_i). \tag{1.3.8}$$

Let $t = (y_1, \cdots, y_n)$ be the order statistics of $x_1, \cdots, x_n$, and $\mathcal{A}_t$ be a sub σ-algebra of $\mathcal{B}^n$ induced by t. Obviously the measure given by (1.3.8) is symmetrical, and all $\mathcal{A}_t$-measurable functions are also symmetrical about the coordinate axes. To verify the completeness of the order statistic t, we need to prove that, for an $\mathcal{A}_t$-measurable function g, if for $\forall P \in \mathcal{P}$ we have

$$\int_{\mathcal{R}^n} g(x_1, \cdots, x_n) \prod_{i=1}^{n} f(x_i) d\nu(x_i) = 0, \tag{1.3.9}$$

then for $\forall B \in \mathcal{A}_t$, we must have

$$\int_B g(x_1, \cdots, x_n) \prod_{i=1}^{n} d\nu(x_i) = 0. \tag{1.3.10}$$

In fact, we only need to prove the situation of $B = B_1 \times \cdots \times B_n$, where $B_j \in \mathcal{B}$ satisfies $\nu(B_j) > 0, j = 1, \cdots, n$. Define

$$f_j(x) = \frac{I_{B_j}(x)}{\nu(B_j)}.$$

This is also a continuous distribution. Let $f(x) = \sum \alpha_j f_j(x)$, where $\alpha_j > 0$ and $\sum \alpha_j = 1$. From (1.3.9) and the symmetry of g, we have

$$\int_B g(x_1, \cdots, x_n) \prod_{i=1}^{n} f(x_i) d\nu(x_i)$$
$$= \int_{\mathcal{R}^n} g(x_1, \cdots, x_n) \prod_{i=1}^{n} \Big[\sum_{j=1}^{n} \alpha_j f_j(x_i)\Big] d\nu(x_i) = 0.$$

This is a homogeneous polynomial of α_j, and from the arbitrariness of α_j and the symmetry of g, we have

$$\int_{\mathcal{R}^n} g(x_1, \cdots, x_n) \prod_{i=1}^{n} f_i(x_i) d\nu(x_i) = 0$$

i.e. Eq. (1.3.10) holds.

Example 1.3.6. Let $x_1, \cdots, x_n$ be a set of samples drawn from the uniform distribution $U(0, \theta)$, from Example 1.2.8, $t(\boldsymbol{x}) = \max\{x_i\}$ is a sufficient statistic for θ. We will prove t is also a complete statistic. From Example 1.2.8, we know the density function of t is

$$h_\theta(t) = \begin{cases} \dfrac{1}{\theta^n} n t^{n-1}, & \text{if} \quad 0 < t < \theta, \\ 0, & \text{otherwise.} \end{cases}$$

Let $g(t)$ be a measurable function satisfying $E_\theta(g(T)) = 0$ for any $\theta > 0$. Taking derivatives on both sides of the equation:

$$\begin{aligned}
0 &= \frac{\partial}{\partial\theta}\int_0^\theta g(t)\frac{1}{\theta^n}nt^{n-1}dt \\
&= \frac{1}{\theta^n}\frac{\partial}{\partial\theta}\int_0^\theta g(t)nt^{n-1}dt + \left(\frac{\partial}{\partial\theta}\left(\frac{1}{\theta^n}\right)\right)\int_0^\theta g(t)nt^{n-1}dt \\
&= \frac{\theta^{n-1}}{\theta^n}\, ng(\theta) + 0 \\
&= \frac{1}{\theta}ng(\theta).
\end{aligned}$$

Since $n/\theta \neq 0$, $g(\theta) = 0$. Thus $t(\boldsymbol{x}) = \max\{x_i\}$ is a complete sufficient statistic.

Now we will discuss the relation between the completeness and the minimal sufficient statistic.

Theorem 1.3.3. *Let t be a statistic in the statistical space $(\mathcal{X}, \mathcal{A}, \mathcal{P})$. If the sufficient statistic t is complete, then it must be a minimal sufficient statistic.*

Proof. Let $\mathcal{A}_t$ be a sub σ-algebra of $\mathcal{A}$ induced by t, and $\mathcal{A}_1$ be any sufficient region. From the definition of minimal sufficient statistic in (1.2.32), we need to prove for $\forall A_t \in \mathcal{A}_t$, there exists $A_1 \in \mathcal{A}_1$, s.t. we must have $P_\theta(A_t \triangle A_1) = 0$ for $\forall\theta \in \Theta$, where the definition has been given in Example 1.1.12.

Since both $\mathcal{A}_t$ and $\mathcal{A}_1$ are sufficient regions, then there exist the following conditional means independent of θ:

$$f(x) = E(I_{A_t}|\mathcal{A}_1, x)$$

$$g(x) = E(f(X)|\mathcal{A}_t, x),$$

where $f(x)$ and $g(x)$ are $\mathcal{A}_1$ and $\mathcal{A}_t$-measurable functions with the rang of [0,1], respectively. From Theorem 1.2.1, for $\forall\theta \in \Theta$

$$E_\theta I_{A_t}(X) = E_\theta f(X) = E_\theta g(X). \tag{1.3.11}$$

Since t is a complete statistic, from $E_\theta[I_{A_t}(X) - g(X)] = 0$, we have

$$I_{A_t}(x) = g(x)$$

a.e. From Theorem 1.2.2, we have

$$I_{A_t}(x) = I_{A_t}(x)\cdot I_{A_t}(x) = I_{A_t}(x)g(x) = E(I_{A_t}(X)f(X)|\mathcal{A}_t, x)$$

almost everywhere. From Theorem 1.2.1 again, we have

$$E_\theta I_{A_t}(X) = E_\theta I_{A_t}(X) f(X) \tag{1.3.12}$$

for $\forall \theta \in \Theta$. Combining (1.3.11) with (1.3.12), we have

$$E_\theta I_{A_t}(X)(1 - f(X)) = E_\theta f(X)(1 - I_{A_t}(X)) = 0.$$

Since the above integrands are nonnegative, then we have

$$I_{A_t}(x)(1 - f(x)) = f(x)(1 - I_{A_t}(x))$$

a.e., *i.e.* we have $I_{A_t}(x) = f(x)$ a.e. It suffices to let

$$A_1 = \{x;\ f(x) = 1\}$$

then we have $P_\theta(A_t \triangle A_1) = 0$ for $\forall \theta \in \Theta$. □

1.3.3 *Sufficiency and Completeness*

From Theorem 1.3.3, we can easily judge the minimal sufficiency of a statistic, which is favorable to make statistical inferences. Now we analyze the exponential family.

Let $x_1, \cdots, x_n$ be i.i.d. from a natural exponential family with the pdf form as (1.3.3). The statistical space is $(\mathcal{X}^n, \mathcal{A}^n, \mathcal{P}^n)$, where $\mathcal{P}^n \ll \nu^n$, and for $P_\theta^n \in \mathcal{P}^n$ we have

$$\frac{dP_\theta^n}{d\nu^n} = c^n(\theta) \exp\left\{\sum_{i=1}^k \theta_i \sum_{j=1}^n t_i(x_j)\right\} \prod_{j=1}^n h(x_j). \tag{1.3.13}$$

Theorem 1.3.4. *In Eq. (1.3.13), let*

$$\boldsymbol{t} = \Big(\sum_{j=1}^n t_1(x_j), \cdots, \sum_{j=1}^n t_k(x_j)\Big), \tag{1.3.14}$$

then the statistic $\mathbf{t}$ *is sufficient for* θ*, and is complete with respect to* $\mathcal{P}^n$*.*

Proof. From the Neyman's Factorization Theorem, the statistic $\boldsymbol{t}$ given by (1.3.14) is sufficient. Considering that an $\mathcal{A}^n$-measurable function g satisfying for $\forall \theta \in \Theta$, we have

$$\int_{\mathcal{X}^n} g(\boldsymbol{x}) c^n(\theta) \exp\left\{\sum_{i=1}^k \theta_i t_i^*(\boldsymbol{x})\right\} h^*(\boldsymbol{x}) d\nu^n(\boldsymbol{x}) = 0,$$

where $\boldsymbol{x} = (x_1, \cdots, x_n)$, $t_i^*(\boldsymbol{x}) = \sum_{j=1}^n t_i(x_j)$, and $h^*(\boldsymbol{x}) = \prod_{j=1}^n h(x_j)$. Compared with Eq. (1.3.1), the equation above is also a generalized Laplace transform, then we have $g(\boldsymbol{x}) = 0$ a.e. Thus $\boldsymbol{t}$ is a complete statistic. □

Applying Theorems 1.3.4 and 1.3.3, it is easy to get the minimal statistics of a series of probability distributions.

1.3.4 *Ancillary Statistics*

We have known that sufficient statistics are quite important, since they include all the information of the parameters. On the contrary, there exist some statistics with distributions independent of parameters, called ancillary statistics. Let $s(x)$ be a statistic from $(\mathcal{X}, \mathcal{A}, \mathcal{P})$ to $(\mathcal{S}, \mathcal{B}, \mathcal{Q})$, where $\mathcal{P} = \{P_\theta;\ \theta \in \Theta\}$, and $\mathcal{Q} = \{Q_\theta;\ \theta \in \Theta\}$. From the construction of statistical space, we know that $Q_\theta(B) = P_\theta(s^{-1}(B))$ for $\forall B \in \mathcal{B}$, $\forall\theta \in \Theta$. If for any θ and $B \in \mathcal{B}$, $Q_\theta(B)$ is independent of θ, then $s(x)$ is called an **ancillary statistic**. Although an ancillary statistic contains no information about the parameter, it has at least two advantages, one is its invariance to the parameter, and the other is its independence from sufficient statistics.

Example 1.3.7. (Ancillary statistics of location parameter) For given probability measure P_θ, its corresponding **Cumulative Distribution Function** is defined as $F_\theta(x) = P_\theta(X \le x)$. If $\mathcal{P} \ll \nu$, and $dP_\theta/d\nu = f_\theta$, then

$$F_\theta(x) = \int_{-\infty}^{x} f_\theta(y)d\nu(y). \tag{1.3.15}$$

Let $F(x)$ be a cdf, if for $\forall\theta \in \Theta$ we have $F_\theta(x) = F(x-\theta)$, then $\mathcal{P}$ is called a **location distribution family** and θ a **location parameter**.

Let $x_1, \cdots, x_n$ be a set of samples from $F(x-\theta)$, we will prove that $s(x) = \max\{x_i\} - \min\{x_i\}$ is an ancillary statistic. Let $z_1, \cdots, z_n$ denote a set of samples from $F(x)$, then $x_1 = z_1 + \theta, \cdots, x_n = z_n + \theta$. For any s we have

$$\begin{aligned} P_\theta(S(X) \le s) &= P_\theta(\max\{X_i\} - \min\{X_i\} \le s) \\ &= P_\theta(\max\{Z_i + \theta\} - \min\{Z_i + \theta\} \le s) \\ &= P_\theta(\max\{Z_i\} - \min\{Z_i\} \le s). \end{aligned}$$

Obviously the probability distribution does not depend on the parameter θ. From the arbitrariness of s, we know that $s(x)$ is an ancillary statistic. Usually, $s(x)$ is called the **sample range**.

Example 1.3.8. (Ancillary statistics of scale parameter) $\mathcal{P}$ with cdf in the form of $F(x/\sigma)$ is called a **scale distribution family**, and σ a **scale parameter**, where $\sigma > 0$. Let $x_1, \cdots, x_n$ be a set of samples, we will prove that $\boldsymbol{s}(\boldsymbol{x}) = (x_1/x_n, \cdots, x_{n-1}/x_n)$ is an ancillary statistic.

Let $z_1, \cdots, z_n$ denote a set of samples from $F(x)$, then we have $x_1 = \sigma z_1, \cdots, x_n = \sigma z_n$. For any $\boldsymbol{s} = (s_1, \cdots, s_{n-1})$ we have

$$\begin{aligned} P_\sigma(\boldsymbol{S}(\boldsymbol{X}) \le \boldsymbol{s}) &= P_\sigma(X_1/X_n \le s_1, \cdots, X_{n-1}/X_n \le s_{n-1}) \\ &= P_\sigma(Z_1/Z_n \le s_1, \cdots, Z_{n-1}/Z_n \le s_{n-1}), \end{aligned}$$

which is independent of σ.

The theorem below shows the relationship between ancillary statistics and complete sufficient statistics.

Theorem 1.3.5. *Let $(\mathcal{X}, \mathcal{A}, \mathcal{P})$ be a statistical space, where $\mathcal{P} = \{P_\theta; \theta \in \Theta\}$. Let $T(x)$ be a complete sufficient statistic, and $S(x)$ be an ancillary statistic, then $T(X)$ and $S(X)$ are mutually independent (denoted as $T(X) \perp\!\!\!\perp S(X)$).*

Proof. Let $\mathcal{A}_t$ and $\mathcal{A}_s$ be sub σ-algebras of $\mathcal{A}$ induced by $T(x)$ and $S(x)$, respectively. We need to prove that for any $\theta \in \Theta, A_t \in \mathcal{A}_t$, and $A_s \in \mathcal{A}_s$ we have

$$P_\theta(A_t \cap A_s) = P_\theta(A_t)P_\theta(A_s). \tag{1.3.16}$$

From the condition, $P_\theta(A_s)$ is independent of θ, let $P_\theta(A_s) = a$. Let

$$f(x) = P(A_s|\mathcal{A}_t, x) = E(I_{A_s}(X)|\mathcal{A}_t, x).$$

Since $T(x)$ is a sufficient statistic, and the above conditional probability is independent of the parameter θ. From Theorem 1.2.1, for $\forall \theta \in \Theta$,

$$E_\theta f(X) = E_\theta E(I_{A_s}(X)|\mathcal{A}_t, X) = E_\theta(I_{A_s}(X)) = a.$$

In other words, we have $E(f(X)-a) = 0$ for $\forall \theta \in \Theta$. From the completeness of $\mathcal{A}_t$, we have $f(x) = a$ a.e. with respect to $\mathcal{A}_t$ and $\mathcal{P}$. Then from (1.2.4)

$$P_\theta(A_t \cap A_s) = \int_{A_t} P(A_s|\mathcal{A}_t, x)dP_\theta(x) = aP_\theta(A_t) = P_\theta(A_s)P_\theta(A_t).$$

□

Theorem 1.3.5 is usually called as Basu's Lemma. For further discussion about ancillary statistics and Basu's Lemma, see Basu (1958, 1959), Koehn and Thomas (1975), and Lehmann (1980, 1986). In some situations, it is convenient to prove the independence by using Basu's Lemma, here is an example.

Example 1.3.9. (Independence of the normal sample mean, sample range and sample variance) Let $x_1, \cdots, x_n$ be a set of samples from $N(\mu, \sigma^2)$. Let $\bar{X}$, L and S^2 respectively denote the sample mean, the sample range and the sample variance. For any given σ_0^2, $\bar{X}$ is a complete sufficient statistic for μ, and both L and S^2 are ancillary statistics for μ, by Basu's Lemma, $\bar{X}$, L, and S^2 are mutually independent. By the arbitrariness, we know the result holds for any σ^2.

1.4 Estimation Methods Based on Statistical Space

Although we mainly focus on the study of theory and methods of parameter tests, we will review some estimation methods at first, since the methods of tests are usually based on the estimation of parameter. Furthermore, we will analyze some basic properties of estimation procedures.

Essentially, estimation methods can be classified into two categories: one is that the form of distribution is partly or completely unknown, thus we can only estimate some typical characteristics of the distributions (such as the mean, the variance, and the median), or estimate the cdf itself and so on; the other is that the form of distribution is known, thus we can estimate the parameters in the distribution. Generally speaking, there is no advantage without disadvantage. In principle, we may grasp more information in the second category, and thus we can obtain more accurate estimates. However, if the estimation method depends on the distribution too heavily, then the estimate may behave badly once the information about the the distribution is unreliable (*i.e.*, there are some deviations in the form of distribution). In other words, the estimation method is not robust enough. We will discuss the problem in this section.

During the following discussion, it is usual to suppose that $x_1, \cdots, x_n$ is an i.i.d. sample from $(\mathcal{X}, \mathcal{A}, \mathcal{P})$. If we let $x = (x_1, \cdots, x_n)$, then the statistical space has the form $(\mathcal{X}^n, \mathcal{A}^n, \mathcal{P}^n)$, where $\mathcal{P}^n = \{P^n; P \in \mathcal{P}\}$. Now, we will discuss the first category of estimation methods.

1.4.1 *Moment Estimation and Median Estimation*

Since moments are very important tools for judging the characteristics of distribution of random variable, it is very reasonable to use the sample moments to estimate population moments when the information about the distribution of random variable is not sufficient. Let $x_1, \cdots, x_n$ be a set of samples, and M_s and m_s denote the s-th population moments and sample moments, respectively, *i.e.*

$$M_s = EX_1^s, \quad \text{and} \quad m_s = \frac{1}{n}\sum_{i=1}^{n} x_i^s. \tag{1.4.1}$$

m_s is called a **moment estimator** of M_s. More generally, if some parameter θ can be denoted as a function of moments, say

$$\theta = g(M_1, \cdots, M_k), \tag{1.4.2}$$

then $\hat{\theta} = g(m_1, \cdots, m_k)$ is called a moment estimator of θ.

Example 1.4.1. The moment estimator of the mean μ is

$$\widehat{\mu} = m_1 = \bar{x} = \frac{1}{n}\sum_{i=1}^{n} x_i. \tag{1.4.3}$$

Since the variance σ^2 can be denoted as $\sigma^2 = M_2 - M_1^2$, from (1.4.2), the moment estimator of the variance is

$$\hat{\sigma}^2 = m_2 - m_1^2 = \frac{1}{n}\sum_{i=1}^{n} x_i^2 - \bar{x}^2 = \frac{1}{n}\sum_{i=1}^{n}(x_i - \bar{x})^2. \tag{1.4.4}$$

Since the moment estimator is also a random variable, we usually need to calculate the mean and variance of the moment estimator to test the effect of the estimator. Recall (1.1.5) and (1.1.6), we have

$$E\bar{X} = \frac{1}{n}\sum_{i=1}^{n} EX_i = \mu, \tag{1.4.5}$$

and

$$V\bar{X} = E(\bar{X} - \mu)^2 = \frac{1}{n^2}\sum_{i=1}^{n} E(X_i - \mu)^2 = \frac{\sigma^2}{n}. \tag{1.4.6}$$

Compared with Example 1.2.1, the above result coincides with that of normal distribution. From Eq. (1.4.5), we know it is always reasonable to estimate the population mean by the sample mean whatever the distribution is, since the mean of the estimator is consistent to the population mean. This property is called **unbiasedness**, which will be discussed further later. Equation (1.4.6) shows that it is more accurate to estimate the population mean by the sample mean than by using each observation value alone.

Example 1.4.2. Let $x_1, \cdots, x_n$ be i.i.d. Bernoulli trials with success probability p. From Example 1.3.4, we know that $x = x_1 + \cdots + x_n$ has the binomial distribution $Bi(n,p)$. From Example 1.1.1, $E_pX = np$, the moment estimator of p is $\hat{p} = x/n$. In many practical problems, we usually need to consider the ratio of success to failure, *i.e.*

$$\omega = \frac{p}{1-p}, \tag{1.4.7}$$

which is called **odds**. From Eq. (1.4.2), the estimator of ω is

$$\hat{\omega} = \frac{\hat{p}}{1-\hat{p}} = \frac{x}{n-x}.$$

If we let $y_1, \cdots, y_m$ be i.i.d. Bernoulli trials with success probability q, then $y = y_1 + \cdots + y_m$ has the binomial distribution $Bi(m, q)$. Its odds is

$$\nu = \frac{q}{1-q},$$

and the estimator of ν is $\hat{\nu} = y/(m-y)$. $\theta = \omega/\nu$, the ratio of one odds to another, is a powerful statistic to measure the size of the two success probabilities, which will be called **odds ratio**. From Eq. (1.4.2), its estimator can be written as

$$\hat{\theta} = \frac{\hat{\omega}}{\hat{\nu}} = \frac{x(m-y)}{(n-x)y}. \tag{1.4.8}$$

Obviously we can claim that $q = p$ when $\hat{\theta} = 1$, and that $q > p$ when $\hat{\theta} > 1$. However, since $\hat{\theta}$ is a random variable, the probability of the event $\{\hat{\theta} = 1\}$ occurring is very small even if the true parameter is exactly 1. Then how to compare the size relation of p to q by using $\hat{\theta}$? This is a problem that will be discussed in the next chapter.

Besides the mean and the variance, the median is also an important statistical index of centrality. Let $(\mathcal{X}, \mathcal{A}, \mathcal{P})$ be a statistical space, where $\mathcal{P} = \{P_\theta; \theta \in \Theta\}$ is a continuous distribution, *i.e.* $\mathcal{P} \ll \nu$ with ν a Lebesgue measure. If θ satisfies the following equation

$$F_\theta(\theta) = P_\theta(X \leq \theta) = \frac{1}{2}, \tag{1.4.9}$$

then θ is called a **median**, in other words, the median is the solution of $F_\theta(\theta) = 1/2$. Since it is a continuous distribution, the solution exists uniquely with probability one. Especially when the distribution is symmetrical, the median coincides with the mean.

Let $\boldsymbol{x} = (x_1, \cdots, x_n)$ be an i.i.d. sample from the statistical space $(\mathcal{X}, \mathcal{A}, \mathcal{P})$, and $x_{(1)} < \cdots < x_{(n)}$ denote the order statistics, and the **sample median** is defined as

$$m_{\boldsymbol{x}} = \begin{cases} x_{(n+1)/2}, & \text{when } n \text{ is odd,} \\ \frac{1}{2}\left[x_{(n/2)} + x_{(n/2+1)}\right], & \text{when } n \text{ is even.} \end{cases} \tag{1.4.10}$$

We can estimate the population median by the sample median, *i.e.* $\hat{\theta} = m_{\boldsymbol{x}}$.

For the given $\theta \in \Theta$, let $S(\theta) = \#(x_i \leq \theta) = \#(x_{(i)} \leq \theta)$ denote the the number of x_i's that are less than or equal to θ, which is a non-decreasing step function of θ. Since θ is the median, based on the idea of moment estimation, we can regard the estimator of θ as the solution of

$$S(\theta) = n - S(\theta), \tag{1.4.11}$$

i.e. $S(\hat{\theta}) = n/2$. We can get that $\hat{\theta} = x_{((n+1)/2)}$ when n is odd, and that $\hat{\theta}$ can be any number between $x_{(n/2)}$ and $x_{(n/2+1)}$ when n is even, which coincides with the sample median m_x defined by (1.4.10).

We have known that the mean coincides with the median when the distribution is symmetrical, such as normal distribution. So, logically, we can estimate the mean by the sample median. And then which estimator is better? We will discuss this problem in the last part of this chapter.

1.4.2 *Maximum Likelihood Estimators*

Essentially, the maximum likelihood method discusses the problem of parameter estimation when the pdf is given. The method is generally credited to Fisher, although its roots date back as far as Lambert, Daniel Bernoulli, and Lagrange in the eighteenth century (Scholz, 1985). It is by far the most popular general method of estimation in statistics.

Let $(\mathcal{X}, \mathcal{A}, \mathcal{P})$ be a statistical space, where $\mathcal{P} = \{P_\theta; \theta \in \Theta\}$ and $\mathcal{P} \ll \nu$. For $\theta \in \Theta$, let $dP_\theta/d\nu = f_\theta$. Furthermore, let $X_1, \cdots, X_n$ be an i.i.d. sample from the distribution with pdf $f_\theta(x)$, and its corresponding observation is $x = (x_1, \cdots, x_n)$. Thus we can get the values of $f_\theta(x_1), \cdots, f_\theta(x_n)$, where θ is unknown. The basic idea of maximum likelihood method is to find the parameter θ such that the values of the density functions attain their maxima. Here, the joint pdf of $\boldsymbol{x} = (x_1, \cdots, x_n)$ is

$$L(\boldsymbol{x}; \theta) = \prod_{i=1}^{n} f_\theta(x_i), \tag{1.4.12}$$

or by the monotonicity of log function,

$$l(\boldsymbol{x}; \theta) = \ln L(\boldsymbol{x}; \theta) = \sum_{i=1}^{n} \ln f_\theta(x_i). \tag{1.4.13}$$

$L(\boldsymbol{x}; \theta)$ and $l(\boldsymbol{x}; \theta)$ are called a **likelihood function** and **log-likelihood Function** of the sample respectively. $\hat{\theta}$ is called a **maximum likelihood estimator** (MLE)of the parameter θ, if $\hat{\theta} \in \Theta$ and satisfies

$$L(\boldsymbol{x}; \hat{\theta}) = \sup_{\theta \in \Theta} L(\boldsymbol{x}; \theta) \quad \text{or} \quad l(\boldsymbol{x}; \hat{\theta}) = \sup_{\theta \in \Theta} l(\boldsymbol{x}; \theta). \tag{1.4.14}$$

Obviously, if the pdf is derivable with respect to θ_i, and the solution of

$$\frac{\partial}{\partial \theta_i} l(\boldsymbol{x}; \theta) = 0 \tag{1.4.15}$$

lies in the parameter space Θ, then the MLE of θ is the solution of Eq. (1.4.15).

Example 1.4.3. Let $x_1, \cdots, x_n$ be a set of samples from the normal distribution $N(\mu, \sigma^2)$ to calculate the MLE of the parameters. It is easy to get that

$$l(\boldsymbol{x};\boldsymbol{\theta}) = -\frac{n}{2}\ln(2\pi\sigma^2) - \frac{1}{2\sigma^2}\sum_{i=1}^{n}(x_i-\mu)^2,$$

where $\boldsymbol{\theta} = (\mu, \sigma^2)$. Taking partial derivatives with respect to μ and σ^2, we have

$$\frac{\partial}{\partial\mu}l(\boldsymbol{x};\mu,\sigma^2) = \frac{1}{\sigma^2}\sum_{i=1}^{n}(x_i-\mu) = 0, \tag{1.4.16}$$

and

$$\frac{\partial}{\partial\sigma^2}l(\boldsymbol{x};\mu,\sigma^2) = -\frac{n}{2\sigma^2} + \frac{1}{2\sigma^4}\sum_{i=1}^{n}(x_i-\mu)^2 = 0. \tag{1.4.17}$$

Notice that Eq. (1.4.16) implies that we always have $\hat{\mu} = \bar{x}$ irrespective of the variance. Substituting it into Eq. (1.4.17) yields $\hat{\sigma}^2 = 1/n\sum_{i=1}^{n}(x_i - \bar{x})^2$.

However it needs to prove that $(\bar{x}, \hat{\sigma}^2)$ is the MLE of θ, since the log-likelihood function is not a concave function of θ. Recall Example 1.3.1, we take the following transformation

$$\theta_1 = \frac{\mu}{\sigma^2}, \quad t_1(\boldsymbol{x}) = \sum_{i=1}^{n} x_i;$$

$$\theta_2 = -\frac{1}{2\sigma^2}, \quad t_2(\boldsymbol{x}) = \sum_{i=1}^{n} x_i^2.$$

Then the likelihood function can be written as a natural exponential family,

$$L(\boldsymbol{x};\boldsymbol{\theta}^*) = u(\boldsymbol{\theta}^*)\exp\{\theta_1 t_1(\boldsymbol{x}) + \theta_2 t_2(\boldsymbol{x})\},$$

where $\boldsymbol{\theta}^* = (\theta_1, \theta_2)$ and $u(\boldsymbol{\theta}^*) = (-\theta_2/\pi)^{n/2}\exp\{n\theta_1^2/(4\theta_2)\}$. Then the log-likelihood function is

$$l(\boldsymbol{x};\boldsymbol{\theta}^*) = \ln u(\boldsymbol{\theta}^*) + \theta_1 t_1(\boldsymbol{x}) + \theta_2 t_2(\boldsymbol{x}).$$

From Theorem 1.3.2 we know $l(\boldsymbol{x};\boldsymbol{\theta}^*)$ is a concave function of $\boldsymbol{\theta}^*$, so the MLE of $\boldsymbol{\theta}^*$ exists uniquely, which is the solution of the equations below,

$$\begin{cases} E_{\boldsymbol{\theta}^*}t_1(\boldsymbol{X}) = n\mu = t_1(\boldsymbol{x}), \\ E_{\boldsymbol{\theta}^*}t_2(\boldsymbol{X}) = n(\sigma^2+\mu^2) = t_2(\boldsymbol{x}). \end{cases}$$

We have $\hat{\mu} = \bar{x}$ and $\hat{\sigma}^2 = \sum_{i=1}^{n}(x_i - \bar{x})^2/n$. It can be seen that this coincides with the moment estimators. It is necessary to point out that when parameter spaces are different, estimation methods may be different too. Take the consideration of the parameter space with the restriction of the mean being nonnegative as an example.

$$\Theta_1 = \{(\mu, \sigma^2); 0 \leq \mu < +\infty, 0 < \sigma^2 < \infty\}$$

then computing the MLE of μ is equivalent to computing

$$\min_{\mu \geq 0}(\bar{x} - \mu)^2.$$

Then the MLE of μ is $\hat{\mu}^* = \max\{\bar{x}, 0\}$. Notice that

$$\begin{aligned}(\hat{\mu} - \mu)^2 &= (\bar{x} - \mu)^2 \\ &= (\bar{x} - \hat{\mu}^*)^2 + 2(\bar{x} - \hat{\mu}^*)(\hat{\mu}^* - \mu) + (\hat{\mu}^* - \mu)^2\end{aligned}$$

for $\forall \mu \geq 0$, where the cross-product term satisfies

$$(\bar{x} - \hat{\mu}^*)\hat{\mu}^* = 0, \tag{1.4.18}$$

and

$$(\bar{x} - \hat{\mu}^*)\mu \leq 0. \tag{1.4.19}$$

For $\forall \mu \geq 0$ we have $(\hat{\mu} - \mu)^2 \geq (\hat{\mu}^* - \mu)^2$, thus

$$E_{\boldsymbol{\theta}}(\hat{\mu} - \mu)^2 > E_{\boldsymbol{\theta}}(\hat{\mu}^* - \mu)^2$$

holds for $\forall \boldsymbol{\theta} \in \Theta_1$. This result shows that when the restriction condition is true, the deviation of the MLE is smaller than that without any restriction. Lee (1981) had ever calculated the deviation.

Example 1.4.4. Let $\boldsymbol{x} = (x_1, \cdots, x_k)$ be a set of samples from a multinomial distribution $M(n; p_1, \cdots, p_k)$. There are two methods to get the MLE of the parameter $\boldsymbol{\theta} = (p_1, \cdots, p_k)$.

(i) Let $p_k = 1 - p_1 - \cdots - p_{k-1}$. From Example 1.1.3,

$$l(\boldsymbol{x}; \boldsymbol{\theta}) = \sum_{i=1}^{k-1} x_i \ln p_i + x_k \ln(1 - p_1 - \cdots - p_{k-1}) + c,$$

where c is a constant independent of the parameter. Solving Eq. (1.4.15) yields

$$\frac{x_i}{p_i} = \frac{x_k}{1 - p_1 - \cdots - p_{k-1}}, \quad i = 1, \cdots, k-1.$$

Notice that $x_1 + \cdots + x_k = n$, then the MLE of p_i is

$$\hat{p}_i = \frac{x_i}{n}, i = 1, \cdots, k.$$

(ii) Applying the Lagrange multiplier method. The objective function is

$$H(\boldsymbol{x}; \boldsymbol{\theta}, \lambda) = \sum_{i=1}^{k} x_i \ln p_i + \lambda(1 - p_1 - \cdots - p_k),$$

where λ is the Lagrange multiplier. Our aim is to solve the equations

$$\begin{cases} \dfrac{\partial}{\partial p_i} H(x; \theta, \lambda) = \dfrac{x_i}{p_i} - \lambda = 0, \quad i = 1, \cdots, k, \\ \dfrac{\partial}{\partial \lambda} H(x; \theta, \lambda) = 1 - p_1 - \cdots - p_k = 0. \end{cases}$$

After summation, we have $\sum\limits_{i=1}^{k} x_i = \lambda \sum\limits_{i=1}^{k} p_i$, *i.e.* $\lambda = n$. Then the MLE of p_i is $\hat{p}_i = \dfrac{x_i}{n}$, $i = 1, \cdots, k$.

Then the reasonability about the idea of maximum likelihood method will be discussed. Let $L_n(x; \theta)$ denote the likelihood function when the sample size is n, then we have the following theorem.

Theorem 1.4.1. *Let $(\mathcal{X}, \mathcal{A}, \mathcal{P})$ be a statistical space, $\mathcal{P} \ll \nu$ and $dP_\theta / d\nu = f_\theta$. If θ and f_θ are one-to-one, and the true value of the parameter θ_0 is an interior point of Θ, then for $\forall \theta \neq \theta_0$ we have*

$$\lim_{n \to \infty} P_{\theta_0} \{ L_n(\boldsymbol{X}; \theta_0) > L_n(\boldsymbol{X}; \theta) \} = 1.$$

Proof. From the definition of likelihood function, we know that

$$L_n(\boldsymbol{x}; \theta_0) > L_n(\boldsymbol{x}; \theta) \quad \text{if and only if} \quad \frac{1}{n} \sum_{i=1}^{n} \ln \frac{f_\theta(x_i)}{f_{\theta_0}(x_i)} < 0.$$

Since θ_0 is an interior point of Θ, the above equation can be written as a form of integration when $n \to \infty$, *i.e.*

$$E_{\theta_0} \ln \frac{f_\theta(X)}{f_{\theta_0}(X)} < 0.$$

Since θ and f_θ are one-to-one, and $\ln x$ is a concave function, by Jensen inequality the left side of the above equation is strictly smaller than

$$\ln E_{\theta_0} \frac{f_\theta(X)}{f_{\theta_0}(X)} = 0.$$

Therefore this completes the proof of the theorem. □

Above theorem shows that when n is quite large, the likelihood function attains its maximum at the true value of parameter with probability one. Though we do not know the true parameter in the estimation problem, it is generally reasonable to obtain the MLE based on the sample just like the idea of moment method. For the same estimation method, generally speaking, the bigger the sample size, the better the estimator. We will exemplify the proposition by applying the maximum likelihood method. Let $\hat{\theta}$ be an estimator of θ. For $\theta \in \Theta$, $E_\theta(\hat{\theta} - \theta)^2$ is called the **mean squared error** of $\hat{\theta}$. Obviously the mean squared error is an important index to measure the difference between the estimator and the true value. The smaller the mean squared error, the better the estimator.

Theorem 1.4.2. *Let x_{n+m} denote a set of samples in size $n + m$ from the statistical space $(\mathcal{X}, \mathcal{A}, \mathcal{P})$, which can be divided into two non-empty parts $x_{n+m} = (x_n, x_m)$. If $\mathcal{P}$ belongs to the natural exponential family (cf. (1.3.3)), and $E_\theta t_i(X)$ is a linear function of θ, say $h(\theta)$, then for any interior point θ of Θ we have*

$$E_\theta(\hat{\theta}_{n+m} - \theta)^2 < \min\{E_\theta(\hat{\theta}_n - \theta)^2,\ E_\theta(\hat{\theta}_m - \theta)^2\},$$

where $\hat{\theta}_s$ denotes the MLE of θ when the sample size is s.

Proof. Let $l(x;\theta)$ denote the log-likelihood function of the sample x, then we have

$$l(x_{n+m};\theta) = l(x_n;\theta) + l(x_m;\theta).$$

From Theorem 1.3.2, $\hat{\theta}_s$ is a solution of $E_\theta t_i(X_s) = t_i(x_s)$. Let $h(\theta)$ be the linear function as given in the assumption, then

$$\begin{aligned} h(\hat{\theta}_{n+m}) &= \frac{1}{n+m} E_{\hat{\theta}_{n+m}} t_i(X_{n+m}) \\ &= \frac{1}{n+m} t_i(x_{n+m}) \\ &= \frac{1}{n+m} t_i(x_n) + \frac{1}{n+m} t_i(x_m) \\ &= \frac{1}{n+m} E_{\hat{\theta}_n} t_i(X_n) + \frac{1}{n+m} E_{\hat{\theta}_m} t_i(X_m) \\ &= \frac{n}{n+m} h(\hat{\theta}_n) + \frac{m}{n+m} h(\hat{\theta}_m). \end{aligned}$$

According to the assumption of linearity of h, we have

$$\hat{\theta}_{n+m} = \frac{n}{n+m}\hat{\theta}_n + \frac{m}{n+m}\hat{\theta}_m. \tag{1.4.20}$$

Without the loss of generality, suppose that $\hat{\theta}_1$ is not a constant. Then we will apply the mathematical induction method to prove

$$E_\theta(\hat{\theta}_s - \theta)^2 = \frac{1}{s} E_\theta(\hat{\theta}_1 - \theta)^2 + \frac{s-1}{s}[E_\theta(\hat{\theta}_1 - \theta)]^2. \tag{1.4.21}$$

Notice that when $E_\theta \hat{\theta}_s = \theta$, the last term in the above equation is zero. At first, we will prove that (1.4.21) holds for $s = 2$.

$$\begin{aligned} E_\theta(\hat{\theta}_2 - \theta)^2 &= E_\theta\left(\frac{1}{2}\hat{\theta}_1 + \frac{1}{2}\hat{\theta}_1^* - \theta\right)^2 \\ &= E_\theta\left[\frac{1}{4}(\hat{\theta}_1 - \theta)^2 + \frac{1}{2}(\hat{\theta}_1 - \theta)(\hat{\theta}_1^* - \theta) + \frac{1}{4}(\hat{\theta}_1^* - \theta)^2\right] \\ &= \frac{1}{2}E_\theta(\hat{\theta}_1 - \theta)^2 + \frac{1}{2}[E_\theta(\hat{\theta}_1 - \theta)]^2, \end{aligned}$$

where $\hat{\theta}_1^*$ and $\hat{\theta}_1$ are the two MLEs using different samples, obviously they are i.i.d. And hence when $s = 2$, Eq. (1.4.21) holds. We will prove the case for $s = p + 1$ assuming that Eq. (1.4.21) holds for $s = p$. By Eq. (1.4.20) and the induction assumption, we have

$$\begin{aligned} E_\theta(\hat{\theta}_{p+1} - \theta)^2 &= E_\theta\left(\frac{p}{p+1}\hat{\theta}_p + \frac{1}{p+1}\hat{\theta}_1\right)^2 \\ &= \left(\frac{p}{p+1}\right)^2 E_\theta(\hat{\theta}_p - \theta)^2 + \frac{2p}{(p+1)^2}E_\theta(\hat{\theta}_p - \theta)E_\theta(\hat{\theta}_1 - \theta) \\ &\quad + \left(\frac{1}{p+1}\right)^2 E_\theta(\hat{\theta}_1 - \theta)^2 \\ &= \frac{p+1}{(p+1)^2}E_\theta(\hat{\theta}_1 - \theta)^2 + \left(\frac{p}{p+1}\right)^2 \cdot \frac{p-1}{p}[E_\theta(\hat{\theta}_1 - \theta)]^2 \\ &\quad + \frac{2p}{(p+1)^2}[E_\theta(\hat{\theta}_1 - \theta)]^2 \\ &= \frac{1}{p+1}E_\theta(\hat{\theta}_1 - \theta)^2 + \frac{p}{p+1}[E_\theta(\hat{\theta}_1 - \theta)]^2. \end{aligned}$$

Then Eq. (1.4.21) holds. From Eq. (1.4.21) we can get

$$E_\theta(\hat{\theta}_p - \theta)^2 - E_\theta(\hat{\theta}_{p+1} - \theta)^2 = \frac{1}{p(p+1)}[E_\theta(\hat{\theta}_1 - \theta)^2 - (E_\theta(\hat{\theta}_1 - \theta))^2].$$

By the Cauchy-Schwarz Inequality, the proof is completed. □

Remark 1.4.1. Let $x_1, \cdots, x_n$ be an independent and identically distributed random sample with density $f(\cdot, \theta)$, where θ is an unknown parameter, and let $\hat{\theta}_n$ denote the maximum likelihood estimator of θ based on the sample. Now, suppose that we obtain an additional observation, say x_{n+1}, from the same distribution. In this case, $\hat{\theta}_{n+1}$ denotes the maximum likelihood estimator of θ based on the $n+1$ observations. Theorem 1.4.2 tells us that $\hat{\theta}_{n+1}$ is preferred to $\hat{\theta}_n$ as estimation of θ under some given conditions. Here, we conjecture that this result still holds for *any* maximum likelihood estimator under some regular conditions such as MSE($\hat{\theta}$) is finite (Shi, 2008).

1.4.3 *Quality of Estimators*

From the above discussion, in principle, we can construct many estimators, then what properties should a good estimator have? Furthermore, how to measure the quality of estimator? Obviously, a good estimator must be a function of sufficient statistics to guarantee that no information about the parameters is lost. Besides that, it should also possess the following properties.

(1) **Unbiasedness.** Let $\boldsymbol{x} = (x_1, \cdots, x_n)$ be a set of samples from the statistical space $(\mathcal{X}, \mathcal{A}, \mathcal{P})$, and $T_n(\boldsymbol{x})$ an estimator of θ. We know that the probability of $T_n(\boldsymbol{x}) = \theta$ is quite small almost for any distribution, which motivates us to consider the average situations. If for $\forall\theta \in \Theta$ we have

$$E_\theta T_n(\boldsymbol{X}) = \theta,$$

then $T_n(\boldsymbol{x})$ is called an **unbiased estimator** of θ. For a function of parameter, we can give a similar definition. Let g be a function of θ, if $E_\theta T_n(\boldsymbol{X}) = g(\theta)$ for $\forall\theta \in \Theta$, then $T_n(\boldsymbol{x})$ is called an unbiased estimator of $g(\theta)$. Consider the limiting behavior of the mean, if for $\forall\theta \in \Theta$ we have

$$\lim_{n\to\infty} E_\theta T_n(\boldsymbol{X}) = \theta,$$

then $T_n(\boldsymbol{x})$ is called an **asymptotic unbiased estimator** of θ.

(2) **Consistency.** As we have mentioned above, the probability of $T_n(\boldsymbol{x}) = \theta$ is quite small. However, we may hope that $T_n(\boldsymbol{x})$ approximates to θ with a higher probability when n is large enough, otherwise the estimator will become meaningless. For this we introduce the following criterion based on the convergence in probability: if for $\forall\varepsilon > 0$, we have

$$\lim_{n\to\infty} P_\theta(|T_n(\boldsymbol{X}) - \theta| \geq \varepsilon) = 0, \quad \forall\theta \in \Theta,$$

then $T_n(\boldsymbol{x})$ is called a **consistent estimator** of θ, and denoted as $T_n \xrightarrow{P_\theta} \theta$. Furthermore, we have the following theorem.

Theorem 1.4.3. *Let $T_n(\boldsymbol{x})$ be a consistent estimator of θ. If g is a continuous function of θ, then $g(T_n)$ is a consistent estimator of $g(\theta)$.*

Proof. By the continuity of function, we know that for $\forall \varepsilon > 0$, there exists $\delta > 0$, s.t. $|g(T_n) - g(\theta)| < \varepsilon$ when $|T_n(\boldsymbol{x}) - \theta| < \delta$. Then when $n \to \infty$,

$$\begin{aligned}1 &\geq P_\theta(|g(T_n) - g(\theta)| \leq \varepsilon) \\ &\geq P_\theta(|T_n(X) - \theta| \leq \delta) \to 1, \quad \forall \theta \in \Theta,\end{aligned}$$

or equivalently, $g(T_n) \xrightarrow{P_\theta} g(\theta)$. □

(3) **Asymptotic normality.** For an estimator $T_n(\boldsymbol{x})$, besides the consideration of its approximation to the true parameter θ, we should take the distribution of the deviation $T_n(\boldsymbol{x}) - \theta$ into account as a basic idea in statistics. There are two main reasons. One is that we can determine the speed of convergence about the consistent estimator, and the other is that the limiting distribution should be a normal with mean zero if the deviate is induced by random errors (refer to Problem 1.2). Therefore the property of convergence in distribution is called asymptotic normality. If for any sample size n, there exists a function $\sigma_n^2(\theta)$ of θ, s.t. for any x when $n \to \infty$ we have

$$F_n(x) \to \Phi(x), \tag{1.4.22}$$

where $\Phi(x)$ is the cdf of the standard normal distribution $N(0,1)$ and $F_n(x)$ is the cdf of the following r.v.

$$Z_n = \frac{T_n - \theta}{\sigma_n(\theta)},$$

then $T_n(\boldsymbol{x})$ is called an **asymptotic normal estimator** of θ, and $\sigma_n^2(\theta)$ the **asymptotic variance** of $T_n(\boldsymbol{x})$. If Z denotes an r.v. from $N(0,1)$, then (1.4.22) can also be denoted as $Z_n \xrightarrow{L} Z$.

Example 1.4.5. (Discussing the properties of moment estimators) From Example 1.4.1 we know that the moment estimators of the mean μ and the variance σ^2 are given by $\bar{x}$ in (1.4.3) and $\hat{\sigma}^2$ in (1.4.4) respectively. Let $\boldsymbol{\theta} = (\mu, \sigma^2)$ and $\hat{\boldsymbol{\theta}} = (\bar{x}, \hat{\sigma}^2)$. From the Law of Large Numbers, $\hat{\boldsymbol{\theta}}$ is a consistent estimator of $\boldsymbol{\theta}$. From (1.4.5), (1.4.6) and the Central Limit Theorem we know that, for $\forall \boldsymbol{\theta} \in \Theta$, when $n \to \infty$ we have

$$P_{\boldsymbol{\theta}}\left\{\frac{\sqrt{n}(\bar{X} - \mu)}{\sigma} \leq x\right\} \to \Phi(x).$$

Thus $\bar{x}$ is an asymptotic normal estimator of μ. When σ^2 is unknown, since $\hat{\sigma}^2/\sigma^2 \to 1$ when $n \to \infty$, the above equation can be written as,

$$P_{\boldsymbol{\theta}}\left\{\frac{\sqrt{n}(\bar{X} - \mu)}{\hat{\sigma}} \leq x\right\} \to \Phi(x). \tag{1.4.23}$$

Since $E_{\boldsymbol{\theta}}\hat{\sigma}^2 = (n-1)\sigma^2/n$ for $\forall\theta$, then $\hat{\sigma}^2$ is not an unbiased estimator of σ^2. An unbiased estimator of σ^2 should be

$$\frac{n}{n-1}\hat{\sigma}^2 = \frac{1}{n-1}\sum_{i=1}^{n}(x_i-\bar{x})^2. \tag{1.4.24}$$

Example 1.4.6. (Discussing the properties of median estimators) We will utilize the sample median $\hat{\theta}_n$ to estimate the median θ. Since $\hat{\theta}_n$ is a function of order statistics, from Example 1.3.5, it is a function of sufficient statistics. From the Law of Large Numbers we know that, $\hat{\theta}_n$ is a consistent estimator of θ. Then we will discuss its asymptotic normality.

Let $Z_n = \sqrt{n}(\hat{\theta}_n - \theta)$. For $\forall x \in \mathcal{X}$ we have

$$\begin{aligned} P_\theta(Z_n \le x) &= P_\theta\left\{\sqrt{n}(\hat{\theta}_n-\theta)\le x\right\} \\ &= P_\theta\left\{\hat{\theta}_n \le \theta + \frac{x}{\sqrt{n}}\right\} \\ &= P_\theta\left\{S\left(\theta+\frac{x}{\sqrt{n}}\right)\ge\frac{n}{2}\right\}, \end{aligned}$$

where $S(y) = \#(x_i \le y)$ is given by (1.4.11). Notice that $S(\theta + x/\sqrt{n}) = \sum\limits_{i=1}^{n} I_{\{x_i\le\theta+x/\sqrt{n}\}}$, let

$$Y_{n_i} = I_{\{x_i\le\theta+x/\sqrt{n}\}} - F\left(\theta+\frac{x}{\sqrt{n}}\right),$$

and

$$t_n = \frac{1}{\sqrt{n}}\left[\frac{n}{2} - nF\left(\theta+\frac{x}{\sqrt{n}}\right)\right],$$

then we have

$$\begin{aligned} S\left(\theta+\frac{x}{\sqrt{n}}\right)\ge\frac{n}{2} &\Leftrightarrow \sum_{i=1}^{n} I_{\left\{x_i\le\theta+\frac{x}{\sqrt{n}}\right\}} - nF\left(\theta+\frac{x}{\sqrt{n}}\right) \ge \frac{n}{2} - nF\left(\theta+\frac{x}{\sqrt{n}}\right) \\ &\Leftrightarrow \sum_{i=1}^{n} Y_{n_i} \ge \sqrt{n}t_n. \end{aligned}$$

Notice that $S(y)$ has a binomial distribution $Bi(n,p)$ with $p = F(y)$, it is easy to verify that

$$E_\theta Y_{n_i} = F(\theta+\frac{x}{\sqrt{n}}) - F(\theta+\frac{x}{\sqrt{n}}) = 0,$$

and

$$\begin{aligned} V_\theta Y_{n_i} &= E_\theta Y_{n_i}^2 \\ &= F\left(\theta + \frac{x}{\sqrt{n}}\right)\left[1 - F\left(\theta + \frac{x}{\sqrt{n}}\right)\right] \\ &= F(\theta)(1 - F(\theta)) + o\left(\frac{1}{\sqrt{n}}\right) \\ &= \frac{1}{4} + o\left(\frac{1}{\sqrt{n}}\right), \end{aligned}$$

and

$$\begin{aligned} t_n &= \frac{1}{\sqrt{n}}\left\{\frac{n}{2} - n\left[F(\theta) + F'(\theta)\frac{x}{\sqrt{n}} + o\left(\frac{1}{\sqrt{n}}\right)\right]\right\} \\ &= \frac{1}{\sqrt{n}}\left\{\frac{n}{2} - \frac{n}{2} - \sqrt{n}xf(\theta) - n \cdot o\left(\frac{1}{\sqrt{n}}\right)\right\} \\ &= -xf(\theta) + o(1). \end{aligned}$$

From the Central Limit Theorem, we have

$$\begin{aligned} \lim_{n\to\infty} P_\theta(Z_n \le x) &= \lim_{n\to\infty} P_\theta\left\{\frac{1}{\sqrt{n/4}}\sum_{i=1}^n Y_{n_i} \ge \frac{1}{\sqrt{1/4}}t_n\right\} \\ &\doteq 1 - \Phi(-2xf(\theta)) \\ &= \Phi(2xf(\theta)). \end{aligned}$$

Therefore $\sqrt{n}(\hat{\theta}_n - \theta) \xrightarrow{L} N\left(0, \frac{1}{4f^2(\theta)}\right)$.

Example 1.4.7. (Discussing the properties of MLE) From the Neyman's Factorization Theorem we know that the MLE is a function of sufficient statistics. Especially for an exponential family, from Theorem 1.3.4, the MLE is a function of complete sufficient statistics. From Theorem 1.4.1, we may infer that the MLE $\hat{\theta}_n$ is a consistent estimator of the parameter θ. Now we discuss its asymptotic normality. For $\theta \in \Theta$, let

$$I(\theta) = E_\theta\left(\frac{\partial \ln f_\theta(X)}{\partial\theta}\right)^2. \tag{1.4.25}$$

Usually $I(\theta)$ is called the Fisher information, since it is related to the variance of statistics. From Eq. (1.4.13), we can calculate that

$$\begin{aligned} E_\theta \frac{\partial \ln L(\boldsymbol{X};\theta)}{\partial\theta} &= \sum_{i=1}^n \int \frac{\partial \ln f_\theta(x_i)}{\partial\theta} f_\theta(x_i) d\mu(\boldsymbol{x}) \\ &= \sum_{i=1}^n \int \frac{\partial f_\theta(x_i)}{\partial\theta} d\mu(\boldsymbol{x}) \\ &= \sum_{i=1}^n \frac{\partial}{\partial\theta} \int f_\theta(x_i) d\mu(\boldsymbol{x}) = 0. \end{aligned}$$

From the independence we can get,

$$\begin{aligned} V_\theta \frac{\partial \ln L(\boldsymbol{X};\theta)}{\partial\theta} &= E_\theta \left(\frac{\partial \ln L(\boldsymbol{X};\theta)}{\partial\theta} \right)^2 \\ &= E_\theta \left(\sum_{i=1}^n \frac{\partial \ln f_\theta(X_i)}{\partial\theta} \right)^2 = nI(\theta). \end{aligned} \tag{1.4.26}$$

If $f_\theta(x)$ has the third-order derivative with respect to θ, then we have

$$\sqrt{n}(\hat{\theta} - \theta_0) \xrightarrow{L} N(0, I^{-1}(\theta_0)), \tag{1.4.27}$$

where θ_0, denoting the true value of parameter, is an interior point of Θ. Then we will prove Eq. (1.4.27). Taking a Taylor series expansion of $h(\boldsymbol{x};\theta) = \dfrac{\partial}{\partial\theta} \ln L(\boldsymbol{x};\theta)$ at θ_0, we have

$$h(\boldsymbol{x};\theta) = h(\boldsymbol{x};\theta_0) + (\theta - \theta_0)\frac{\partial}{\partial\theta} h(\boldsymbol{x};\theta_0) + o(|\theta - \theta_0|).$$

Substituting $\hat{\theta}_n$ for θ in the above equation yields

$$0 = h(\boldsymbol{x};\hat{\theta}_n) = h(\boldsymbol{x};\theta_0) + (\hat{\theta}_n - \theta_0)\frac{\partial}{\partial\theta} h(\boldsymbol{x};\theta_0) + o(|\hat{\theta}_n - \theta_0|).$$

Since $\hat{\theta}_n$ is a consistent estimator of θ_0, by neglecting the higher-order infinitesimal term, and solving the above equation, we can get

$$\sqrt{n}(\hat{\theta}_n - \theta_0) = -\sqrt{n} \left[\frac{\partial}{\partial\theta} h(\boldsymbol{x};\theta_0) \right]^{-1} h(\boldsymbol{x};\theta_0).$$

Then we will calculate the limiting distribution of the right side of the above equation. Since

$$\begin{aligned} \frac{1}{n}\left[\frac{\partial}{\partial\theta} h(\boldsymbol{x};\theta)\right] &= \frac{\partial^2}{\partial\theta^2} \left[\frac{1}{n} \sum_{i=1}^n \ln f_\theta(x_i) \right] \\ &\xrightarrow{P_\theta} \frac{\partial^2}{\partial\theta^2} E_\theta \ln f_\theta(X) \\ &= E_\theta \frac{\partial^2 \ln f_\theta(X)}{\partial\theta^2} \\ &= -E_\theta \left(\frac{\partial \ln f_\theta(X)}{\partial\theta} \right)^2, \end{aligned}$$

by Eq. (1.4.25), we have

$$-\frac{1}{n}\left[\frac{\partial}{\partial\theta}h(\boldsymbol{x};\theta_0)\right] \xrightarrow{P_{\theta_0}} I(\theta_0). \tag{1.4.28}$$

Similar to Eq. (1.4.26), we can get that

$$E_{\theta_0}h(X;\theta_0)=0,$$

and

$$V_{\theta_0}h(X;\theta_0)=nI(\theta_0).$$

By the Central Limit Theorem,

$$\sqrt{n}\cdot\frac{1}{n}h(\boldsymbol{x};\theta_0) \xrightarrow{L} N(0,I(\theta_0)).$$

Then we complete the proof of (1.4.27) by (1.4.28).

Similarly, we can get the corresponding result when the parameter is a multidimensional vector. Let the parameter space Θ be a subset of $\boldsymbol{R}^k$, and the true value $\boldsymbol{\theta}_0$ is an interior point of Θ, then we also have

$$\sqrt{n}(\hat{\boldsymbol{\theta}}-\boldsymbol{\theta}_0) \xrightarrow{L} N(\boldsymbol{0},\boldsymbol{I}_k^{-1}(\boldsymbol{\theta}_0)), \tag{1.4.29}$$

where $\boldsymbol{I}_k(\boldsymbol{\theta}_0)$ is a $k\times k$ positive definite matrix with (i,j)-element

$$I_{ij}(\boldsymbol{\theta}_0)=E_{\boldsymbol{\theta}_0}\left(\frac{\partial\ln f_{\boldsymbol{\theta}}(X)}{\partial\theta_i}\right)\left(\frac{\partial\ln f_{\boldsymbol{\theta}}(X)}{\partial\theta_j}\right), \quad i,j=1,\cdots,k.$$

Usually $\boldsymbol{I}_k(\boldsymbol{\theta}_0)$ is called the **Fisher information matrix**.

1.4.4 *Comparison of Estimators*

In previous subsection, we have discussed three properties that a good estimator should have. In fact, many estimators satisfy the three properties. Consequently, it is important to make a comparison of estimators. Essentially, the properties are mainly related to the means of estimators. Now, we will pay much attention to the variances of estimators when we make a comparison among them.

Let $\mathcal{T}$ denote the set of all unbiased estimators. For T_1 and $T_2\in\mathcal{T}$, if $V_\theta T_1(X)\le V_\theta T_2(X)$ for $\forall\theta\in\Theta$, then we say that T_1 is superior to T_2. If $V_\theta T^*(X)\le V_\theta T(X)$ for $\forall T\in\mathcal{T}$ and $\forall\theta\in\Theta$, then $T^*\in\mathcal{T}$ is called a **uniformly minimal variance unbiased estimator (UMVUE)**.

Theorem 1.4.4. *If the UMVUE exists, then it uniquely exists with probability one.*

Proof. Let T_1 and T_2 be UMVUEs, then for $\forall\theta \in \Theta$, we have

$$E_\theta T_1 = E_\theta T_2 = \theta,$$

$$V_\theta T_1 = V_\theta T_2.$$

For $\forall\theta \in \Theta$, we have $E_\theta(T_1 - T_2) = 0$. By the Chebychev inequality, it suffices to prove that

$$\begin{aligned} 0 = V_\theta(T_1 - T_2) &= E_\theta(T_1 - T_2)^2 \\ &= E_\theta T_1(T_1 - T_2) - E_\theta T_2(T_1 - T_2). \end{aligned}$$

Thus, it suffices to prove that for $\forall\theta \in \Theta$, we have $E_\theta T_1(T_1 - T_2) = 0$. If there exists $\theta_0 \in \Theta$ s.t. $E_{\theta_0} T_1(T_1 - T_2) \neq 0$, let $\lambda = E_{\theta_0} T_1(T_1 - T_2)/E_{\theta_0}(T_1 - T_2)^2$ and $T_\lambda = T_1 - \lambda(T_1 - T_2)$. It is easy to verify that $T_\lambda \in \mathcal{T}$, and

$$\begin{aligned} E_{\theta_0} T_\lambda^2 &= E_{\theta_0} T_1^2 - 2\lambda E_{\theta_0} T_1(T_1 - T_2) + \lambda^2 E_{\theta_0}(T_1 - T_2)^2 \\ &< E_{\theta_0} {T_1}^2. \end{aligned}$$

This contradicts with that T_1 is a UMVUE. □

We have ever discussed that when constructing estimators, sufficient statistics, especially complete sufficient statistics should be taken into consideration. The following two theorems will explain that from another point of view.

Theorem 1.4.5. *Let $S(x)$ be a sufficient statistic for θ. For $\forall T \in \mathcal{T}$, let*

$$T^*(x) = E(T(X)|S(x)), \tag{1.4.30}$$

then $T^* \in \mathcal{T}$, and $V_\theta T^* \leq V_\theta T$ for $\forall\theta \in \Theta$.

Proof. Since $S(x)$ is a sufficient statistic for θ, the conditional mean in (1.4.30) is independent of the parameter θ. From Theorem 1.2.1, for $\forall\theta \in \Theta$ we have

$$E_\theta T^*(X) = E_\theta E(T(X)|S(X)) = E_\theta T(X) = \theta.$$

Thus $T^* \in \mathcal{T}$. From Theorems 1.2.1 and 1.2.2, for $\forall\theta \in \Theta$

$$\begin{aligned} &E_\theta(T(X) - T^*(X))(T^*(X) - \theta) \\ &\qquad = E_\theta E_\theta[(T(X) - T^*(X))(T^*(X) - \theta)|S(X)] \\ &\qquad = E_\theta\{(T^*(X) - \theta)E[(T(X) - T^*(X))|S(X)]\} \\ &\qquad = 0. \end{aligned}$$

Then

$$\begin{aligned} V_\theta T(X) &= E_\theta(T(X) - \theta)^2 \\ &= E_\theta(T^*(X) - \theta)^2 + E_\theta((T(X) - T^*(X))^2 \\ &\geq V_\theta T^*(X). \end{aligned}$$

□

Theorem 1.4.6. *If a complete sufficient statistic exists for θ, then there exists a UMVUE for θ that is a function of the complete sufficient statistic.*

Proof. Let $S(x)$ be a complete sufficient statistic for θ. For $T \in \mathcal{T}$, let

$$T^*(x) = E(T(X)|S(x)).$$

For $\forall T_1 \in \mathcal{T}$, from Theorem 1.4.4, it suffices to prove that $V_\theta T^*(X) \leq V_\theta T_1^*(X)$ for $\forall \theta \in \Theta$, where $T_1^*(x) = E(T_1(X)|S(x))$.

Since $E_\theta[T^*(X) - T_1^*(X)] = \theta - \theta = 0$, by the completeness of statistical space, we have

$$T^*(x) = T_1^*(x)$$

almost everywhere (a.e.). This completes the proof of the theorem. □

From the above theorem and Theorem 1.3.4, there always exists a UMVUE for the exponential family, and it is a function of complete sufficient statistics. We will discuss a more general class of comparison problems of estimators. Let $\mathcal{T}^*$ denote the set of all the asymptotic unbiased estimators, *i.e.*

$$\mathcal{T}^* = \{T_n(x);\ \lim_{n\to\infty} E_\theta T_n(x) = \theta, V_\theta T_n(X) < \infty, \forall \theta \in \Theta\}.$$

Let $\sigma_n^2(\theta) = V_\theta T_n(X)$. From the Central Limit Theorem,

$$\frac{1}{\sigma_n(\theta)}(T_n - \theta) \xrightarrow{L} N(0,1). \tag{1.4.31}$$

Thus $\sigma_n^2(\theta)$ is called an asymptotic variance, and T_n an asymptotic normal estimator of θ. We know from the discussion in the previous section that, $\sigma_n(\theta)$ satisfying (1.4.31) may be not unique, but if $\sigma_n'(\theta)$ also satisfies (1.4.31), we must have

$$\frac{\sigma_n(\theta)}{\sigma_n'(\theta)} \to 1 \quad \text{as} \quad n \to \infty. \tag{1.4.32}$$

Thus, in the sense of (1.4.32), we can regard the asymptotic variance of the asymptotic normal estimator to be unique. So, we can compare estimators in virtue of the asymptotic variance. Let $T_{1n}(x)$ and $T_{2n}(x)$ be two asymptotic normal estimators for θ, and their asymptotic variances be σ_{1n}^2 and σ_{2n}^2 respectively. Then

$$e = e(T_{1n}, T_{2n}) = \lim_{n\to\infty} \frac{1/\sigma_{1n}^2}{1/\sigma_{2n}^2} = \lim_{n\to\infty} \frac{\sigma_{2n}^2}{\sigma_{1n}^2} \tag{1.4.33}$$

is called the **asymptotic relative efficiency** of T_{1n} with respect to T_{2n}. Obviously, when $e(T_{1n}, T_{2n}) > 1$, the estimator T_{1n} is superior to T_{2n}. If $x_1, \cdots, x_n$ is an i.i.d. sample satisfying $E_\theta X_i = \theta$ and $V_\theta X_i = \sigma^2(\theta) < \infty$. From the Central Limit Theorem,

$$\frac{\sqrt{n}(\overline{X} - \theta)}{\sigma(\theta)} \xrightarrow{L} N(0, 1).$$

This shows that the asymptotic variance has the order $1/n$, *i.e.*

$$n\sigma_n^2(\theta) \to \sigma^2(\theta), \quad \text{as} \quad n \to \infty.$$

Therefore, when the sample sizes of the two estimators are different (say, n_1 and n_2, respectively), the relation of the two sample sizes is approximately

$$e = \frac{n_1}{n_2} \Longrightarrow n_1 = en_2$$

in order to reach the same asymptotic relative efficiency. When $e < 1$, the sample size of T_{2n} should be larger than that of T_{1n} in order to reach the same efficiency.

Example 1.4.8. Let $x_1, \cdots, x_n$ be a set of samples from $N(\theta, 1)$ to estimate the parameter θ. It can be seen that both the sample mean $\bar{x}_n$ and the sample median m_n are reasonable estimators for θ, and belong to $\mathcal{T}^*$. The asymptotic variance of $\bar{x}_n$ is $1/n$. From Example 1.4.6, when $\theta = 0$, the asymptotic variance of m_n is

$$\frac{1}{4nf^2(0)} = \frac{2\pi}{4n} = \frac{\pi}{2n}.$$

Then $e = e(\bar{x}_n, m_n) = 2/\pi < 1$, and thus $\bar{x}_n$ is superior to m_n. To reach the same efficiency, we must have

$$n_1 = \frac{2}{\pi} n_2,$$

i.e., the sample size of m_n is $\pi/2 \approx 1.57$ times that of $\bar{x}_n$. This shows that $\bar{x}_n$ is superior to m_n as far as estimation of the mean θ is concerned. But when considering from the point of view of **robustness**, we may get a different result. Consider

$$X_1, \cdots, X_n \overset{\text{i.i.d.}}{\sim} (1 - \alpha)\Phi(x - \theta) + \alpha\Phi\left(\frac{x - \theta}{3}\right),$$

where $\alpha \in (0, 1)$. This is a mixture normal distribution, which is still symmetrical about θ. Its pdf is

$$f_\alpha(x) = (1 - \alpha)\phi(x - \theta) + \alpha\frac{1}{3}\phi\left(\frac{x - \theta}{3}\right),$$

where $\phi(x)$ is the pdf of the standard normal distribution. The asymptotic variance of $\bar{x}_n$ is

$$\sigma_{1n}^2 = \frac{1}{n}[(1-\alpha)+9\alpha] = \frac{1}{n}(1+8\alpha),$$

and the asymptotic variance of m_n is

$$\begin{aligned}\sigma_{2n}^2 &= \frac{1}{4nf_\alpha^2(0)} \\ &= \frac{1}{4n}\left[\frac{1-\alpha}{\sqrt{2\pi}} + \frac{\alpha}{9\sqrt{2\pi}}\right]^{-2} \\ &= \frac{\pi}{2n}\left[\frac{9}{9-8\alpha}\right]^2.\end{aligned}$$

It is easy to verify that, when $\alpha = 1/8$,

$$\sigma_{1n}^2 = \frac{2}{n} > \sigma_{2n}^2 \approx \frac{1.9}{n}.$$

This shows that m_n is more robust than $\bar{x}_n$ when the sample distribution has some fluctuations.

In the discussion about the parameter estimation in a statistical space, we can see that it is usual to analyze and compare estimators in different angles. Especially in practical application, we should make a concrete analysis of concrete problems.

1.4.5 *Nonparametric MLE for Population cdf*

Now we will study how to estimate a cdf when its form is completely unknown. Recall the definition of cdf in Example 1.3.7, that is, for the given probability measure P, the corresponding cdf is $F(x) = P(X \leq x)$. Obviously, the cdf is a nondecreasing right-continuous function, *i.e.* the left-hand limit $F(x-0)$ at x may not be equal to the function value $F(x)$ itself. Let $\mathcal{F}$ denote the set of all the cdfs. Let $x_1, \cdots, x_n$ be a set of samples from F, where $F \in \mathcal{F}$, we would like to estimate F by the sample. Recall the definition of likelihood function in (1.4.12), we may define a nonparametric likelihood function as

$$L(F) = \prod_{i=1}^{n}[F(x_i) - F(x_i - 0)]. \tag{1.4.34}$$

If $\hat{F} \in \mathcal{F}$ and satisfies

$$L(\hat{F}) = \sup_{F\in\mathcal{F}} L(F), \tag{1.4.35}$$

then $\hat{F}$ is called a **nonparametric maximum likelihood estimator (NMLE)** of F.

Now we discuss how to solve Eq. (1.4.35). Recall the discussion about the order statistics in Examples 1.2.5 and 1.3.5, let $y_1 \le \cdots \le y_n$ be the order statistics of $x_1, \cdots, x_n$, define

$$F_n(x) = \begin{cases} 0, & \text{if } \ x < y_1, \\ \dfrac{k}{n}, & \text{if } \ y_k \le x < y_{k+1}, \\ 1, & \text{if } \ x \ge y_n. \end{cases} \tag{1.4.36}$$

Obviously, $0 \le F_n(x) \le 1$, and it is a non-decreasing right-continuous function about x, hence $F_n \in \mathcal{F}$. F_n is called an **empirical distribution function**, which is a function of complete sufficient statistics. The following theorem shows that F_n is the NMLE of F, *i.e.* F_n is a solution of Eq. (1.4.35).

Theorem 1.4.7. *For any $F \in \mathcal{F}$, if $F \ne F_n$, then*

$$L(F) < L(F_n). \tag{1.4.37}$$

Proof. Suppose that there are m different values in the sample $\{x_1, \cdots, x_n\}$, and their corresponding order statistics are $y_1^* < \cdots < y_m^*$. Let n_j denote the number of the samples that are equal to y_j^*, *i.e.* $n_j = \#(x_i = y_j^*)$, $i = 1, \cdots, n$ and $j = 1, \cdots, m$. Let $p_j = F(y_j^*) - F(y_j^* - 0)$ and $\hat{p}_j = n_j/n$. If there exists j s.t. $p_j = 0$, by the definition of Eq. (1.4.34), Equation (1.4.37) holds. So we can assume that all $p_j > 0$, then obviously $\sum_{j=1}^m p_j \le 1$.

Since for any $z \ge 0$ we have $\ln z \le z - 1$, where the equality holds if and only if $z = 1$. From the given conditions we know that, there exists at least one j s.t. $p_j \ne \hat{p}_j$, thus we can get

$$\begin{aligned} \ln[L(F)/L(F_n)] &= \sum_{j=1}^m n_j \ln(p_j/\hat{p}_j) \\ &< \sum_{j=1}^m n_j (p_j/\hat{p}_j - 1) \\ &= n \sum_{j=1}^m \hat{p}_j (p_j/\hat{p}_j - 1) \\ &\le 0. \end{aligned}$$

This completes the proof of the theorem. □

By the definition we can see that the empirical distribution function $F_n(x)$ denotes the frequency of the samples not exceeding x, and hence for the given x, $nF_n(x)$ has a binomial distribution $Bi(n,p)$, where $p = F(x)$ and F denotes the true cdf. Thus

$$EF_n(x) = F(x), \tag{1.4.38}$$

$$VF_n(x) = \frac{1}{n}F(x)[1 - F(x)]. \tag{1.4.39}$$

Notice that Eq. (1.4.38) shows that the NMLE of F is unbiased. From the Law of Large Numbers and the Central Limit Theorem, we can obtain the consistency and asymptotic normality of F_n, *i.e.*

$$P\left\{\lim_{n\to\infty} F_n(x) = F(x)\right\} = 1, \tag{1.4.40}$$

and

$$\frac{\sqrt{n}[F_n(x) - F(x)]}{\sqrt{F(x)[1 - F(x)]}} \longrightarrow N(0,1). \tag{1.4.41}$$

In fact, we can obtain a stronger version than Eq. (1.4.40). Let

$$D_n = \sup_x |F_n(x) - F(x)|$$

then we have

$$P\{\lim_{n\to\infty} D_n = 0\} = 1. \tag{1.4.42}$$

Notice that Eq. (1.4.42) states that the empirical distribution function $F_n(x)$ converges to the true cdf $F(x)$ with probability one. Usually, the result is called the Glivenko-Cantelli Theorem. For more detailed discussion, see Loève (1963).

By the empirical distribution function, we can make a further analysis of the moment estimators and the median estimators. Recall Eq. (1.4.1), by substituting the empirical distribution function for the cdf, we get

$$\int x^s dF_n(x) = \frac{1}{n}\sum_{i=1}^{n} x_i^s = m_s.$$

So the moment estimator can also be regarded as an estimator obtained by the empirical distribution function. For the median, the function in Eq. (1.4.11) is equivalent to

$$S(x) = nF_n(x).$$

Therefore the solution of $S(\theta) = n/2$ is exactly the sample median.

1.5 Problems

1.1. For a given statistical space $(\mathcal{X}, \mathcal{A}, \mathcal{P})$ with $\mathcal{P} = \{P_\theta; \theta \in \Theta\}$, let $\mathcal{A}_1$ and $\mathcal{A}_2$ be two sub σ-algebras of $\mathcal{A}$ satisfying $\mathcal{A}_1 \subset \mathcal{A}_2$, and $h(x)$ an $\mathcal{A}$-measurable function. Prove that for $\forall\theta \in \Theta$ we have

$$E_\theta[E_\theta(h(X)|\mathcal{A}_2, x)|\mathcal{A}_1, x] = E_\theta(h(X)|\mathcal{A}_1, x)$$

almost everywhere. Furthermore, explain the minimal sufficient statistics using the above result.

1.2. **(Theory of errors based on the normal distribution)** Let x denote the true but known length of an object. Let x_i denote the i-th measurement result with measurement error $\varepsilon_i = x_i - x$, $i = 1, 2, \cdots, n$. Here, both the ε_i's and the x_i's are r.v.'s. Assume that

(a) There has no systematic error, *i.e.*, the mean of the n measurement results, $\bar{x} = \sum_{i=1}^{n} x_i/n$, is equal to the true length x;

(b) The ε_i's are mutually independent;

(c) The ε_i's have a common distribution with density function $f(\cdot)$.

Verify that ε_i has a normal distribution with mean 0 based on the idea the maximum likelihood method.

1.3. **(The distribution of shoot deviation)** Consider a shooting contest where one aims a bullet at a target. The coordinate system is set up on the target plane where the bull's-eye is the origin O. Let the point where the bullet hits the target be (X, Y). Here, the deviations X and Y are two random variables. Suppose that the following conditions are satisfied:

(a) The pdfs $p(x)$ and $q(y)$ of X and Y are continuous, and $p(0)q(0) > 0$;

(b) X and Y are mutually independent;

(c) The value of joint distribution of X and Y at (x, y) depends only on the distance $r = \sqrt{(x^2 + y^2)}$ between this point and the origin point.

Then both X and Y have a normal distribution with standard deviation $\sigma > 0$, *i.e.* they have the same pdf as follows:

$$p(x) = \frac{1}{\sigma\sqrt{2\pi}} \exp\left\{-\frac{x^2}{2\sigma^2}\right\}.$$

1.4. **(Poisson process)** Consider the number of radioactive particles emitted in a unit of time. Assume that it satisfies the following three properties:

(a) **(Stationary increments)** The number of particles emitted in $[t_0, t_0 + t)$ depends only on the length t but is independent of the starting time t_0. If $P_k(t)$ denotes the probability that there are k particles emitted in a given interval of length t, then

$$\sum_{k=0}^{\infty} P_k(t) = 1$$

holds for any t. This property shows the probability distribution does not change with time.

(b) **(Independent increments or without after-effects)** The event of k particles arriving at the given region in $[t_0, t_0 + t)$ is independent of the event occurring before t_0. The property shows that the numbers of particles in two disjoint intervals are independent. Independent Increment indicates the processes are independent in the disjoint time intervals.

(c) **(Orderliness)** In a sufficiently small interval, exactly one particle arrives at the given region at most. If

$$\psi(t) = \sum_{k=2}^{\infty} P_k(t) = 1 - P_0(t) - P_1(t)$$

then we must have $\psi(t) = o(t)$, *i.e.*

$$\lim_{t \to 0} \frac{\psi(t)}{t} = 0.$$

This property shows that, in practice, two or more particles are impossible to arrive at the given region simultaneously.

Prove that $P_k(t)$ has the Poisson distribution for the given t.

1.5. (The relationship among normal distribution, Poisson distribution, and noncentral χ^2 distribution with 1 degree of freedom) Prove that the noncentral χ^2 distribution with 1 degree of freedom and non-centrality parameter θ^2 can be factorized into an infinite sum of the product of Poisson distribution with parameter $\theta^2/2$ and the χ^2 with $2i + 1$ degrees of freedom, *i.e.*, if the random variable $X - \theta$ has the standard normal distribution, then the pdf of $Y = X^2$ is as follows

$$p_Y(y) = \frac{1}{2\sqrt{2\pi y}} \exp\left\{-\frac{y + \theta^2}{2}\right\}(e^{\sqrt{y}\theta} + e^{-\sqrt{y}\theta}), \quad y > 0,$$

and the above pdf can be transformed into

$$p_Y(y) = \sum_{i=0}^{\infty} P(R = i) f_{2i+1}(y),$$

where R has the Poisson distribution with parameter $\theta^2/2$, and f_m is the pdf of the χ^2_m distribution.

1.6. Let $x_1, \cdots, x_n$ be a set of samples, and let $\bar{x}$ and s^2 denote the sample mean and the variance respectively. Prove that

$$s^2 = \frac{1}{2n(n-1)} \sum_{i=1}^{n} \sum_{j=1}^{n} (x_i - x_j)^2.$$

1.7. Exemplify the following three situations respectively:

(a) MLE exists and is unique;
(b) MLE exists but is not unique;
(c) MLE does not exist.

1.8. Verify the Lagrange's identity: for real numbers $a_1, \cdots, a_n$ and $b_1, \cdots, b_n$ we have

$$\left(\sum_{i=1}^{n} a_i^2\right)\left(\sum_{i=1}^{n} b_i^2\right) - \left(\sum_{i=1}^{n} a_i b_i\right)^2 = \sum_{i=1}^{n-1} \sum_{j=i+1}^{n} (a_i b_j - a_j b_i)^2.$$

Furthermore, prove that the correlation coefficient is equal to 1 if and only if all the sample points lies in a straight line (Wright, 1992).

1.9. Let $X_1, \cdots, X_n$ be a set of samples from the uniform distribution $U(0, \theta)$, where $\Theta = \{\theta : \theta > 0\}$. Let Y_n be the largest order statistic (cf. Example 1.2.5). Prove that Y_n is complete.

1.10. Let $X_1, \cdots, X_n$ be a set of samples from the uniform distribution $U(\theta, \theta + 1]$. For the given $0 \le p < 1$, let $Z = g(X_1 - p) + p$, where the function $g(y)$ is defined as the maximal integer less than or equal to y. Prove that Z is an unbiased estimator for θ, but the UMVUE does not exist.

1.11. Let $X_1, \cdots, X_n$ be i.i.d r.v.s from $U(0, 1)$. For $1 \le i \le n$, let Y_i be the product of the first i variables, *i.e.* $Y_i = X_1 \cdots X_i$. Prove that the distribution of X_{k+1} given that $X_1 = x_1, \cdots, X_k = x_k$ is a uniform distribution $U(0, x_k)$. Furthermore, prove that $E(Y_n) = 1/2^n$.

1.12. Let Y_2 and Y_4 denote the second and the 4-th order statistics of a random sample of size 5 from a distribution of the continuous type having distribution $F(x)$. Compute $P[F(Y_4) - F(Y_2) \ge 0.5]$.

1.13. Let $X_1, X_2, \cdots, X_n$ be a random sample from a uniform distribution $U(0, \theta)$, $\theta > 0$.

(a) Find the MLE of θ.
(b) Find the UMVUE of θ.
(c) Find the method of moments estimate for $\theta(1 - \theta)$.

1.14. $X_1, X_2, \cdots, X_n$ is a random sample from $f(x,\theta) = \theta e^{-\theta x} I(x > 0)$.

(a) Use the factorization theorem to find a sufficient statistic for $\theta \in (0, \infty)$.

(b) Find the unbiased minimum variance estimator of $1/\theta$.

1.15. $X = X_1, X_2, \cdots, X_n$ is a random sample from $U(0, \theta)$.

(a) Find the moment estimator and the maximum likelihood estimator for θ.

(b) Show that one of these estimators is sufficient for θ.

(c) Calculate both estimates for the sample $x = 0.1, 0.2, 0.4, 0.9$, and comment.

1.16. What is meant by an (m, m) exponential family of distributions? What is a curved exponential family?
Write a brief account of data reduction by sufficiency in exponential families.
What is meant by an ancillary statistics? What is the conditionality principle of statistical inference?
Let $Y_1, \cdots, Y_n$ be independent, identically distributed $N(\mu, \mu^2)$, $\mu > 0$.
Show that $(T_1, T_2) = (\Sigma_{i=1}^n Y_i/n, \sqrt{\sum_{i=1}^n Y_i^2/n})$ is minimal sufficient and $Z = T_1/T_2$ is ancillary. Explain why inference about μ should be based on the conditional distribution of $V = \sqrt{n}\, T_2$, given Z, and show that this conditional distribution has density

$$f(\nu|z;\mu) = \frac{k}{\mu^n}\nu^{n-1} \exp\left\{-\frac{1}{2}\left[\frac{\nu}{\mu} - z\sqrt{n}\right]^2\right\},$$

for a constant k, not depending on μ.

1.17. Suppose we have the year 2001 results for tennis matches between the 5 top women players. Let r_{ij} be the number of matches in which player i beat player j and let n_{ij} be the number of matches of player i against player j, for $1 \le i < j \le 5$.
Assume that the (r_{ij}) are independent random variables, and assume

$$r_{ij} \sim Bi(n_{ij}, p_{ij}),$$

and

$$\ln(p_{ij}/(1 - p_{ij})) = \alpha_i - \alpha_j, \quad 1 \le i < j \le 5,$$

with $\alpha_5 = 0$.

(a) Write down the log-likelihood for the unknown parameters, and explain why we need a constraint on $(\alpha_1, \cdots, \alpha_5)$.

(b) How would you find a confidence interval for the probability that player 1 beats player 5?

(c) How would you find a confidence interval for the probability that player 2 beats player 3?

(d) How might you extend the model to allow for a grass court/clay court factor?

[You are not expected to find explicit expressions for the maximum likelihood estimators $\hat{\alpha}_i$.]

1.18. (a) Suppose $(Y|U = u)$ has a Poisson distribution, with mean μu, and U has probability density function $f(u)$, where

$$f(u) = \theta^\theta u^{\theta-1} e^{-\theta u}/\Gamma(\theta), \quad \text{for } u \geq 0.$$

Show that

i. $E(Y) = \mu, \quad V(Y) = \mu + \mu^2/\theta,$

ii. Y has frequency function

$$g(y|\mu) = \frac{\Gamma(\theta + y)\mu^y \theta^\theta}{\Gamma(\theta) y! (\mu + \theta)^{\theta+y}}, \quad \text{for } y = 0, 1, 2, \cdots.$$

(b) If $(Y_1, \cdots, Y_n)$ are independent observations, and Y_i has frequency function $g(y_i|\mu_i)$, where $\ln \mu_i = \beta x_i$ and $x_1, \cdots, x_n$ are given, describe how to estimate β in the case where θ is a known parameter, and derive the asymptotic distribution of your estimator.

1.19. Suppose $x_1, \cdots, x_n$ are drawn independently from a mixture normal distribution with the pdf

$$f(x|\boldsymbol{\theta}) = \alpha f_1(x) + (1 - \alpha) f_2(x),$$

where $f_j(x)$ denotes the density for a normal distribution with mean μ_j and common variance σ^2, and $\boldsymbol{\theta} = (\alpha, \mu_1, \mu_2, \sigma^2)$.
Suppose that we now introduce auxiliary variables Z_{ij} such that

$$Z_{ij} = \begin{cases} 1, & \text{if } x_i \sim N(\mu_j, \sigma^2), \\ 0, & \text{otherwise.} \end{cases}$$

Show that the likelihood function can be written as

$$L(\boldsymbol{x}|\boldsymbol{\theta}, \boldsymbol{Z}) = \prod_{i=1}^{n} \prod_{j=1}^{2} (\alpha_j f_j(x_i))^{z_{ij}}.$$

1.20. Let $(\Omega, \mathcal{A})$ be a measurable space, and let μ be a σ-finite measure on $\mathcal{A}$. Try to show that there must exist a probability measure P on $\mathcal{A}$, such that $\mu \ll P$ and $P \ll \mu$.

1.21. Let the distribution function $F(x)$ of an r.v. X be right-continuous. Try to show that $E(F(X)) \geq 1/2$, with equality if and only if $F(x)$ is continuous everywhere.

1.22. Let $X_1, X_2, \cdots, X_n$ be i.i.d. r.v.s. Let $X_{(1)} \leq X_{(2)} \leq \cdots \leq X_{(n)}$ denote the order statistics of $X_1, X_2, \cdots, X_n$. Suppose that $\varphi(y)$ is a Borel measurable function on the real line, and $E(\varphi(X_1))$ is finite. Prove that

$$E(\varphi(X_1) \mid X_{(1)}) = \frac{1}{n}\varphi(X_{(1)}) + \frac{1}{n}\sum_{i=2}^{n} E(\varphi(X_{(i)}) \mid X_{(1)}).$$

1.23. Let the joint density function of X and Y be

$$[\Gamma(\alpha_1)\Gamma(\alpha_2)\theta_1^{\alpha_1}\theta_2^{\alpha_2}]^{-1}x^{\alpha_1-1}y^{\alpha_2-1}\exp\{-\theta_1^{-1}x - \theta_2^{-1}y\}$$

for $x > 0,\ y > 0,\ \alpha_1 > 0,\ \alpha_2 > 0,\ \theta_1 > 0$, and $\theta_2 > 0$, where α_1 and α_2 are known, θ_1 and θ_2 are parameters.

(a) Find the UMVUE of $\theta_2^2 - \theta_1$.

(b) For $\alpha_1 > 1$, find the UMVUE of θ_1^{-1}.

1.24. (a) Let $X_1, X_2, \ldots, X_n$ be i.i.d. r.v.s ($n \geq 2$) with the common pdf $\sigma^{-1}\exp\{-\sigma^{-1}(x-\mu)\}$ for $x \geq \mu,\ -\infty < \mu < \infty$, and $\sigma > 0$, where μ and σ are parameters. Find the MLEs of μ and σ.

(b) Let $(\mathcal{X}, \mathcal{B}_\mathcal{X}, \mathcal{P})$ be the statistical space of X with $\mathcal{P} = \{P_\theta;\ \theta \in \Theta\}$. Both μ_1 and μ_2 are σ-finite measures on $\mathcal{B}_\mathcal{X}$, satisfying $dP_\theta/d\mu_1 = f_1(x,\theta)$ and $dP_\theta/d\mu_2 = f_2(x,\theta)$. Try to show that the two MLEs of θ based on $f_1(x,\theta)$ and $f_2(x,\theta)$, respectively, are the same.

1.25. Suppose that the joint probability density function of (X, Y) is given by

$$P(X = m, Y = n) = \binom{n}{m} p^m(1-p)^{n-m}\lambda^n e^{-\lambda}/n!$$

for $m = 0, 1, \cdots, n$ and $n = 0, 1, 2, \cdots$, where $0 < p < 1$ and $\lambda > 0$. Find the marginal probability density functions of X and Y.

1.26. Suppose that the r.v. X is symmetric about the zero point, *i.e.*, X and $-X$ have the same distribution with the cdf $F(x)$. Furthermore, suppose that the variance of X is finite. Prove that the variance of X is

$$\int_0^{+\infty} 4x[1 - F(x)]dx.$$

1.27. Suppose that the joint pdf of (X,Y) is given by

$$[2\pi(1-\rho^2)^{1/2}]^{-1}\exp\{-[2(1-\rho^2)]^{-1}(x^2-2\rho xy+y^2)\} \quad \text{for } -1<\rho<1.$$

Let $T=X+Y$. Find the conditional expectation of X given $T=t$, $E(X|T=t)$.

1.28. Suppose that $X_1,\cdots,X_n$ are i.i.d. r.v.s with the pdf $e^{-x}(x>0)$. Let $X_{(1)}=\min\{X_i;\ 1\le i\le n\}$, and $T_n=\sum_{i=1}^n X_i/n - X_{(1)}$. Show that $X_{(1)}$ is independent of T_n.

1.29. Suppose $X_1,\cdots,X_n$ are i.i.d. r.v.s with the Weibull pdf

$$m\eta^{-m}x^{m-1}\exp\{-(x/\eta)^m\} \qquad \text{for } x>0,$$

where $m>0$ and $\eta>0$ are parameters.

(a) Find the pdf of $\ln X_1$.

(b) Find the moment estimator of $p\hat{=}P(X_1<L)$ ($L>0$ is known).

Chapter 2

Methods of Statistical Tests

2.1 Introduction: Background of Problem

Let $(\mathcal{X}, \mathcal{A}, \mathcal{P})$ be a statistical space, where $\mathcal{P} = \{P_\theta;\ \theta \in \Theta\}$, and Θ is a parameter space. Let Θ_0 be a proper subset of Θ. Sometimes, we may be interested in the problem that whether a parameter θ lies in Θ_0 or not. The basic idea for solving this problem is: First, set up the hypothesis

$$H_0 : \theta \in \Theta_0, \tag{2.1.1}$$

then construct a statistic, utilizing samples from the population, and decide whether the hypothesis is true or false based on the statistic. (2.1.1) is usually called as **null hypothesis**, and the statistic which is used to test the hypothesis is called as a **test statistic**.

Now we will explore the basic idea of hypothesis testing based on an example. To evaluate whether a new drug is efficient or not, we consider a clinical trial. Suppose that there are 50 patients taking a new medicine, and 35 patients are cured after a period of time. Then, is this new medicine efficient or not? Depending only on these information, we cannot draw a conclusion. So we discuss this problem in the following two situations according to the difference in information.

(I) Situation with empirical information

Suppose we have a kind of traditional medicine to treat this disease, and the cure rate is 60%. Now suppose testing is to judge whether a new medicine is better than the traditional medicine.

First, we need to construct a model. Let $n = 50$ and X_i represent the response of the i-th patient when one takes the new medicine. Suppose that X_i takes 1 if the i-th patient is cured and 0 otherwise. The quantity θ represents the probability of cure. Then we have

$$P_\theta(X_i = 1) = \theta,\ P_\theta(X_i = 0) = 1 - \theta,\ \text{for}\ \ i = 1, \cdots, n.$$

This is a Bernoulli trial. As noted in Example 1.3.4 of Chapter 1, let $Y = X_1 + \cdots + X_n$, and then $Y \sim Bi(n, \theta)$. From Theorem 1.3.3 of Chapter 1, we can see that Y is the minimal sufficient statistic of θ. Hence we can expand our analysis based on Y.

We are interested in whether the new medicine is better than the traditional medicine or not, hence the parameter space of θ should be

$$\Theta = \{\theta;\ 0.6 \le \theta \le 1\}.$$

Generally, it will tend to be conservative to construct null hypothesis, corresponding to Equation (2.1.1), let

$$H_0 : \theta = \theta_0 = 0.6. \tag{2.1.2}$$

Here $\Theta_0 = \{\theta_0\}$. A hypothesis which contains only one point is called a **simple null hypothesis**, and **composite null hypothesis** otherwise. If the null hypothesis is rejected, then the new medicine is regarded to be better than the traditional medicine; otherwise, the new medicine is regarded to be consistent with the traditional medicine.

Next, we will see the null hypothesis which adopts such conservative assumption is both beneficial for the analysis and helpful to calculate.

For the sake of test, an intuitive idea is to seek a good estimate $\hat{\theta}$ of parameter θ, and then judge whether $\hat{\theta}$ is far from θ_0 or not. If the distance between $\hat{\theta}$ and θ_0 is very far, we will reject the null hypothesis (H_0), otherwise we accept the null hypothesis (H_0). As noted in Theorem 1.4.5 and Example 1.4.2 of Chapter 1, we know that $\hat{\theta}(Y) = Y/n$ is the minimal variance unbiased estimate of θ, and it is a good estimate. In this example, the estimated value of the sample is $\hat{\theta}(y) = y/n = 35/50 = 0.7$. Now, we have to judge whether 0.7 is far from $\theta_0 = 0.6$ or not. How to judge? **The basic idea of testing theory is to decide the difference between $\hat{\theta}(y)$ and θ_0 by comparing the magnitude of probability.**

Let p represent the probability of event $\{Y;\ \hat{\theta}(Y) \ge \hat{\theta}(y)\}$ when the null hypothesis H_0 is true, which will be called as the **p-value**, then

$$p = P_{\theta_0}\{Y;\ \hat{\theta}(Y) \ge \hat{\theta}(y)\}. \tag{2.1.3}$$

Obviously, a smaller p-value means that the possibility of occurrence is less. Hence when the p-value is small, say, $p \le \alpha \in (0, 1)$, we regard that $\hat{\theta}(y)$ is far from θ_0, and the null hypothesis H_0 is not true. Here α is called the **level probability**, usually take $\alpha = 0.05$. In our example,

$$p = P_{\theta_0}\{Y;\ \hat{\theta}(Y) \ge \hat{\theta}(y)\} = \sum_{y \ge 35} \binom{n}{y} {\theta_0}^y (1 - \theta_0)^{n-y}, \tag{2.1.4}$$

here $n = 50$, $\theta_0 = 0.6$. The probability computation is quite complicated, recalling Example 1.4.5, we can apply a similar method.

Let

$$U_n(Y) = \frac{Y - n\theta_0}{\sqrt{n\theta_0(1-\theta_0)}},$$

then when $n \to \infty$, $U_n(Y) \xrightarrow{L} N(0,1)$. Since

$$U_n(y) = \frac{y - n\theta_0}{\sqrt{n\theta_0(1-\theta_0)}} = \frac{35 - 50 \times 0.6}{\sqrt{50 \times 0.6 \times 0.4}} = 1.4434, \tag{2.1.5}$$

then Eq. (2.1.4) is approximately turned to be

$$\begin{aligned} p &= P_{\theta_0}\{Y;\ \hat{\theta}(Y) \geq \hat{\theta}(y)\} \\ &= P_{\theta_0}\{Y;\ U_n(Y) \geq U_n(y)\} \\ &\approx 1 - \Phi(1.4434) = 0.075. \end{aligned}$$

Since $p > \alpha = 0.05$, we cannot reject the null hypothesis H_0, hence, we can only regard that there is no significant difference between the new medicine and the traditional medicine.

However, when the sample size n is different, the problem will change fundamentally. If we let $n = 100, y = 70$, that $\hat{\theta}(y) = 0.7$ seems to be unchanged, however, in Equation (2.1.5), $U_n(y) = 2.04$, and the p-value is $p = 1 - \Phi(2.04) = 0.02 < \alpha$. In the case, we have sufficient reasons to reject the null hypothesis, and regard the new medicine is superior to the traditional medicine. **Although the estimated values are the same, the larger the sample size, the higher reliability in both the estimation theory and testing theory, which is the same as the Law of Large Numbers.**

As discussed above, we can see that for a testing problem, first we need to construct a model, and then choose a test statistic, finally compute p-value. The following methods are equivalent. Corresponding to Eq. (2.1.3), let θ_α satisfy

$$P_{\theta_0}\{Y;\ \hat{\theta}(Y) \geq \theta_\alpha\} = \alpha, \tag{2.1.6}$$

then when $\hat{\theta}(y) \geq \theta_\alpha$, we reject the null hypothesis H_0, otherwise we cannot. For the composite null hypothesis, corresponding to Eqs. (2.1.3) and (2.1.6) are

$$p = \sup_{\theta \in \Theta_0} P_\theta\{Y;\ \hat{\theta}(Y) \geq \hat{\theta}(y)\};$$

$$\sup_{\theta \in \Theta_0} P_\theta\{Y;\ \hat{\theta}(Y) \geq \theta_\alpha\} = \alpha.$$

Usually we call the above θ_α as **α-critical value.**

(II) Situation without empirical information

For the same question, in the absence of empirical information, we must arrange for a comparative clinical trial. Patients can take the traditional medicine, they can also take the placebo without any effect, and we call it control treatment. Assuming that the number of patients is 40, and 23 patients are cured at the same time. We can get Table 2.1.1, which is called a **contingency table.**

Table 2.1.1 Comparison of drug efficacy

	Cured (B_1)	Failed (B_2)	Total
Taking new drug (A_1)	$35(n_{11})$	$15(n_{12})$	$50(n_{1+})$
Taking placebo (A_2)	$23(n_{21})$	$17(n_{22})$	$40(n_{2+})$
Total	$58(n_{+1})$	$32(n_{+2})$	$90(n)$

As can be seen from Table 2.1.1, A_i and B_j represent random events whether patient is cured by taking the new medicine, respectively. Using N_{ij} to represent the number of random events $A_i \bigcap B_j$ which take place at the same time, $i,j = 1,2$. Let $p_{ij} = P\{A_i \cap B_j\}$, let $N_{i+} = N_{i1} + N_{i2}, N_{+j} = N_{1j} + N_{2j}, N = N_{1+} + N_{2+} = N_{+1} + N_{+2}$, corresponding sample value $n_{ij}, n_{i+}, n_{+j}, n$ are given as Table 2.1.1 shown. Now, we can construct several models to solve the problem.

Poisson distribution model. Suppose that there are no any limitations about the data. Let N_{ij} obey the Poisson distribution independently, then

$$P\{N_{ij} = n_{ij}\} = e^{-p_{ij}} \frac{{p_{ij}}^{n_{ij}}}{n_{ij}!}, \quad i,j = 1,2.$$

It is easy to verify that $EN_{ij} = p_{ij}$. To consider the effectiveness of new medicine, we can make a testing in the parameter space $\{p_{ij};\ p_{11} \geq p_{21}\}$

$$H_0 : p_{11} = p_{21}, \tag{2.1.7}$$

or in the parameter space $\{p_{ij};\ p_{22} \geq p_{12}\}$

$$H_0 : p_{12} = p_{22}. \tag{2.1.8}$$

The usual method is to combine the above two equations, let

$$\theta = \frac{p_{11}p_{22}}{p_{12}p_{21}}, \tag{2.1.9}$$

then the parameter space is $\{\theta;\ \theta \geq 1\}$, and we want to test

$$H_0 : \theta = 1. \tag{2.1.10}$$

If the null hypothesis is rejected, the new medicine is regarded to be effective. Recalling the discussion in Example 1.4.2, the above θ is called an **odds ratio**.

Multinomial distribution model. Suppose that $N = n$ is given. Then $N_{ij} \sim M(n; p_{ij})$, and we have

$$P\{N_{ij} = n_{ij} \text{ for } i, j = 1, 2\} = \frac{n!}{n_{11}!n_{12}!n_{21}!n_{22}!} \sum_{i=1}^{2} \sum_{j=1}^{2} {p_{ij}}^{n_{ij}}.$$

As noted in Example 1.1.3 of Chapter 1, $EN_{ij} = np_{ij}$. For testing problem, the null hypothesis can also be represented by Eq. (2.1.10).

Binomial distribution model. Suppose $N_{1+} = n_{1+}$ and $N_{2+} = n_{2+}$ are given. The data in Table 2.1.1 obey two independent binomial distributions, then

$$P\{N_{11} = n_{11}\} = \binom{n_{1+}}{n_{11}} {p_{11}}^{n_{11}} (1 - p_{11})^{n_{1+} - n_{11}},$$

$$P\{N_{21} = n_{21}\} = \binom{n_{2+}}{n_{21}} {p_{21}}^{n_{21}} (1 - p_{21})^{n_{2+} - n_{21}}.$$

As noted in Example 1.1.1 of Chapter 1, $EN_{11} = n_{1+}p_{11}, EN_{21} = n_{2+}p_{21}$. Hence testing problem can be represented by (2.1.7), (2.1.8) or (2.1.10), here $p_{12} = 1 - p_{11}, p_{22} = 1 - p_{21}$.

Hypergeometric distribution model. Suppose $N_{i+} = n_{i+}, N_{+j} = n_{+j}, i, j = 1, 2$ are given. There is only a random variable, it may be set N_{11}, then N_{11} obeys the noncentral hypergeometric distribution, then

$$P\{N_{11} = n_{11}\} = \frac{\binom{n_{+1}}{n_{11}} \binom{n_{+2}}{n_{1+} - n_{11}} \theta^{n_{11}}}{\sum\limits_{i=\max\{0, n_{1+}+n_{+1}-n\}}^{\min\{n_{+1}, n_{+1}\}} \binom{n_{+1}}{i} \binom{n_{+2}}{n_{1+} - n_{11}} \theta^{i}},$$

here θ is the odds ratio, given by Eq. (2.1.9). Now testing problem is given by (2.1.10), especially when $\theta = 1$, N_{11} is distributed as the hypergeometric distribution, then

$$P\{N_{11} = n_{11}\} = \frac{\binom{n_{+1}}{n_{11}} \binom{n_{+2}}{n_{1+} - n_{11}}}{\binom{n}{n_{1+}}}. \tag{2.1.11}$$

From the discussion above, we can see that for a statistical testing, we can construct various models to study in different angles. In many cases, study results will be different due to various models. As to the question that which model is more suitable, we have to analyze specific issues according to the background of the practical problems. In the discussion of several models above, for the null hypothesis (2.1.10), testing problem on odds ratio is common. It is interesting that MLE of odds ratio θ in any model is

$$\hat{\theta}(n_{ij}) = \frac{n_{11}n_{22}}{n_{12}n_{21}}. \tag{2.1.12}$$

Let $\hat{\theta}(N_{ij})$ denote the corresponding random variable $\hat{\theta}(n_{ij})$. According to the above-mentioned idea, it is necessary to calculate the p-value, *i.e.* the probability when $H_0 : \theta = 1$ is true

$$p = P\{\hat{\theta}(N_{ij}) \geq \hat{\theta}(n_{ij})\}. \tag{2.1.13}$$

The probability can be computed by Eq. (2.1.11), but it requires a large amount of computation. We need to consider an approximate distribution of $\hat{\theta}(N_{ij})$, let $\hat{\psi}(N_{ij}) = \ln \hat{\theta}(N_{ij})$, then when H_0 is true, the approximate distribution of $\hat{\psi}$ is

$$\hat{\psi} \sim N(0, \nu). \tag{2.1.14}$$

Here $\nu = 1/n_{11} + 1/n_{12} + 1/n_{21} + 1/n_{22}$ (see Problem 2.1). For the above example,

$$\hat{\theta}(n_{ij}) = \frac{35 \times 17}{15 \times 23} = 1.7246,$$

$$\hat{\psi}(n_{ij}) = \ln 1.7246 = 0.5450,$$

$$\nu = 0.1976, \sqrt{\nu} = 0.4445.$$

Let Z be a random variable which has the standard normal distribution, then the p-value in Equation (2.1.13) can be calculated approximately as

$$\begin{aligned} p &= P\{\hat{\theta}(N_{ij}) \geq \hat{\theta}(n_{ij})\} \\ &\approx P\{Z \geq \hat{\psi}(n_{ij})/\sqrt{\nu}\} \\ &= P\{Z \geq 1.2261\} = 0.11. \end{aligned}$$

The probability is quite great, we cannot say that new medicine is effective. In fact, from the table of data analysis, we can conclude that the new medicine has no obvious advantage in cure.

Test statistic and test function. We will summarize and systematize the issues discussed above in this section. Our purpose is to discuss the statistical testing presented in (2.1.1). Let $X_1, \cdots, X_n$ be a set of samples from the statistical space $(\mathcal{X}, \mathcal{A}, \mathcal{P})$, and let $X = (X_1, \cdots, X_n)$. $T(X)$ is presented for testing (2.1.1), and it is a statistic from the space $(\mathcal{X}^n, \mathcal{A}^n, \mathcal{P}^n)$ to $(\mathcal{T}, \mathcal{B}, \mathcal{P}^{\mathcal{T}})$, here $\mathcal{P}^{\mathcal{T}} = \{P_\theta^T; \theta \in \Theta\}$. Usually such $T(X)$ is called a **test statistic** or **test**. Let $t(x)$ be the sample value of $T(X)$, then

$$p = \sup_{\theta \in \Theta_0} P_\theta\{T(X) \geq t(x)\} = \sup_{\theta \in \Theta_0} P_\theta^T\{T \geq t\} \tag{2.1.15}$$

is called a p-**value**. When the p-value is small, say, $p \leq \alpha = 0.05$, we reject null hypothesis. We can also use a simple function to denote the event in Eq. (2.1.15). Let

$$\phi(x) = \begin{cases} 1, & \text{if } \ T(X) \geq t(x), \\ 0, & \text{otherwise}, \end{cases} \tag{2.1.16}$$

or

$$\phi(t) = \begin{cases} 1, & \text{if } \ T \geq t, \\ 0, & \text{otherwise}. \end{cases} \tag{2.1.17}$$

Then the probability in Eq. (2.1.15) can be expressed as in the form of mathematical expectation

$$p = \sup_{\theta \in \Theta_0} E_\theta \phi(X) = \sup_{\theta \in \Theta_0} E_\theta^T \phi(T). \tag{2.1.18}$$

Clearly the function given by (2.1.16) or (2.1.17) is measurable with the range $[0, 1]$, and it is called the **test function**. From Eq. (2.1.18) we can see that the judgement for the testing depends on mathematical expectation of the test function. For $\forall \theta \in \Theta$, $E_\theta \phi(X)$ is called the power in point θ of the test function $\phi(x)$, and

$$\beta_\phi(\theta) = E_\theta \phi(X)$$

is called the **power function** of $\phi(x)$. The following theorem indicates that a good test statistic should be a function of sufficient statistics.

Theorem 2.1.1. *Let $T(X)$ be a sufficient statistic, for any test function $\phi(x)$, there exists a test function which only depends on $T(X)$ and has the same power function with $\phi(x)$.*

Proof. Given $\phi(x)$, since $T(X)$ is a sufficient statistic, we can let

$$\begin{aligned} \psi(t) &= E_\theta(\phi(X)|T(X) = t) \\ &= E(\phi(X)|T(X) = t), \end{aligned}$$

thus for $\forall \theta \in \Theta$, as noted in Theorem 1.2.1 in Chapter 1,

$$\begin{aligned} E_\theta \psi(T) &= E_\theta[E\phi(X)|T] \\ &= E_\theta \phi(X). \end{aligned}$$

□

2.2 Likelihood Ratio Tests

We have discussed the basic idea of maximum likelihood procedure and proved the fact that the maximum likelihood estimator (MLE) has some optimal statistical properties under many circumstances in the last section of Chapter 1. In this section, we will discuss the testing problem based on MLEs.

2.2.1 *The Likelihood Ratio Methods*

Let $x_1, \cdots, x_n$ be a set of samples from $(\mathcal{X}, \mathcal{A}, \mathcal{P})$, where $\mathcal{P} = \{P_\theta; \theta \in \Theta\}$. Let $x = (x_1, \cdots, x_n)$. Suppose that there exists a σ-finite measure υ in the sample space such that $\mathcal{P} \ll \upsilon$, let $dP_\theta/d\upsilon = f_\theta$. Now we also express the density function as $f(x;\theta)$ for convenience. Let Θ_0 be a proper subset of Θ. Consider the testing problem (2.1.1), *i.e.*,

$$H_0 : \theta \in \Theta_0. \tag{2.2.1}$$

Here the likelihood function and the log-likelihood function are

$$L(x;\theta) = \prod_{i=1}^{n} f(x_i;\theta),$$

and

$$l(x;\theta) = \ln L(x;\theta) = \sum_{i=1}^{n} \ln f(x_i;\theta),$$

respectively. Let $\hat{\theta}_0$ and $\hat{\theta}$ be the MLEs when the parameter is restricted on Θ_0 and Θ, respectively, *i.e.*,

$$L(x;\hat{\theta}_0) = \sup_{\theta\in\Theta_0} L(x;\theta),$$

and

$$L(x;\hat{\theta}) = \sup_{\theta\in\Theta} L(x;\theta).$$

So the ratio

$$\Lambda(x) = \frac{L(x;\hat{\theta}_0)}{L(x;\hat{\theta})}$$

is called as the **likelihood ratio test** of the testing problem (2.1.1) (LRT for short). As the likelihood function is a nonnegative and unimodal function in many situations, by Theorem 1.4.1 in Chapter 1, we can get the following results:

(1) $0 \le \Lambda(x) \le 1$;
(2) When $\hat{\theta}$ is far away from $\hat{\theta}_0$, $\Lambda(x)$ is small; otherwise, $\Lambda(x)$ is large;
(3) If the null is true in the testing problem (2.2.1), then $\Lambda(x) \to 1$ when $n \to \infty$.

Based on the discussion in the last section, the null hypothesis, H_0, should be rejected when $\Lambda(x)$ is small. Corresponding to Eq. (2.1.15), we can denote the p-value as

$$p = \sup_{\theta \in \Theta_0} P_\theta\{\Lambda(X) \le \Lambda(x)\}. \tag{2.2.2}$$

For calculation convenience, we use the log-likelihood form of LRT, *i.e.*,

$$\begin{aligned}\lambda(x) &= -2\ln\Lambda(x)\\ &= 2[\ln L(x;\hat{\theta}) - \ln L(x;\hat{\theta}_0)]\\ &= 2\sum_{i=1}^{n}[\ln f(x_i;\hat{\theta}) - \ln f(x_i;\hat{\theta}_0)],\end{aligned} \tag{2.2.3}$$

so, corresponding to Eq. (2.2.2), the p-value can be denoted as

$$p = \sup_{\theta \in \Theta_0} P_\theta\{\lambda(X) \ge \lambda(x)\}. \tag{2.2.4}$$

Example 2.2.1. (Binomial parameter test) Consider the testing problem proposed by (2.1.2). We first discuss it in a general way. Let $Y \sim Bi(n,\theta)$, $\Theta = \{\theta; \theta_0 \le \theta \le 1\}$ for a given $\theta_0 \in (0,1)$. The null hypothesis is

$$H_0 : \theta = \theta_0.$$

To get the LRT, we should calculate the MLE of θ under Θ_0 and Θ, respectively. As it is obvious that $\hat{\theta}_0 = \theta_0$, we need to compute the $\hat{\theta}$ only. Since $\theta^y(1-\theta)^{n-y}$ is a unimodal function of θ and reaches its maximum value at y/n, so

$$\hat{\theta} = \begin{cases}\theta_0, & \text{if} \quad y/n \le \theta_0,\\ y/n, & \text{if} \quad y/n > \theta_0.\end{cases}$$

Thus the LRT is

$$\Lambda(y) = \begin{cases}1, & \text{if} \quad y/n \le \theta_0,\\ \dfrac{\theta_0^y(1-\theta_0)^{n-y}}{(y/n)^y(1-y/n)^{n-y}}, & \text{if} \quad y/n > \theta_0.\end{cases} \tag{2.2.5}$$

Since $\Lambda(y)$ is a decreasing function of y when $\theta_0 < y/n$, so the p-value is

$$\begin{aligned}p &= P_{\theta_0}\{\Lambda(Y) \le \Lambda(y)\}\\ &= P_{\theta_0}\{Y \ge y\}.\end{aligned} \tag{2.2.6}$$

It is obvious that the above p-value coincides with both Eq. (2.1.4) and our intuition.

Note that it is a common method for statistical testing in (2.2.7), that is, according to the likelihood ratio test criterion, the null hypothesis can be rejected when $\Lambda(y) \leq c$ for some constant c. Since $\Lambda(y)$ is a monotone decreasing function of y, this is equivalence of that if there exists a constant, say c', the null hypothesis should be rejected when $y \geq c'$. Because it is very convenient to calculate the probability by Y, so Y is also called a LRT.

Let's consider a test with a composite hypothesis. Let $\Theta = \{\theta;\ 0 \leq \theta \leq 1\}$, the null hypothesis is

$$H_0 : \theta \leq \theta_0, \tag{2.2.7}$$

i.e., $\Theta_0 = \{\theta;\ 0 \leq \theta \leq \theta_0\}$. Under Θ, the MLE of θ is $\hat{\theta} = y/n$, under Θ_0, the MLE of θ is

$$\hat{\theta}_0 = \begin{cases} y/n, & \text{if } \ y/n \leq \theta_0, \\ \theta_0, & \text{if } \ y/n > \theta_0. \end{cases}$$

It is obvious that LRT is the same as (2.2.6) completely. Computing the p-value, by monotonicity (see Example 3.6.3 in Chapter 3 for details),

$$\begin{aligned} p &= \sup_{\theta \in \Theta_0} P_\theta\{\Lambda(Y) \leq \Lambda(y)\} \\ &= \sup_{\theta \in \Theta_0} P_\theta\{Y \geq y\} \\ &= P_{\theta_0}\{Y \geq y\}, \end{aligned}$$

which is consistent with (2.2.7). In the next chapter, we will discuss that for many statistics, the p-value of a composite null hypothesis can be achieved on the boundary. We will call the distribution of parameter on the boundary as the **least favorable distribution**.

Example 2.2.2. Normal mean distribution Let $x_1, \cdots, x_n$ be a set of samples from a normal distribution $N(\mu, \sigma^2)$. For a given μ_0 we can consider the following two testing problems:

$$H_{01} : \mu = \mu_0, \tag{2.2.8}$$

$$H_{02} : \mu \leq \mu_0. \tag{2.2.9}$$

Usually, the null hypothesis denoted by (2.2.8) is called as **two-sided testing** and the null hypothesis denoted by (2.2.9) as **one-sided testing**. In the following chapters, we will discuss these problems in detail.

1) Assume that σ^2 is known. For H_{01}, the parameter space $\Theta = \{\mu; -\infty < \mu < +\infty\}$, and $\Theta_{01} = \{\mu_0\}$. Under Θ_{01}, the MLE of μ is

$\hat{\mu}_0 = \mu_0$ obviously. Under Θ, from Example 1.4.3 in Chapter 1, $\hat{\mu} = \bar{x}$, and from Eq. (2.2.3)

$$\begin{aligned}\lambda(x) &= 2[\ln L(x;\hat{\mu}) - \ln L(x;\hat{\mu_0})]\\ &= \frac{n}{\sigma^2}(\bar{x}-\mu_0)^2.\end{aligned}$$

Let $y = \sqrt{n}(\bar{x}-\mu_0)/\sigma$. Obviously $\lambda(X) \geq \lambda(x)$ is equivalent to $|Y| \geq |y|$. If H_{01} is true and $Y \sim N(0,1)$, the p-value is

$$\begin{aligned}p &= P_{\mu_0}\{|Y| \geq |y|\}\\ &= 2(1-\Phi(|y|)),\end{aligned}$$

where Φ is the standard normal cumulative distribution function (cdf). For H_{02}, the parameter space $\Theta_{02} = \{\mu; -\infty < \mu \leq \mu_0\}$, so

$$\hat{\mu}_0 = \begin{cases}\bar{x}, & \text{if } \bar{x} \leq \mu_0,\\ \mu_0, & \text{if } \bar{x} \geq \mu_0.\end{cases}$$

Thus, the LRT is

$$\Lambda(x) = \begin{cases}0, & \text{if } \bar{x} < \mu_0,\\ \dfrac{n}{\sigma^2}(\bar{x}-\mu_0)^2, & \text{if } \bar{x} \geq \mu_0.\end{cases} \tag{2.2.10}$$

Let $y = \sqrt{n}(\bar{x}-\mu_0)/\sigma$, from Eq. (2.2.10), it suffices to consider the situation that $y > 0$. Here the p-value is

$$\begin{aligned}p &= P_{\mu_0}\{Y \geq y\}\\ &= 1-\Phi(y).\end{aligned}$$

2) Assume that σ^2 is unknown. There are two unknown parameters, *i.e.*, $\theta = (\mu, \sigma^2)$, where σ^2 is called a **nuisance parameter** since it does not present in Eqs. (2.2.8) and (2.2.9). The parameter spaces are

$$\begin{aligned}\Theta &= \{(\mu,\sigma^2);\ -\infty < \mu < \infty, \sigma^2 > 0\},\\ \Theta_{01} &= \{(\mu,\sigma^2);\ \mu = \mu_0, \sigma^2 > 0\},\\ \Theta_{02} &= \{(\mu,\sigma^2);\ \mu \leq \mu_0, \sigma^2 > 0\}.\end{aligned}$$

It is easy to compute that under Θ and Θ_{01}, the MLEs of θ are $\hat{\theta} = (\bar{x}, \sum(x_i-\bar{x})^2/n)$ and $\hat{\theta}_0 = (\mu_0, \sum(x_i-\mu_0)^2/n)$ respectively. For the testing problem (2.2.9), the LRT is

$$\begin{aligned}\Lambda(x) &= \left[\frac{\sum_{i=1}^n(x_i-\bar{x})^2}{\sum_{i=1}^n(x_i-\mu_0)^2}\right]^{n/2}\\ &= \left[\frac{n-1}{n-1+t^2}\right]^{n/2},\end{aligned}$$

where

$$t = t(x) = \frac{\sqrt{n}(\bar{x} - \mu_0)}{s}, \tag{2.2.11}$$

and

$$s^2 = \frac{1}{n-1} \sum_{i=1}^{n} (x_i - \bar{x})^2.$$

Note that $\Lambda(x)$ is a monotone decreasing function of t, so reject the null hypothesis when $\Lambda(x)$ is small $\Longleftrightarrow$ reject it when $t(x)$ is large, thus the p-value is

$$p = P\{|t(X)| \geq |t(x)|\}.$$

Under H_0, S^2/σ^2 has a χ^2 distribution with $(n-1)$ degrees of freedom (see Problem 2.2), however, $(x - \mu_0)/\sigma$ has a standard normal distribution. From Example 1.1.9 in Chapter 1, $t(X)$ has a Student's t-distribution with $n-1$ degrees of freedom.

For the testing problem (2.2.9) we also only consider the case that $t(x) > 0$. Here the p-value is

$$p = P\{t(X) \geq t(x)\}.$$

The $t(x)$ given by (2.2.11) is called a **Student's t-test statistic**, or **t-test**, usually used to deal with the one-sided or two-sided mean test when the variance is unknown. If the sample is not drawn from the normal distribution, we can draw the conclusion that $t(X)$ has an asymptotic normal distribution used by Example 1.4.5 in the last chapter.

Example 2.2.3. (Testing problems about two normal populations) Let $x_1, \cdots, x_n$ and $y_1, \cdots, y_m$ be two sets of samples from the normal distributions $N(\mu_1, \sigma_1^2)$ and $N(\mu_2, \sigma_2^2)$, respectively. We want to compare these two normal distributions based on them.

Comparing means. Consider the following one-sided and two-sided testing problems

$$H_{01} : \mu_1 = \mu_2, \tag{2.2.12}$$

$$H_{02} : \mu_1 \leq \mu_2. \tag{2.2.13}$$

Two variances are known. Two sample means $\bar{x}$ and $\bar{y}$ are the sufficient statistics for the two population means. Because $\bar{X} \sim N(\mu_1, \sigma_1^2/n)$ and $\bar{Y} \sim N(\mu_2, \sigma_2^2/m)$, so $Z = \bar{X} - \bar{Y} \sim N(\mu, \sigma^2)$, in which $\mu =$

$\mu_1 - \mu_2, \sigma^2 = (m\sigma_1^2 + n\sigma_2^2)/mn$. The problem becomes into the case which has been discussed in Example 2.2.2 with $\mu_0 = 0$.

Two variances are equal but unknown. The log-likelihood function can be written as

$$l(x, y; \theta) = \frac{m+n}{2} \ln \sigma^2 - \frac{1}{2\sigma^2} \left[\sum_{i=1}^{n} (x_i - \mu_1)^2 + \sum_{i=1}^{m} (y_i - \mu_2)^2 \right] + c,$$

where $\theta = (\mu_1, \mu_2, \sigma^2)$. The MLE of μ is $\hat{\mu} = (n\bar{x} + m\bar{y})/(n+m)$ when $\mu_1 = \mu_2 = \mu$; when there is no restriction, $\hat{\mu}_1 = \bar{x}$ and $\hat{\mu}_2 = \bar{y}$. Thus the MLEs of σ^2 are

$$\hat{\sigma}_0^2 = \frac{1}{n+m} \left[\sum_{i=1}^{n} (x_i - \hat{\mu})^2 + \sum_{i=1}^{m} (y_i - \hat{\mu})^2 \right],$$

and

$$\hat{\sigma}^2 = \frac{1}{n+m} \left[\sum_{i=1}^{n} (x_i - \bar{x})^2 + \sum_{i=1}^{m} (y_i - \bar{y})^2 \right],$$

respectively. Through computing, the LRT should reject H_0 if

$$t(x, y) = \frac{(\bar{x} - \bar{y}) \Big/ \sqrt{\dfrac{1}{m} + \dfrac{1}{n}}}{\sqrt{\left[\sum\limits_{i=1}^{n} (x_i - \hat{x})^2 + \sum\limits_{i=1}^{m} (y_i - \hat{y})^2 \right] /(m+n-2)}}$$

takes a large value. It is easy to get that $t(X, Y) \sim t(m+n-2)$ under H_0 (see Problem 2.3).

Two variances are unknown. The problem becomes very complex. It is the famous Behrens-Fisher problem. Some related research results can be found in Stuart *et al.* (1999).

Comparing variances. Consider the following two testing problems:

$$H_{01} : \sigma_1^2 = \sigma_2^2, \tag{2.2.14}$$

$$H_{02} : \sigma_1^2 \geq \sigma_2^2. \tag{2.2.15}$$

The parameter spaces are

$$\Theta = \{(\mu_1, \sigma_1^2, \mu_2, \sigma_2^2);\ -\infty < \mu_i < \infty, \sigma_i^2 > 0 \text{ for } i = 1, 2\},$$

$$\Theta_{01} = \{(\mu_1, \sigma_1^2, \mu_2, \sigma_2^2);\ -\infty < \mu_i < \infty, \sigma_1^2 = \sigma_2^2 \text{ for } i = 1, 2\},$$

$$\Theta_{02} = \{(\mu_1, \sigma_1^2, \mu_2, \sigma_2^2);\ -\infty < \mu_i < \infty, \sigma_1^2 \geq \sigma_2^2 \text{ for } i = 1, 2\}.$$

The MLEs are always equal to $\hat{\mu}_1 = \bar{x}$, $\hat{\mu}_2 = \bar{y}$ in which μ_1 and μ_2 are nuisance parameters. Under Θ, the MLEs of variances are $\hat{\sigma}_1^2 = 1/n\sum_{i=1}^n(x_i-\bar{x})^2$, and $\hat{\sigma}_2^2 = 1/m\sum_{i=1}^m(y_i-\bar{y})^2$, respectively. Under Θ_{01}, let $\sigma_1^2 = \sigma_2^2 = \sigma^2$, so the MLE of σ^2 is

$$\hat{\sigma}^2 = \frac{1}{n+m}\left[\sum_{i=1}^n(x_i-\bar{x})^2 + \sum_{i=1}^m(y_i-\bar{y})^2\right].$$

For Θ_{01}, it is easy to get the LRT

$$\begin{aligned}\Lambda(x,y) &= \left(\frac{\hat{\sigma}_1^2}{\hat{\sigma}^2}\right)^{\frac{n}{2}} \cdot \left(\frac{\hat{\sigma}_2^2}{\hat{\sigma}^2}\right)^{\frac{m}{2}} \\ &= \left(\frac{n}{m+n}\right)^{\frac{n}{2}}\left(\frac{m}{m+n}\right)^{\frac{m}{2}} \\ &\quad \cdot \left[1+\frac{n-1}{m-1}F(x,y)\right]^{-\frac{m}{2}}\left[1+\frac{m-1}{n-1}\frac{1}{F(x,y)}\right]^{-\frac{n}{2}},\end{aligned} \tag{2.2.16}$$

where $F(x,y) = {S_1}^2/{S_2}^2$,

$$S_1^2 = \frac{1}{n-1}\sum_{i=1}^n(x_i-\bar{x})^2, \quad S_2^2 = \frac{1}{m-1}\sum_{i=1}^m(y_i-\bar{y})^2. \tag{2.2.17}$$

As $\Lambda(x,y)$ is a convex function of F by Eq. (2.2.16), reject the null hypothesis when $\Lambda(x,y)$ is small, which is equivalent to reject the null hypothesis when F is too small or too large. By the characteristics of F in Eq. (2.2.17), the p-value is

$$\begin{aligned}p &= P\{\Lambda(X,Y) \le \Lambda(x,y)\} \\ &= 2\cdot\min\{P\{F(X,Y)\le F(x,y)\}, P\{F(X,Y)\ge F(x,y)\}\}.\end{aligned} \tag{2.2.18}$$

Furthermore, ${S_1}^2/{\sigma_1}^2$ and ${S_2}^2/{\sigma_2}^2$ have the χ^2 distribution with $n-1$ and $m-1$ degrees of freedom respectively. From Example 1.1.10, if H_0 is true, *i.e.*, $\sigma_1^2 = \sigma_2^2$, $F(X,Y)$ has an F-distribution with $n-1$ and $m-1$ degrees of freedom. Although the p-value discussed above can be calculated, the computation is very complicated. Next, we will consider the method of α-critical value discussed in the last section.

For a given level α, say $\alpha = 0.05$, the null hypothesis can be rejected if the p-value is small, say $p \le \alpha$. This is equivalent to the fact that there exists a constant Λ_α and reject null hypothesis if $\Lambda(x,y) \le \Lambda_\alpha$, where Λ_α satisfies

$$P\{\Lambda(X,Y) \le \Lambda_\alpha\} = \alpha.$$

By Eq. (2.2.18), it is equivalent to find constants c_1 and c_2, reject the null hypothesis if $F(x, y) \leq c_1$ or $F(x, y) \geq c_2$ when c_1 and c_2 satisfy

$$P\{F(X,Y) \leq c_1\} = P\{F(X,Y) \geq c_2\} = \frac{\alpha}{2},$$

and $F(X, Y)$ is a random variable which has an F-distribution with $n - 1$ and $m - 1$ degrees of freedom.

$F(x, y)$ given by Eq. (2.2.17) is called an **F-test statistic** or **F-test** and is used to compare whether two variances come from the same population or not.

2.2.2 *LRT's Limiting Distribution*

It can be seen from the following chapters that the LRT is the same as MLE that has some optimal statistical properties. However, there are two obvious disadvantages: (i) it depends on the density function in the model strongly. We will discuss this problem for the Student's t-test in the next chapter; (ii) Sometimes it is a hard work to obtain the null distribution. It is important to compute the limiting distribution of LRT with the null hypothesis.

2.2.2.1 *Limiting distribution under simple null hypothesis*

Firstly, we give a lemma which is very common in computing. Then through a sample we sort out the basic thought on computing the limiting distribution. Lastly, we give a theorem.

A matrix $\boldsymbol{B}$ is called an **idempotent matrix** if it is symmetric and satisfies $\boldsymbol{B}^2 = \boldsymbol{B}$. It is obvious that for a given idempotent matrix $\boldsymbol{B}$ there exists an orthogonal matrix $\boldsymbol{Q}$ such that $\boldsymbol{QBQ'}$ is a diagonal matrix and the elements on the diagonal are either 1 or 0. If r denotes the number of 1-element, then r is equal to the rank of $\boldsymbol{B}$, *i.e.*, $r = \text{rank}(\boldsymbol{B})$.

Lemma 2.2.1. *Let $X_1, \cdots, X_n$ be an i.i.d. sample from $N(0, 1)$, and $\boldsymbol{X} = (X_1, \cdots, X_n)'$. $\boldsymbol{X'BX} \sim \chi^2_r$ if and only if $\boldsymbol{B}$ is an idempotent matrix and $r = rank(\boldsymbol{B})$.*

Proof. For any symmetric matrix $\boldsymbol{B}$, there exists an orthogonal matrix $\boldsymbol{Q}$, such that

$$\boldsymbol{QBQ'} = \boldsymbol{\Lambda} = \text{diag}\{\lambda_1, \cdots, \lambda_n\}.$$

So $\boldsymbol{B}^2 = \boldsymbol{B}$ and $\text{rank}(\boldsymbol{B}) = r$ if and only if there are r characteristic roots equal to 1 and others are 0. Now let $\boldsymbol{Y} = \boldsymbol{QX}$, then $\boldsymbol{Y} \sim N(\boldsymbol{0}, \boldsymbol{I}_n)$ and

$\boldsymbol{X} = \boldsymbol{Q}'\boldsymbol{Y}$, so

$$\boldsymbol{X}'\boldsymbol{B}\boldsymbol{X} = \boldsymbol{Y}'\boldsymbol{Q}\boldsymbol{B}\boldsymbol{Q}'\boldsymbol{Y} = \sum_{i=1}^{n} \lambda_i Y_i^2.$$

Since $\sum_{i=1}^{n} \lambda_i Y_i^2 \sim \chi_r^2$ if and only if there are r λ_i's equal to 1 and others are 0, and this completes the proof. □

Example 2.2.4. (Testing multinomial parameter) Let $X \sim M(n; \boldsymbol{p})$, and $\boldsymbol{\pi}$ be a given multinomial parameter, *i.e.*, $\boldsymbol{\pi} = (\pi_1, \cdots, \pi_k)$, $\pi_i > 0$, and $\sum_{i=1}^{k} \pi_i = 1$. Test whether the true parameters are $\boldsymbol{\pi}$ or not, *i.e.* ,

$$H_0 : \boldsymbol{p} = \boldsymbol{\pi}. \tag{2.2.19}$$

Let $\boldsymbol{x} = (x_1, \cdots, x_k)$ be a set of samples. By Example 1.4.4, we can get that the MLE of $\boldsymbol{p}$ is $\hat{\boldsymbol{p}} = (\hat{p}_1, \cdots, \hat{p}_k)$ when $\hat{p}_i = x_i/n$. Under H_0, as the MLE of $\boldsymbol{p}$ is $\boldsymbol{\pi}$, the log-LR test is

$$\lambda(\boldsymbol{x}) = 2n \sum_{i=1}^{k} \hat{p}_i \ln \frac{\hat{p}_i}{\pi_i}. \tag{2.2.20}$$

Note that the second-order Taylor series expansion of $\ln(1+z)$ is

$$\ln(1+z) = z - \frac{1}{2}z^2 + o(z^2).$$

If H_0 is true, then $\hat{\boldsymbol{p}} \to \boldsymbol{\pi}$ by Example 1.4.7. Since

$$\ln \frac{\hat{p}_i}{\pi_i} = \ln\left(1 + \frac{\hat{p}_i - \pi_i}{\pi_i}\right),$$

we can expand Eq. (2.2.20) as

$$\begin{aligned}
\lambda(\boldsymbol{x}) &= 2n\sum_{i=1}^{k} [\pi_i + (\hat{p}_i - \pi_i)] \left[\frac{\hat{p}_i - \pi_i}{\pi_i} - \frac{1}{2}\left(\frac{\hat{p}_i - \pi_i}{\pi_i}\right)^2\right] + o(\|\hat{p} - \pi\|^2) \\
&= 2n\sum_{i=1}^{k} (\hat{p}_i - \pi_i) + n\sum_{i=1}^{k} \frac{(\hat{p}_i - \pi_i)^2}{\pi_i} + o(\|\hat{\boldsymbol{p}} - \boldsymbol{\pi}\|^2).
\end{aligned}$$

The first right-side term is zero, the second term can be written as

$$\chi^2 = \sum_{i=1}^{k} \frac{(x_i - n\pi_i)^2}{n\pi_i}. \tag{2.2.21}$$

By Lemma 2.2.1, we can prove that if H_0 is true, the limiting distribution of χ^2 is a χ^2 distribution with $k-1$ degrees of freedom, *i.e.*, $\chi^2 \xrightarrow{L} \chi_{k-1}^2$. We shall prove this result in the next section.

Now we will discuss a general theorem about the limiting distribution of the LR test. According to Example 2.2.4, the following two conditions are necessary, which are usually called **regularity conditions**.

Let $x_1, \cdots, x_n$ be a set of samples from the statistical space $(\mathcal{X}, \mathcal{A}, \mathcal{P})$ with $\mathcal{P} = \{P_{\boldsymbol{\theta}};\ \boldsymbol{\theta} \in \Theta\}$, let Θ be a set belonged to $\boldsymbol{R}^k$, and let ν be a σ-finite measure on $(\mathcal{X}, \mathcal{A})$, satisfying $\mathcal{P} \ll \nu$. Furthermore, let $dP_{\boldsymbol{\theta}}(\boldsymbol{x})/d\nu(\boldsymbol{x}) = f(\boldsymbol{x};\boldsymbol{\theta})$. Suppose that $\boldsymbol{\theta}$ and $f(\boldsymbol{x};\boldsymbol{\theta})$ are in one-to-one correspondence and that the density function satisfies the following conditions:

(I) The function $f(\boldsymbol{x};\boldsymbol{\theta})$ has third-order partial derivatives with respect to $\boldsymbol{\theta}$ and satisfies

$$\int \frac{\partial f(\boldsymbol{x};\boldsymbol{\theta})}{\partial \theta_i} d\nu(\boldsymbol{x}) = \frac{\partial}{\partial \theta_i} \int f(\boldsymbol{x};\boldsymbol{\theta}) d\nu(\boldsymbol{x}) = 0, \quad i = 1, \cdots, k;$$

$$\int \frac{\partial^2 f(\boldsymbol{x};\boldsymbol{\theta})}{\partial \theta_i \theta_j} d\nu(\boldsymbol{x}) = \frac{\partial^2}{\partial \theta_i \theta_j} \int f(\boldsymbol{x};\boldsymbol{\theta}) d\nu(\boldsymbol{x}) = 0, \quad i, j = 1, \cdots, k;$$

$$\left| \frac{\partial^3 \ln f(\boldsymbol{x};\boldsymbol{\theta})}{\partial \theta_i \partial \theta_j \partial \theta_s} \right| < M(\boldsymbol{x}), \quad i, j, s = 1, \cdots, k,$$

where $M(\boldsymbol{x})$ is an integrable function in $(\mathcal{X}, \mathcal{A}, \mathcal{P})$.

(II) For $\forall \boldsymbol{\theta} \in \Theta$, $\boldsymbol{I}_k(\boldsymbol{\theta}) = (I_{ij}(\boldsymbol{\theta}))_{k \times k}$ is a positive definite matrix where

$$I_{ij}(\boldsymbol{\theta}) = \int \Big(\frac{\partial \ln f(\boldsymbol{x};\boldsymbol{\theta})}{\partial \theta_i} \Big) \Big(\frac{\partial \ln f(\boldsymbol{x};\boldsymbol{\theta})}{\partial \theta_j} \Big) f(\boldsymbol{x};\boldsymbol{\theta}) d\nu(\boldsymbol{x}).$$

Recall Example 1.4.7, the above conditions are exactly the ones that are used to compute the limiting distribution of the MLE $\hat{\boldsymbol{\theta}}$. Especially, $I_k(\boldsymbol{\theta})$ in Condition (II) is the **Fisher information**.

Theorem 2.2.1. *Let $\boldsymbol{\theta}_0$ be an interior point in Θ. If the null $H_0 . \boldsymbol{\theta} = \boldsymbol{\theta}_0$ is true, the log-LRT $\lambda(\boldsymbol{x}) = -2 \ln \Lambda(\boldsymbol{x})$ has a limiting χ^2 distribution with k degrees of freedom, i.e., $\lambda(\boldsymbol{X}) \xrightarrow{L} \chi^2_k$.*

Proof. By Eq. (2.2.2),

$$\lambda(\boldsymbol{x}) = 2[l(\boldsymbol{x};\hat{\boldsymbol{\theta}}) - l(\boldsymbol{x};\boldsymbol{\theta}_0)],$$

where $l(\boldsymbol{x};\boldsymbol{\theta}) = \sum_{i=1}^{n} \ln f(x_i;\boldsymbol{\theta})$ is the log-likelihood function. Expanding $l(\boldsymbol{x};\boldsymbol{\theta}_0)$ at $\hat{\boldsymbol{\theta}}$, we have

$$l(\boldsymbol{x};\boldsymbol{\theta}_0) = l(\boldsymbol{x};\hat{\boldsymbol{\theta}}) + (\boldsymbol{\theta}_0 - \hat{\boldsymbol{\theta}})' l'(\boldsymbol{x};\hat{\boldsymbol{\theta}}) + \frac{1}{2} (\boldsymbol{\theta}_0 - \hat{\boldsymbol{\theta}})' l''(\boldsymbol{x};\boldsymbol{\theta}^*) (\boldsymbol{\theta}_0 - \hat{\boldsymbol{\theta}}).$$

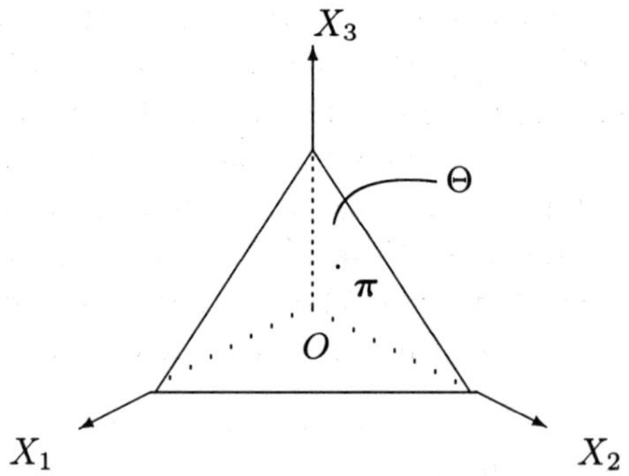

Fig. 2.2.1 Multinomial parameter space

By the MLE's properties, the right-hand second term is zero, and in the third term

$$l''(\boldsymbol{x};\boldsymbol{\theta}^*) = \Big(\sum_{i=1}^n \frac{\partial \ln f(x_i;\boldsymbol{\theta})}{\partial\theta_s\partial\theta_t}\,|_{\boldsymbol{\theta}=\boldsymbol{\theta}^*}\Big)_{k\times k},$$

here, $\boldsymbol{\theta}^*$ satisfies $||\boldsymbol{\theta}^* - \boldsymbol{\theta}_0|| \le ||\hat{\boldsymbol{\theta}} - \boldsymbol{\theta}_0||$. Let $\boldsymbol{D}(\boldsymbol{\theta}_0) = \sqrt{n}(\hat{\boldsymbol{\theta}} - \boldsymbol{\theta}_0)$, and $c(\boldsymbol{\theta}_0) = -\frac{1}{n}l''(\boldsymbol{x};\boldsymbol{\theta}_0)$. By the regularity condition (I), we can get that

$$\lambda(\boldsymbol{x}) = D'(\boldsymbol{\theta}_0)c(\boldsymbol{\theta}_0)\boldsymbol{D}(\boldsymbol{\theta}_0) + o(||\hat{\boldsymbol{\theta}} - \boldsymbol{\theta}_0||^2). \qquad (2.2.22)$$

From Example 1.4.7, we can get that $\boldsymbol{D}(\boldsymbol{\theta}_0) \xrightarrow{L} N(\boldsymbol{0}, \boldsymbol{I}_k^{-1}(\boldsymbol{\theta}_0))$, where $\boldsymbol{I}_k(\boldsymbol{\theta}_0)$ can be obtained by the regularity condition (II). Based on the Law of Large Numbers and Eq. (1.4.28) in Example 1.4.7, $c(\boldsymbol{\theta}_0) \xrightarrow{P_{\boldsymbol{\theta}_0}} \boldsymbol{I}_k(\boldsymbol{\theta}_0)$. Since $\boldsymbol{I}_k^{\frac{1}{2}}(\boldsymbol{\theta}_0)\boldsymbol{D}(\boldsymbol{\theta}_0) \xrightarrow{L} N(\boldsymbol{0}, \boldsymbol{I}_k)$, $\lambda(\boldsymbol{X}) \xrightarrow{L} \chi_k^2$. □

Note that the assumption that $\boldsymbol{\theta}_0$ is an interior point of Θ is important in the theorem above. In Example 2.2.4, since the parameters in multinomial distribution satisfy $p_1+\cdots+p_k = 1$, the number of independent parameters is only $k-1$ and Θ is a set including interior points in $k-1$-dimensional space, see Fig. 2.2.1. But under the null hypothesis (2.2.19), $\boldsymbol{\pi}$ must be an interior point of Θ. If $\boldsymbol{\theta}_0$ is the boundary point of Θ, then the computation of the null distribution will become very difficult. We shall discuss this problem further in Chapter 4 in detail.

2.2.2.2 *Limiting distribution under the composite null hypothesis*

Firstly we will give an expression which is equivalent to the LRT, then sort out the basic idea on computing the limiting distribution through an example and lastly give a general theorem. Let $l(\boldsymbol{x};\boldsymbol{\theta})$ denote the log-likelihood function and let

$$\boldsymbol{V}(\boldsymbol{\theta}) = \frac{1}{\sqrt{n}}\left(\frac{\partial l(\boldsymbol{x};\boldsymbol{\theta})}{\partial\theta_1},\cdots,\frac{\partial l(\boldsymbol{x};\boldsymbol{\theta})}{\partial\theta_k}\right)'. \tag{2.2.23}$$

Lemma 2.2.2. *Suppose that the regularity conditions are satisfied, if the true value $\boldsymbol{\theta}_0$ is an interior point of Θ, then $\boldsymbol{V}(\boldsymbol{\theta}_0) \xrightarrow{L} N(\boldsymbol{0}, \boldsymbol{I}_k(\boldsymbol{\theta}_0))$, where $\boldsymbol{I}_k(\boldsymbol{\theta}_0)$ is given by the regularity condition (II). Furthermore,*

$$||\boldsymbol{D}(\boldsymbol{\theta}_0)\boldsymbol{I}_k(\boldsymbol{\theta}_0) - \boldsymbol{V}(\boldsymbol{\theta}_0)||^2 \xrightarrow{P_{\theta_0}} 0,$$

where $\boldsymbol{D}(\boldsymbol{\theta}_0) = \sqrt{n}(\hat{\boldsymbol{\theta}} - \boldsymbol{\theta}_0)$ and $\hat{\boldsymbol{\theta}}$ is the MLE of $\boldsymbol{\theta}$.

Proof. For simplicity, we only prove the case when $k = 1$. Since $\hat{\theta}$ is the solution of the likelihood equation $l'(x;\theta) = 0$, expanding $l'(x;\hat{\theta})$ at θ_0 yields

$$0 = l'(x;\theta_0) + (\hat{\theta} - \theta_0)l''(x;\theta_0) + \frac{1}{2}(\hat{\theta} - \theta_0)^2 l'''(x;\theta^*), \tag{2.2.24}$$

where $|\theta^* - \theta_0| \leq |\hat{\theta} - \theta_0|$. By the regularity condition (I), the last term can be expressed as $o((\hat{\theta} - \theta_0)^2)$. Recall Eq. (1.4.28) in Example 1.4.7 that

$$-\frac{1}{n}l''(x;\theta_0) \xrightarrow{P_{\theta_0}} I(\theta_0), \tag{2.2.25}$$

so Eq. (2.2.24) can be written as

$$D(\theta_0)\cdot\left(-\frac{1}{n}l''(x;\theta_0)\right) - V(\theta_0) = \frac{1}{\sqrt{n}}\,o((\hat{\theta} - \theta_0)^2).$$

By Eq. (2.2.25) we can draw the lemma's conclusion, and

$$D(\theta_0) \xrightarrow{L} N(0, I^{-1}(\theta_0)) \Rightarrow V(\theta_0) \xrightarrow{L} N(0, I(\theta_0)).$$

□

Example 2.2.5. (Testing multivariate normal mean under linear restriction) Let $x_1,\cdots,x_n$ be a set of samples from k-dimensional normal distribution $N(\boldsymbol{\theta}, \boldsymbol{I}_k)$, and let $\boldsymbol{A} = (a_1,\cdots,a_r)$ be a given $k \times r$ matrix with full rank. The null hypothesis is

$$H_0 : \boldsymbol{A}'\boldsymbol{\theta} = \boldsymbol{0}.$$

It is very obvious that under H_0 the MLE of normal mean $\boldsymbol{\theta}$ is the solution of

$$\min_{\boldsymbol{A}'\boldsymbol{\theta}=\boldsymbol{0}} ||\bar{\boldsymbol{x}} - \boldsymbol{\theta}||^2, \tag{2.2.26}$$

where $\bar{\boldsymbol{x}}$ denotes the sample mean, and $||\bar{\boldsymbol{x}} - \boldsymbol{\theta}||^2 = \sum_{i=1}^{k}(x_i - \theta_i)^2$. Let $L_{\boldsymbol{A}}$ be an r-dimensional linear space which is spanned by $\boldsymbol{a}_1, \cdots, \boldsymbol{a}_r$, and let $\boldsymbol{b}_1, \cdots, \boldsymbol{b}_s$ be a base of $L_{\boldsymbol{A}}^{\perp}$ which is the orthogonal complement space of $L_{\boldsymbol{A}}$, where $s = k - r$.

$$\boldsymbol{A}'\boldsymbol{\theta} = \boldsymbol{0} \Leftrightarrow \exists \boldsymbol{\eta} \in \boldsymbol{R}^s,\ \boldsymbol{\theta} = \eta_1 \boldsymbol{b}_1 + \cdots + \eta_s \boldsymbol{b}_s.$$

Then Eq. (2.2.26) can be rewritten as

$$\min_{\boldsymbol{\eta}} ||\bar{\boldsymbol{x}} - \boldsymbol{B}\boldsymbol{\eta}||^2, \tag{2.2.27}$$

where $\boldsymbol{B} = (\boldsymbol{b}_1, \cdots, \boldsymbol{b}_s)$. This is a projection problem. We can get

$$\hat{\boldsymbol{\eta}} = (\boldsymbol{B}'\boldsymbol{B})^{-1}\boldsymbol{B}'\bar{\boldsymbol{x}}$$

as the solution of Eq. (2.2.27). So the solution of Eq. (2.2.26) is

$$\hat{\boldsymbol{\theta}}_0 = \boldsymbol{B}\hat{\boldsymbol{\eta}} = \boldsymbol{B}(\boldsymbol{B}'\boldsymbol{B})^{-1}\boldsymbol{B}'\bar{\boldsymbol{x}}.$$

Let $\boldsymbol{\theta}_0$ be the true parameter satisfying the null hypothesis. Since $\bar{\boldsymbol{X}} \sim N(\boldsymbol{\theta}_0, \boldsymbol{I}_k/n)$, and $\boldsymbol{B}(\boldsymbol{B}'\boldsymbol{B})^{-1}\boldsymbol{B}'$ is an idempotent matrix, it is easy to get that

$$\sqrt{n}(\hat{\boldsymbol{\theta}}_0 - \boldsymbol{\theta}_0) \sim N(\boldsymbol{0}, \boldsymbol{B}(\boldsymbol{B}'\boldsymbol{B})^{-1}\boldsymbol{B}').$$

As the MLE of $\boldsymbol{\theta}$ without any restriction is $\bar{\boldsymbol{x}}$, the LRT is

$$\lambda(\boldsymbol{x}) = \bar{\boldsymbol{x}}'(\boldsymbol{I}_k - \boldsymbol{B}(\boldsymbol{B}'\boldsymbol{B})^{-1}\boldsymbol{B}')\bar{\boldsymbol{x}}$$

for testing H_0. As $\boldsymbol{I}_k - \boldsymbol{B}(\boldsymbol{B}'\boldsymbol{B})^{-1}\boldsymbol{B}'$ is also an idempotent matrix and its rank is $k - s$, we can get $\lambda(\boldsymbol{X}) \sim \chi^2_{k-s}$ by Lemma 2.2.1.

As enlightened by Example 2.2.5, we can consider a composite null hypothesis with r restricted conditions, *i.e.*, $q_1(\boldsymbol{\theta}) = 0, \cdots, q_r(\boldsymbol{\theta}) = 0$, where q_1 has the first order continuous derivative function. These restricted conditions are equivalent that there exist $s = k - r$ new parameters $\eta_1, \cdots, \eta_s$ such that the original parameters can be expressed as the functions of the new parameters:

$$H_0 : \theta_i = g_i(\eta_1, \cdots, \eta_s), \quad i = 1, \cdots, k, \tag{2.2.28}$$

and g_i has the first order continuous derivative function.

Theorem 2.2.2. *Consider the composite null hypothesis as given by (2.2.28). If the regularity conditions are satisfied and the true parameter is an interior point of the parameter space corresponding to the null hypothesis, then the log-LRT $\lambda(x) = -2\ln\Lambda(x)$ has a limiting χ^2 distribution with $k - s$ degrees of freedom, i.e.,* $\lambda(X) \xrightarrow{L} \chi^2_{k-s}$.

Proof. We firstly discuss the parameter space expressed by (2.2.28). Let

$$\boldsymbol{g}(\boldsymbol{\eta}) = (g_1(\boldsymbol{\eta}), \cdots, g_k(\boldsymbol{\eta}))'.$$

Then there exists an $\boldsymbol{\eta}_0$ such that $\boldsymbol{\theta}_0 = \boldsymbol{g}(\boldsymbol{\eta}_0)$ and $\boldsymbol{\eta}_0$ is also an interior point. As the likelihood equation should be represented as $l(x;\boldsymbol{\theta}) = l(x;\boldsymbol{g}(\boldsymbol{\eta}))$, we can get the MLE of $\boldsymbol{\eta}$, $\hat{\boldsymbol{\eta}}_0$, by solving

$$\frac{\partial l(x;\boldsymbol{g}(\boldsymbol{\eta}))}{\partial \eta_i} = 0, i = 1, \cdots, s.$$

Similar to the discussion in Example 1.4.7, we can get

$$\sqrt{n}(\hat{\boldsymbol{\eta}}_0 - \boldsymbol{\eta}_0) \xrightarrow{L} N(\boldsymbol{0}, \boldsymbol{I}_s^{-1}(\boldsymbol{\eta}_0)), \tag{2.2.29}$$

where $\boldsymbol{I}_s(\boldsymbol{\eta}) = (I_{ij}(\boldsymbol{\eta}))_{s\times s}$,

$$I_{ij}(\boldsymbol{\eta}) = \int \left(\frac{\partial \ln f(x;\boldsymbol{g}(\boldsymbol{\eta}))}{\partial \eta_i}\right)\left(\frac{\partial \ln f(x;\boldsymbol{g}(\boldsymbol{\eta}))}{\partial \eta_j}\right) f(x;\boldsymbol{g}(\boldsymbol{\eta}))d\nu(x).$$

Let

$$\boldsymbol{U}(\boldsymbol{\eta}) = \frac{1}{\sqrt{n}}\left(\frac{\partial l(x;\boldsymbol{g}(\boldsymbol{\eta}))}{\partial \eta_1}, \cdots, \frac{\partial l(x;\boldsymbol{g}(\boldsymbol{\eta}))}{\partial \eta_s}\right)'. \tag{2.2.30}$$

Similar to the proof of Lemma 2.2.2, we can get

$$\boldsymbol{U}(\boldsymbol{\eta}_0) \xrightarrow{L} N(\boldsymbol{0}, \boldsymbol{I}_s(\boldsymbol{\eta}_0)),$$

$$||\boldsymbol{H}(\boldsymbol{\eta}_0)\boldsymbol{I}_s(\boldsymbol{\eta}_0) - \boldsymbol{U}(\boldsymbol{\eta}_0)||^2 \xrightarrow{P_{\boldsymbol{\theta}_0}} 0,$$

where $\boldsymbol{H}(\boldsymbol{\eta}_0) = \sqrt{n}(\hat{\boldsymbol{\eta}}_0 - \boldsymbol{\eta}_0)$. Let $\hat{\boldsymbol{\theta}}_0 = \boldsymbol{g}(\hat{\boldsymbol{\eta}}_0)$, by Eq. (2.2.22) and Lemma 2.2.2, we can get

$$\begin{aligned}\lambda(x) &= 2[l(x;\hat{\boldsymbol{\theta}}) - l(x;\hat{\boldsymbol{\theta}}_0)]\\ &= 2[l(x;\hat{\boldsymbol{\theta}}) - l(x;\boldsymbol{\theta}_0)] - 2[l(x;\hat{\boldsymbol{\theta}}_0) - l(x;\boldsymbol{\theta}_0)]\\ &= \boldsymbol{D}'(\boldsymbol{\theta}_0)\boldsymbol{I}_k(\boldsymbol{\theta}_0)\boldsymbol{D}(\boldsymbol{\theta}_0) - \boldsymbol{H}'(\boldsymbol{\eta}_0)\boldsymbol{I}_s(\boldsymbol{\eta}_0)\boldsymbol{H}(\boldsymbol{\eta}_0)\\ &\quad + o(||\hat{\boldsymbol{\theta}} - \boldsymbol{\theta}_0||^2) + o(||\hat{\boldsymbol{\theta}}_0 - \boldsymbol{\theta}_0||^2)\\ &= \boldsymbol{V}'(\boldsymbol{\theta}_0)\boldsymbol{I}_k^{-1}(\boldsymbol{\theta}_0)\boldsymbol{V}(\boldsymbol{\theta}_0) - \boldsymbol{U}'(\boldsymbol{\eta}_0)\boldsymbol{I}_s^{-1}(\boldsymbol{\eta}_0)\boldsymbol{U}(\boldsymbol{\eta}_0)\\ &\quad + o(||\hat{\boldsymbol{\theta}} - \boldsymbol{\theta}_0||^2) + o(||\hat{\boldsymbol{\theta}}_0 - \boldsymbol{\theta}_0||^2),\end{aligned} \tag{2.2.31}$$

where $\boldsymbol{V}(\boldsymbol{\theta})$ and $\boldsymbol{U}(\boldsymbol{\eta})$ are given by (2.2.23) and (2.2.30) respectively. Notice that $\boldsymbol{\theta} = \boldsymbol{g}(\boldsymbol{\eta})$ and by compound derivative formula, we can get

$$\boldsymbol{U}(\boldsymbol{\eta}_0) = \boldsymbol{A}(\boldsymbol{\eta}_0)\boldsymbol{V}(\boldsymbol{\theta}_0),$$

$$\boldsymbol{I}_s(\boldsymbol{\eta}_0) = \boldsymbol{A}(\boldsymbol{\eta}_0)\boldsymbol{I}_k(\boldsymbol{\theta}_0)\boldsymbol{A}(\boldsymbol{\eta}_0),$$

where $\boldsymbol{A}(\boldsymbol{\eta}) = (a_{ij}(\boldsymbol{\eta}))_{r\times k}$, $a_{ij}(\boldsymbol{\eta}) = \dfrac{\partial g_j(\boldsymbol{\eta})}{\partial \eta_i}$, $i = 1, \cdots, r$, $j = 1, \cdots, k$. Thus, Eq. (2.2.22) can be rewritten as

$$\begin{aligned}\lambda(x) \approx\ & \boldsymbol{V}'(\boldsymbol{\theta}_0)\boldsymbol{I}_k^{-1}(\boldsymbol{\theta}_0)\boldsymbol{V}(\boldsymbol{\theta}_0)\\ & -\boldsymbol{V}'(\boldsymbol{\theta}_0)\boldsymbol{A}'(\boldsymbol{\eta}_0)[\boldsymbol{A}(\boldsymbol{\eta}_0)\boldsymbol{I}_k(\boldsymbol{\theta}_0)\boldsymbol{A}'(\boldsymbol{\eta}_0)]^{-1}\boldsymbol{A}(\boldsymbol{\eta}_0)\boldsymbol{V}'(\boldsymbol{\theta}_0).\end{aligned}$$

Let $\boldsymbol{W}(\boldsymbol{\theta}_0) = \boldsymbol{I}_k^{-\frac{1}{2}}(\boldsymbol{\theta}_0)\boldsymbol{V}(\boldsymbol{\theta}_0)$, the above formula becomes

$$\lambda(x) \approx \boldsymbol{W}'(\boldsymbol{\theta}_0)\boldsymbol{G}(\boldsymbol{\theta}_0)\boldsymbol{W}(\boldsymbol{\theta}_0),$$

where

$$\boldsymbol{G}(\boldsymbol{\theta}_0) = \boldsymbol{I}_k - \boldsymbol{I}_k^{\frac{1}{2}}(\boldsymbol{\theta}_0)\boldsymbol{A}'(\boldsymbol{\eta}_0)[\boldsymbol{A}(\boldsymbol{\eta}_0)\boldsymbol{I}_k(\boldsymbol{\theta}_0)\boldsymbol{A}'(\boldsymbol{\eta}_0)]^{-1}\boldsymbol{A}(\boldsymbol{\eta}_0)\boldsymbol{I}_k^{\frac{1}{2}}(\boldsymbol{\theta}_0).$$

It is easy to verify that $\boldsymbol{G}(\boldsymbol{\theta}_0)$ is an idempotent matrix and $\text{rank}(G(\boldsymbol{\theta}_0)) = k - s$. Since $\boldsymbol{W}(\boldsymbol{\theta}_0) \xrightarrow{L} N(\boldsymbol{0}, \boldsymbol{I}_k)$, by Lemma 2.2.1, we can get $\lambda(X) \xrightarrow{L} \chi^2_{k-s}$. □

There are many authors who have discussed the limiting distribution of LRT, such as Wilks (1938); Rao (1973). One can refer to Shapiro (1988); Robertson *et al.* (1988); Silvapulle and Sen (2004) and so on when the true parameter $\boldsymbol{\theta}_0$ is a boundary point of Θ.

2.3 Tests of Goodness-of-fit

The goodness-of-fit test discussed in this section is mainly used to treat the problem of discrete distribution tests, so it is called sometimes as categorical data analysis (Bishop *et al.*, 1975). The goodness-of-fit test was proposed by Karl Pearson in 1900, which is used to compare a theoretical distribution, such as a normal, Poisson etc., with the observed data from a sample. We will discuss this problem in the later part of this section. Because under most conditions, the limiting distribution of the goodness-of-fit test is a χ^2 distribution, it is also called as Pearson χ^2 test.

Review the discussion in Example 2.2.4. Let $X \sim M(n; \boldsymbol{p})$. For a given multinomial distribution with parameters $\boldsymbol{\pi}$, consider the problem of testing

$$H_0: \ \boldsymbol{p} = \boldsymbol{\pi}. \tag{2.3.1}$$

Let $\boldsymbol{x} = (x_1, \cdots, x_k)$ be a set of samples. By Eq. (2.2.22), the likelihood ratio test is equivalent to that

$$\chi^2 = \sum_{i=1}^{k} \frac{(x_i - n\pi_i)^2}{n\pi_i}. \tag{2.3.2}$$

Since x_i is an observation, we know that $EX_i = n\pi_i$ is the expectation from Example 1.1.3. For intuition, Eq. (2.3.2) can be written as

$$\chi^2 = \sum \frac{(O-E)^2}{E}. \tag{2.3.3}$$

In the subsequent discussion, we will find that the significance of Eq. (2.3.3) is far from its content. Now, let us complete the proof of Example 2.2.4 at first.

2.3.1 *Limiting Distribution*

Let $\hat{\boldsymbol{p}}$ denote the MLE of $\boldsymbol{p}$, that is $\hat{\boldsymbol{p}} = \boldsymbol{x}/n$. We discuss the limiting distribution of $\hat{\boldsymbol{p}}$ first. Let $\boldsymbol{U}_n = \sqrt{n}(\hat{\boldsymbol{p}} - \boldsymbol{p})$, where $\boldsymbol{p}$ is the true value of parameter. From Example 1.1.3, we know the expectation of $\boldsymbol{U}_n$ is zero, and the covariance matrix is $\boldsymbol{D_p} - \boldsymbol{pp}'$, where $\boldsymbol{D_p}$ is the diagonal matrix with diagonal elements $p_1, \cdots, p_k$, that is $\boldsymbol{D_p} = \text{diag}\{p_1, \cdots, p_k\}$.

Theorem 2.3.1. *If* $\boldsymbol{p}$ *is the true value of the parameter, then* $\boldsymbol{U}_n \xrightarrow{L} \boldsymbol{U}$, *where* $\boldsymbol{U} \sim N(\boldsymbol{0}, \boldsymbol{D_p} - \boldsymbol{pp}')$.

Proof. We will prove it with the moment generating function. Recall that when $\boldsymbol{X}$ has a multivariate normal distribution $N(\boldsymbol{\mu}, \boldsymbol{\Sigma})$, the moment generating function of $\boldsymbol{X}$ is

$$M_{\boldsymbol{X}}(\boldsymbol{t}) = Ee^{\boldsymbol{t}'\boldsymbol{X}} = \exp\left\{\frac{1}{2}(\boldsymbol{t}-\boldsymbol{\mu})'\boldsymbol{\Sigma}(\boldsymbol{t}-\boldsymbol{\mu})\right\}. \tag{2.3.4}$$

When $\boldsymbol{X}$ has a multinomial distribution $M(n;\boldsymbol{p})$, the moment generating function of $\boldsymbol{X}$ is:

$$M_{\boldsymbol{X}}(\boldsymbol{t}) = Ee^{\boldsymbol{t}'\boldsymbol{X}} = \left[\sum_{i=1}^{k} p_i e^{t_i}\right]^n. \tag{2.3.5}$$

Now, let $M_{\boldsymbol{U}_n}(\boldsymbol{t})$ be the moment generating function of $\boldsymbol{U}_n$, then we have

$$M_{\boldsymbol{U}_n}(\boldsymbol{t}) = Ee^{\boldsymbol{t}'\boldsymbol{U}_n} = Ee^{\frac{1}{\sqrt{n}}\boldsymbol{t}'\boldsymbol{X} - \sqrt{n}\boldsymbol{t}'\boldsymbol{p}} = e^{-\sqrt{n}\boldsymbol{t}'\boldsymbol{p}} M_{\boldsymbol{X}}\left(\frac{1}{\sqrt{n}}\boldsymbol{t}\right),$$

here, $M_{\boldsymbol{X}}(\boldsymbol{t}/\sqrt{n})$ denotes the moment generating function of the multinomial distribution with parameter $\boldsymbol{t}/\sqrt{n}$. From Eq. (2.3.5)

$$\begin{aligned} M_{\boldsymbol{U}_n}(\boldsymbol{t}) &= e^{-\sqrt{n}\boldsymbol{t}'\boldsymbol{p}}\left[\sum_{i=1}^{k} p_i e^{\frac{1}{\sqrt{n}}t_i}\right]^n \\ &= \left[\sum_{i=1}^{k} p_i e^{\frac{1}{\sqrt{n}}(t_i - \boldsymbol{t}'\boldsymbol{p})}\right]^n. \end{aligned}$$

Note that when z is very small, the Taylor series expansion of e^z is $e^z = 1 + z + z^2/2 + o(z^2)$, then

$$M_{\boldsymbol{U}_n}(\boldsymbol{t}) = \left[\sum_{i=1}^{k} p_i \left(1 + \frac{1}{\sqrt{n}}(t_i - \boldsymbol{t}'\boldsymbol{p}) + \frac{1}{2n}(t_i - \boldsymbol{t}'\boldsymbol{p})^2\right) + o\left(\frac{1}{n}\right)\right]^n$$

$$= \left[1 + \frac{1}{\sqrt{n}}\sum_{i=1}^{k} p_i(t_i - \boldsymbol{t}'\boldsymbol{p}) + \frac{1}{2n}\sum_{i=1}^{k} p_i(t_i - \boldsymbol{t}'\boldsymbol{p})^2 + o\left(\frac{1}{n}\right)\right]^n,$$

the second term in the bracket of the above equation is zero, while the third term is

$$\sum_{i=1}^{k} p_i(t_i - \boldsymbol{t}'\boldsymbol{p})^2 = \boldsymbol{t}'(\boldsymbol{D_p} - \boldsymbol{p}\boldsymbol{p}')\boldsymbol{t}.$$

Therefore

$$M_{\boldsymbol{U}_n}(\boldsymbol{t}) = \left[1 + \frac{1}{2n}\boldsymbol{t}'(\boldsymbol{D_p} - \boldsymbol{p}\boldsymbol{p}')\boldsymbol{t} + o\left(\frac{1}{n}\right)\right]^n,$$

then, when $n \to \infty$, we have

$$M_{\boldsymbol{U}_n}(\boldsymbol{t}) \to \exp\left\{\frac{1}{2}\boldsymbol{t}'(\boldsymbol{D_p} - \boldsymbol{p}\boldsymbol{p}')\boldsymbol{t}\right\}.$$

Comparing with Eq. (2.3.4), the conclusion holds. □

Similar to Lemma 2.2.1, we can also get the following lemma.

Lemma 2.3.1. *Let* $\boldsymbol{X} \sim N(\boldsymbol{0}, \boldsymbol{\Sigma})$. *If* $\boldsymbol{\Sigma}$ *is an idempotent matrix, then,* $\boldsymbol{X}'\boldsymbol{X} \to \chi^2_r$, *where* $r = rank(\boldsymbol{\Sigma})$.

Proof. From Lemma 2.2.1, there exists an orthogonal matrix O such that $\boldsymbol{O\Sigma O}'$ is a diagonal matrix with r diagonal elements equal to 1 (and all other elements equal to zero), which is denoted by $\boldsymbol{A}_r$. Let $\boldsymbol{Y} = \boldsymbol{OX}$, then $\boldsymbol{Y} \sim N(\boldsymbol{0}, \boldsymbol{A}_r)$. It is easy to verify that $\boldsymbol{X}'\boldsymbol{X} = \boldsymbol{Y}'\boldsymbol{Y} = \sum_{i=1}^{k} Y_i^2 \sim \chi^2_r$, and this completes the proof the lemma. □

Theorem 2.3.2. *Suppose that* $\boldsymbol{X} \sim M(n; \boldsymbol{p})$, *and let* $\boldsymbol{x}$ *be the sample of* $\boldsymbol{X}$ *and*

$$\chi^2 = \sum_{i=1}^{k} \frac{(x_i - n\pi_i)^2}{n\pi_i}, \tag{2.3.6}$$

then, when $H_0 : p = \boldsymbol{\pi}$ *is true,* $\chi^2 \xrightarrow{L} \chi^2_{k-1}$.

Proof. Let $\boldsymbol{U}_n = \sqrt{n}(\hat{\boldsymbol{p}} - \boldsymbol{\pi})$, where $\hat{\boldsymbol{p}} = (\hat{p}_1, \cdots, \hat{p}_k)', \hat{p}_i = x_i/n$. From Eq. (2.3.6), we can get

$$\chi^2 = \boldsymbol{U}_n' \boldsymbol{D}_{\boldsymbol{\pi}}^{-1} \boldsymbol{U}_n,$$

where $\boldsymbol{D}_{\boldsymbol{\pi}} = \text{diag}\{\pi_1, \cdots, \pi_k\}$. From Theorem 2.3.1, we know that $\boldsymbol{U}_n \xrightarrow{L} \boldsymbol{U}$, where $\boldsymbol{U} \sim N(\boldsymbol{0}, \boldsymbol{D}_{\boldsymbol{\pi}} - \boldsymbol{\pi}\boldsymbol{\pi}')$. Let $\boldsymbol{V}_n = \boldsymbol{D}_{\boldsymbol{\pi}}^{-\frac{1}{2}} \boldsymbol{U}_n$, then $\boldsymbol{V}_n \xrightarrow{L} \boldsymbol{V}$, where $\boldsymbol{V} \sim N(\boldsymbol{0}, \boldsymbol{I}_k - \sqrt{\boldsymbol{\pi}}\sqrt{\boldsymbol{\pi}}')$ and $\sqrt{\boldsymbol{\pi}} = (\sqrt{\pi_1}, \cdots, \sqrt{\pi_k})'$. Furthermore, since $\boldsymbol{U}_n = \boldsymbol{D}_{\boldsymbol{\pi}}^{\frac{1}{2}} \boldsymbol{V}_n$, then

$$\chi^2 = \boldsymbol{U}_n' \boldsymbol{D}_{\boldsymbol{\pi}}^{-1} \boldsymbol{U}_n = \boldsymbol{V}_n' \boldsymbol{V}_n.$$

Note that $\sum_{i=1}^{k} \pi_i = 1$, $\boldsymbol{I}_k - \sqrt{\boldsymbol{\pi}}\sqrt{\boldsymbol{\pi}}'$ is an idempotent matrix. From Lemma 2.3.1, the conclusion holds. □

Example 2.3.1. (Heredity experiments) Mendel proposed the theory of inheritance around 1866. He supported his theory by the result from the pea crossbreed experiment. He studied two physical characteristics of peas: Color and shape. That is, Y(Yellow), G(Green) and R(Round), W(Wrinkled). Mendel crossbred pure yellow and pure green peas. He predicted the second generation would have four kinds of results: YY, YG, GY, GG. The experimental results have shown that if the genotype includes Y the pea must be yellow; The G is recessive gene, the pea are green with genotype GG. If we use θ_1 to denote the probability of yellow, then $\theta_1 = 3/4$, that is, the probability of green is $1 - \theta_1 = 1/4$. Similarly, if we use θ_2 to denote the probability of round, then $\theta_2 = 3/4$, while the probability of wrinkled is $1 - \theta_2 = 1/4$.

Now crossbreed the yellow and round peas with the green and wrinkled peas, there are 4 possible outcomes. According to the independence or Mendel's second law (see Feller, 1968, §5.5), their associated probabilities are respectively

$$\begin{aligned} P_1 &= P\{\text{YR}\} = \theta_1\theta_2 = 9/16, \\ P_2 &= P\{\text{YW}\} = \theta_1(1-\theta_2) = 3/16, \\ P_3 &= P\{\text{GR}\} = (1-\theta_1)\theta_2 = 3/16, \\ P_4 &= P\{\text{GW}\} = (1-\theta_1)(1-\theta_2) = 1/16. \end{aligned} \tag{2.3.7}$$

After his experiments, Mendel obtained 556 grains of peas. The corresponding numbers of the 4 gene-pairs were

YR	YW	GR	GW
315	101	108	32

Now we need to verify Mendel's hypothesis, *i.e.*, whether H_0 denoted by (2.3.7) is true or false. Using the χ^2 statistic given in Eq. (2.3.6), we have

O :	315	101	108	32
E :	312.75	104.25	104.25	34.75

then we have $\chi^2 = 0.47$. Here, χ^2 has the chi-square distribution with 3 degrees of freedom. Calculating the p-value, we get $P\{\chi_3^2 \geq 0.4\} \approx 0.39$. The probability is too great to reject H_0. Therefore we believe that the experimental results support Mendel's hypothesis.

2.3.2 *Test of Independence in Two-Way Contingency Table*

The goodness-of-fit test is mainly used to test contingency tables. Reconsider the analysis of the effect of the drug shown in Table 2.1.1. Let X and Y be two random variables denoting whether the patients take the medicine or not and whether the patients are cured or not. Logically, it is easy to establish the following equivalent relation:

•Whether the medicine is efficient or not,
•Whether X is independent of Y or not. (2.3.8)

Let $p_{ij} = P\{X \in A_i, Y \in B_j\}, p_{i+} = P\{X \in A_i\}$, and $p_{+j} = P\{Y \in B_j\}$, then we have

$$\begin{aligned} p_{i+} &= P\{X \in A_i\} \\ &= P\{X \in A_i, Y \in B_1\} + P\{X \in A_i, Y \in B_2\} \\ &= p_{i1} + p_{i2}; \\ p_{+j} &= P\{Y \in B_j\} \\ &= P\{X \in A_1, Y \in B_j\} + P\{X \in A_2, Y \in B_j\} \\ &= p_{1j} + p_{2j}, \end{aligned}$$

and

$$p_{1+} + p_{2+} = p_{+1} + p_{+2} = 1.$$

The independence between X and Y implies $p_{ij} = P\{X \in A_i, Y \in B_j\} = P\{X \in A_i\}P\{Y \in B_j\} = p_{i+}p_{+j}$. Therefore, the null hypothesis is defined as

$$H_0: \ p_{ij} = p_{i+}p_{+j}, \quad i = 1, 2; j = 1, 2. \tag{2.3.9}$$

It is easy to know that the MLEs of p_{i+} and p_{+j} under H_0 are $\hat{p}_{i+} = n_{i+}/n$ and $\hat{p}_{+j} = n_{+j}/n$, respectively. Given the marginal data, that is, $N_{i+} = n_{i+}, N_{+j} = n_{+j}$, if H_0 is true, we have

$$EN_{ij} = n\hat{p}_{i+}\hat{p}_{+j}.$$

From Eq. (2.3.3), we can obtain the goodness-of-fit test

$$\chi^2 = \sum_{i=1}^{2}\sum_{j=1}^{2} \frac{(n_{ij} - n\hat{p}_{i+}\hat{p}_{+j})^2}{n\hat{p}_{i+}\hat{p}_{+j}}. \tag{2.3.10}$$

From Eq. (2.3.9), the number of independent parameters is $(2-1)+(2-1)$ in the parameter space corresponding to H_0. However, in the whole parameter space, the number of independent parameters is $2 \times 2 - 1$. Since $(2 \times 2 - 1) - (2-1) - (2-1) = 1$, by Theorem 2.2.2, we have $\chi^2 \xrightarrow{L} \chi_1^2$.

Using this method to calculate the data given by Table 2.1.1, we can get

O :	35	15	23	17
E :	32.22	17.78	25.78	22

Thus we have $\chi^2 = 1.52$, and the p-value is $P\{\chi_r^2 \geq 1.52\} \approx 0.19$. This probability is too great to reject the null hypothesis. Therefore we may deem that the new medicine is independent of cure rate, that is, we cannot believe that the new medicine is efficient.

Usually, Table 2.1.1 is called a two-way 2×2 contingency table. It is easy to generalize the above discussion to the situations of two-way $I \times J$ contingency table. Suppose that there are I disjoint events $A_1, \cdots, A_I$ and J disjoint events $B_1, \cdots, B_J$. Corresponding to Eq. (2.3.9), we need to test the null hypothesis

$$H_0: \ p_{ij} = p_{i+}p_{+j} \quad i = 1, \cdots, I; j = 1, \cdots, J. \tag{2.3.11}$$

The goodness-of-fit test here is given by Eq. (2.3.10). When H_0 is true, its limiting distribution is a χ^2 with $(I \times J - 1) - (I-1) - (J-1) = (I-1)(J-1)$ degrees of freedom. We can give the following theorem.

Theorem 2.3.3. *For a given $I \times J$ two-way contingency table, the null hypothesis of independence test is given by (2.3.11), and the goodness-of-fit test is*

$$\chi^2 = \sum_{i=1}^{I}\sum_{j=1}^{J} \frac{(n_{ij} - n\hat{p}_{i+}\hat{p}_{+j})^2}{n\hat{p}_{i+}\hat{p}_{+j}},$$

Table 2.3.1 Death penalty verdict (Ignoring victim's race)

Defendant's race	Death penalty Yes	No	Total
White	53	430	483
Black	15	176	191
Total	68	606	674

where $\hat{p}_{i+} = n_{i+}/n, \hat{p}_{+j} = n_{+j}/n$*, then, when the null hypothesis is true, we have*

$$\chi^2 \xrightarrow{L} \chi^2_{(I-1)(J-1)}.$$

For discussion convenience, sometimes we also call the intersection event that A_i and B_j occur simultaneously as cell (i, j) or equivalently, $A_i \bigcap B_j \iff (i, j)$. Let $\{X = i\}$ and $\{Y = j\}$ denote $\{X \in A_i\}$ and $\{Y \in B_j\}$, respectively.

2.3.3 *Test of Independence in Three-Way Contingency Table*

To strengthen the comprehension of the problem, we analyze an example given by Agresti (2002), which is related to whether or not there exists racial discrimination in the death penalty verdict in USA. The 647 subjects were the defendants in indictments involving cases with multiple murders in Florida between 1976 and 1987. The data is given in Table 2.3.1, which is a 2×2 two-way contingency table. Now we want to test the independence between death penalty and race of defendant. Based on the data given in Table 2.3.1, we have

O :	53	430	15	176
E :	48.73	434.27	19.27	171.73

then we have $\chi^2 = 1.47$ by calculating Eq. (2.3.10). The p-value is $P\{\chi^2_1 \geq 1.47\} \approx 0.18$. Obviously, the probability is too great to reject the null hypothesis. Therefore we can believe that death penalty verdict is independent of the defendant's race. From Table 2.3.1 we can see that the sample proportion receiving the death penalty is higher for white defendants than that for black defendants:

- White victim 11.9%;

Table 2.3.2 Death penalty verdict (Considering victim's race)

(a) Victim's race is white ($Z = 1$)

Defendant's race	Death penalty Yes	No	Total
White	53	414	467
Black	11	37	48
Total	64	451	515

(b) Victim's race is black ($Z = 2$)

Defendant's race	Death penalty Yes	No	Total
White	0	16	16
Black	4	139	143
Total	4	155	159

- Black victim 10.2%.

However it is necessary to analyze this problem further. In addition, if the victim's race is put into consideration, we can obtain Table 2.3.2.

We can see that Table 2.3.1 can be obtained by combining the data in Tables 2.3.2(a) and 2.3.2(b). As given in Table 2.3.1, let X denote the defendant's race, and Y denote the death penalty verdict. Furthermore, let Z denote the victim's race. We can obtain Tables 2.3.2(a) and 2.3.2(b) when $Z = 1$ and $Z = 2$, respectively. The core of testing is the problem of conditional independence when Z has been given. Corresponding to the case of two-way, let n_{ijk} denote the number of observations in cell (i, j, k). Let

$$P_{ijk} = P\{X = i, Y = j, Z = k\},$$

and

$$P_{i+k} = P\{X = i|Z = k\}, P_{+jk} = P\{Y = j|Z = k\}.$$

Based on Table 2.3.2, we consider the following three testing problems:

$$H_0 : P_{ijk} = P_{i+k} \cdot P_{+jk}, \quad i = 1, 2; j = 1, 2; k = 1, 2,$$

$$H_{01} : P_{ij1} = P_{i+1} \cdot P_{+j1}, \quad i = 1, 2; j = 1, 2,$$

$$H_{02} : P_{ij2} = P_{i+2} \cdot P_{+j2}, \quad i = 1, 2, j = 1, 2. \tag{2.3.12}$$

For the last two testing problems, let $X^2(k)$ denote the goodness-of-fit test given $Z = k$, then

$$X^2(k) = \sum_{i=1}^{2}\sum_{j=1}^{2} \frac{(n_{ijk} - n_{++k}\hat{p}_{i+k}\hat{p}_{+jk})^2}{n_{++k}\hat{p}_{i+k}\hat{p}_{+jk}}, \tag{2.3.13}$$

where $\hat{p}_{i+k} = n_{i+k}/n_{++k} = (n_{i1k} + n_{i2k})/n_{++k}$, $\hat{p}_{+jk} = n_{+jk}/n_{++k} = (n_{1jk} + n_{2jk})/n_{++k}$, $n_{++k} = n_{1+k} + n_{2+k} = n_{+1k} + n_{+2k}$. The goodness-of-fit test for H_0 is

$$X_2^2 = X^2(1) + X^2(2).$$

By independence, we know that the degrees of freedom of X_2^2 are $2 \times 1 \times 1 = 2$, thus we have $X_2^2 \to \chi_2^2$.

From Table 2.3.2, $X^2(1) = 5.34, X^2(2) = 0.46, X_2^2 = X^2(1) + X^2(2) = 5.8$. Check the χ^2 distribution table, $P\{\chi_1^2 \geq 5.34\} \approx 0.02$, $P\{\chi_1^2 \geq 0.46\} \approx 0.5$, $P\{\chi_2^2 \geq 5.8\} \approx 0.05$. In this way we may draw the following conclusion: When the victim is white, the death penalty verdict is related to the defendant's race. we can also see this based on proportions, one is 22.9% and the other is 11.3%; When the victim is black, the death penalty verdict is independent of the defendant's race. While considering the two circumstances simultaneously, we cannot accept the hypothesis that the death penalty verdict is independent of the defendant's race. As a result, there exists racial discrimination when the victim's race is considered.

It is very easy to generalize the three-way contingency table to general situation as $I \times J \times K$. Similar to Eq. (2.3.12), we consider testing the independence among the K conditions

$$\begin{array}{l} H_0 : P_{ijk} = P_{i+k} \cdot P_{+jk}, i = 1, \cdots, I; j = 1, \cdots, J; k = 1, \cdots, K. \\ H_{0k} : P_{ijk} = P_{i+k} \cdot P_{+jk}, i = 1, \cdots, I; j = 1, \cdots, J. \end{array} \tag{2.3.14}$$

Theorem 2.3.4. *For a given $I \times J \times K$ three-way contingency table, the testing problem of conditional independence is given as Eq. (2.3.14). The goodness-of-fit test $X^2(k)$ for the null hypothesis H_{0k} can be given by Eq. (2.3.13). The goodness-of-fit test for H_0 is*

$$X_K^2 = X^2(1) + \cdots + X^2(K). \tag{2.3.15}$$

When the null hypothesis is true, we have $X^2(k) \xrightarrow{L} \chi^2_{(I-1)(J-1)}$, and $X_K^2 \xrightarrow{L} \chi^2_{K(I-1)(J-1)}$.

As Table 2.3.1 can be combined by the two contingency tables, *i.e.* Tables 2.3.2(a) and 2.3.2(b), a three-way $I \times J \times K$ contingency table can be combined by K two-way $I \times J$ contingency table. To analyze a three-way or two two-way contingency table separately, on the one hand, we can use the goodness-of-fit test; on the other hand, we can use the odds ratio test discussed in Section 2.1. However, the conclusion may be different in these two ways, even contradict. When one constructs a three-way contingency table, the third variable usually denotes some kind of attribute, such as race, sex, or age. Sometimes it is unimaginable that the conclusion is different just because of the difference of the attribute. As an example, let us analyze the efficacy of a medical treatment. The third variable denotes sex: male or female. The medical treatment can be associated with a higher recovery rate for treated patients compared with the recovery rate for untreated patients; yet, treated male patients and treated female patients can each have lower recovery rates when compared with untreated male patients and untreated female patients. Conversely, higher recovery rates for treated patients in each subpopulation. Simpson (1951) first discussed this problem. Subsequently, there are many statisticians studying the compressibility of multi-way contingency tables. Further, it has been stimulating many people to study the latent variable which has been considered by Fisher. For overview results, see Bishop *et al.* (1975); Breslow and Day (1980); Agresti (2002).

2.3.4 *Goodness-of-Fit Test of Distribution Function*

Now return to the problem that has been originally proposed by Karl Pearson when he proposed the goodness-of-fit test. Let $x_1, \cdots, x_n$ be a set of samples from $F(x)$, and let $F_0(x)$ be a given distribution function. Now we want to test

$$H_0: \quad F(x) = F_0(x). \tag{2.3.16}$$

For example, we want to test whether the scores of some students are distributed as $N(\mu, \sigma^2)$ or not. In this situation, we are more interested in testing the distribution. Therefore, we usually substitute MLEs for unknown parameters, namely $\mu = \bar{x}, \sigma^2 = \sum_{i=1}^n (x_i - \bar{x})^2/n$.

Let (a, b) denote the interval from which samples are drawn, a and b may be negative or positive infinity. Taking $k-1$ real numbers, partitioning (a, b) into k intervals: $(a, y_1], (y_1, y_2], \cdots, (y_{k-1}, b)$. Let n_i denote the number of samples which fall in the interval $(y_{i-1}, y_i]$. Here, $y_0 = a, y_k = b$. Let

$$p_{0i} = F_0(y_i) - F_0(y_{i-1}), \quad i = 1, \cdots, k.$$

By Eq. (2.3.3), the goodness-of-fit test about the distribution function is

$$\chi^2 = \sum_{i=1}^{k} \frac{(n_i - np_{0i})^2}{np_{0i}},$$

where $n = n_1 + \cdots + n_k$. When the null hypothesis (2.3.16) is true, the limiting distribution of χ^2 is χ^2_{k-1}.

In the above testing process, the problem of determining the number k of the intervals and the real number y_i is very important and very difficult. No matter how to partition, it is inadmissible when the samples are too small in any interval. Otherwise, the result will be affected significantly (see Simpson and Margolin, 1986).

2.4 Sign Tests and Rank Tests

This section mainly discusses a testing problem concerning medians. In the discussion, we can see that the proposed methods need not to make any hypothesis about the special forms of distributions. So these methods can be called to be **distribution-free**, or **nonparametric**. For more detailed discussions relating to the contents in this section, see Hettmansperger (1984).

First, let us recall the definition of median in Section 1.4. Let $(\mathcal{X}, \mathcal{A}, \mathcal{P})$ denote a statistical space, where $\mathcal{P} = \{P_\theta; \theta \in \Theta\}$ is a continuous distribution, namely $\mathcal{P} \ll \nu$, and ν is the Lebesgue measure. θ is called a median if

$$P_\theta\{X \leq \theta\} = P_\theta\{X \geq \theta\} = \frac{1}{2}. \tag{2.4.1}$$

Let $F(x) = P_\theta\{X \leq x\}$. So the median is the solution of $F(\theta) = 1/2$. Let $x_1, \cdots, x_n$ be a set of samples from the statistical space $(\mathcal{X}, \mathcal{A}, \mathcal{P})$, and then we consider testing

$$H_0 : \theta = 0. \tag{2.4.2}$$

Here, we make some explanations of the above-mentioned testing:

(1) Suppose that the parameter space is $\Theta = \{\theta;\ \theta \geq 0\}$, namely considering one-sided test. This is only for the sake of discussion convenience, and is easy to be generalized into the situation of two-sided test. Recalling the discussion in Section 2.1, we can see that Eq. (2.4.2) is consistent with $H_0 : \theta \leq 0$.

(2) H_0 has universality. For example, if we test the hypothesis $H_0' : \theta = \theta_0$, let $y_i = x_i - \theta_0$, $i = 1, \cdots, n$, so when H_0' is true, $P_0\{Y \leq 0\} = P_{\theta_0}\{X - \theta_0 \leq 0\} = 1/2$, then it can be transformed into the form of Eq. (2.4.2).

(3) H_0 is a composite null hypothesis, because it includes all continuous distributions with zero medians.

2.4.1 *Sign Tests*

This is one of the most ancient methods of statistics, traced back to Arbuthnott (1710) who made an investigation for the birth rate of male infants in London. Although the sign testing method is simple, it possesses many favorable statistical properties, especially constructing the basic idea of the nonparametric method. Let

$$G = \sum_{i=1}^{n} g(X_i), \tag{2.4.3}$$

here, $g(x) = 1$ when $x > 0$; $g(x) = 0$ otherwise. Then G is a random variable denoting the number of positive ones among $X_1, \cdots, X_n$. Let $g = \sum g(x_i)$ denote the sample value of G. Obviously, when g is larger, we can reject H_0 as given in (2.4.2), and regard that the median $\theta > 0$. The G given in Eq. (2.4.3) is called a **sign test**.

Now we discuss the distribution of G if H_0 is true. For $\theta \in \Theta$, let

$$p_\theta = P_\theta\{X > 0\} = 1 - F(-\theta).$$

According to Example 1.3.4, $g(X_1), \cdots, g(X_n)$ are a set of samples from Bernoulli trials with success probability p_θ, thus we have $G \sim Bi(n, p_\theta)$. Note that the following result is independent of the sample distribution F:

$$H_0 : \theta = 0 \Leftrightarrow H_0 : p_0 = \frac{1}{2}. \tag{2.4.4}$$

However, if $\theta > 0$, p_θ is relevant to the sample distribution F. We can easily get

$$E_\theta G = n(1 - F(-\theta)),$$

$$V_\theta G = n(1 - F(-\theta))F(-\theta).$$

According to the Central Limit Theorem, when θ is the true value,

$$\frac{G - E_\theta G}{\sqrt{V_\theta G}} \xrightarrow{L} Z \sim N(0, 1).$$

Especially when H_0 is true, $E_0 G = n/2, V_0 G = n/4$. So the p-value is

$$\begin{aligned} P\{G \geq g\} &= P\left\{\frac{G - E_0 G}{\sqrt{V_0 G}} \geq \left(g - \frac{n}{2}\right) \Big/ \sqrt{\frac{n}{4}}\right\} \\ &\approx 1 - \Phi\left(\left(g - \frac{n}{2}\right) \Big/ \sqrt{\frac{n}{4}}\right), \end{aligned} \tag{2.4.5}$$

Table 2.4.1 Continuity correction for the binomial distribution $Bi(5, 1/2)$.

	g=0	g=1	g=3
$P(G \leq g)$	0.0313	0.1875	0.5
$\Phi(Z_g)$	0.0127	0.0901	0.6736
$\Phi(Z_g^*)$	0.0367	0.1867	0.5

where $\Phi(\cdot)$ denotes the distribution function of the standard normal distribution. Moreover, let

$$Z_g = \left(g - \frac{n}{2}\right) \Big/ \sqrt{\frac{n}{4}}, \tag{2.4.6}$$

by Eq. (2.4.5), the p-value can be approximated by $1 - \Phi(Z_g)$. Because the binomial distribution is discrete, then when n is smaller, the approximation based on the normal distribution given in Eq. (2.4.5) has lower precision. Therefore, in application, we usually make 0.5 correction about Z_g in Eq. (2.4.6). The intent of the correction is to make Z_g focus on the origin: when $g - n/2 < 0$, let

$$Z_g^* = \left(g + 0.5 - \frac{n}{2}\right) \Big/ (\sqrt{\frac{n}{4}}); \tag{2.4.7}$$

When $g - n/2 \geq 0$, let

$$Z_g^* = \left(g - 0.5 - \frac{n}{2}\right) \Big/ \left(\sqrt{\frac{n}{4}}\right). \tag{2.4.8}$$

Table 2.4.1 illustrates the effect of the continuity correction for the binomial distribution $Bi(5, 1/2)$. Here, when $g = 0$ or $g = 1$ we can use Eq. (2.4.7); When $g = 3$ we can use Eq. (2.4.8). It can be seen that the correction is quite effective.

2.4.2 *Wilcoxon Signed Rank Tests*

In the next chapter we will see that sign test has lots of statistical properties, such as uniformly most powerful. However, it only utilizes sign information of observations, but no information about the distance between each observation and the origin, which is not enough especially for symmetrical distribution. Now the testing method discussed will make use of this information.

Let $F(x)$ be a continuous distribution function, and **symmetrical** about the point 0, namely

$$F(x) = 1 - F(-x). \tag{2.4.9}$$

Then the point 0 is both the median and the mean (if the mean exists). We assume the samples $x_1, \cdots, x_n$ are drawn randomly from $F(x-\theta)$, then θ is both the median and the mean. Suppose that the parameter space is $\Theta = \{\theta; \theta \geq 0\}$, the testing problem is

$$H_0 : \theta = 0. \tag{2.4.10}$$

Many random variables have symmetrical distribution, especially those related to error and random effect. In addition, we can also use the symmetrical distribution to study paired-data. For example, when we analyze a clinical trial, let T denote the treatment (such as drug) effect, C denote the control effect. We are interested in whether there exists significant difference between the treatment and the control or not. The null hypothesis is that there is no significant difference. Therefore, if the null hypothesis is true,

$$P\{T - C \leq x\} = P\{C - T \leq x\} = P\{-(T - C) \leq x\}.$$

If let $X = T - C$, then both X and $-X$ have the same distribution, namely the distribution of X is symmetrical about the point 0.

We can discuss the characteristics of the random sample from the symmetrical distribution about the point 0:

(1) The amount of positive numbers is almost the same as that of negative numbers among $x_1, \cdots, x_n$;
(2) The sum of positive numbers corresponding to $|x_1|, \cdots, |x_n|$ is almost the same as that of the negative numbers.

The above two characteristics are the base of constructing test statistics. In order to characterize the second one, we need to introduce the definition of rank.

In general, among $z_1, \cdots, z_n$, the rank of z_i represents the number in $z_1, \cdots, z_n$ which is not larger than z_i. So, let $z_{(1)} < \cdots < z_{(n)}$ be order statistics, if $z_i = z_{(j)}$, then the **rank** of z_i is j, denoted by $r(z_i) = j$.

Now, let $|x|_{(1)}, \cdots, |x|_{(n)}$ be the order statistics of $|x_1|, \cdots, |x_n|$, and r_i denote the rank of $|x_i|$, namely $|x_i| = |x|_{(r_i)}$, then the **Wilcoxon sign test** is

$$T = \sum_{i=1}^{n} iW_i = \sum_{i=1}^{n} r_i g(x_i), \tag{2.4.11}$$

where $W_i = 1$ if $|x|_{(i)}$ is corresponding to the positive observation; $W_i = 0$ if $|x|_{(i)}$ is corresponding to the negative observation, and $g(x)$ is given by the sign test (2.4.3). When T is larger, we should reject the null hypothesis. So, this statistic possesses two characteristics that samples should have.

Now we will discuss the distribution of T when the null hypothesis H_0 is true. Let R_i denote the random variable corresponding to r_i.

Theorem 2.4.1. *When H_0 is true, $g(X_1), \cdots, g(X_n)$ are independent of $(R_1, \cdots, R_n)$.*

Proof. Because $(g(X_1), |X_1|), \cdots, (g(X_n), |X_n|)$ are mutually independent, and $(R_1, \cdots, R_n)$ is the function of $(|X_1|, \cdots, |X_n|)$, so it suffices to prove that $g(X_i)$ and $|X_i|$ are independent. In fact,

$$\begin{aligned} P\{g(X_i) = 1, |X_i| \leq x\} &= P\{0 \leq X_i \leq x\} \\ &= F(x) - \frac{1}{2} \\ &= \frac{1}{2}(2F(x) - 1) \\ &= P\{g(X_i) = 1\}P\{|X_i| \leq x\}. \end{aligned}$$

Similarly, we have $P\{g(X_i) = 0, |X_i| \leq x\} = P\{g(X_i) = 0\}P\{|X_i| \leq x\}$.

Let d_i satisfy $|x_{d_i}| = |x|_{(i)}$, so W_i in Eq. (2.4.11) satisfies

$$W_i = g(x_{d_i}), \tag{2.4.12}$$

here, d_i is called as **anti-rank**. Let D_i denote a random variable corresponding to d_i. Similar to the proof of Theorem 2.4.1, we can get that $g(X_1), \cdots, g(X_n)$ and $(D_1, \cdots, D_n)$ are independent. □

Theorem 2.4.2. *If the null hypothesis H_0 is true, then $W_1, \cdots, W_n$ are i.i.d., and*

$$P\{W_i = 0\} = P\{W_i = 1\} = \frac{1}{2}. \tag{2.4.13}$$

Proof. Let $D = (D_1, \cdots, D_n)$, and $d = (d_1, \cdots, d_n)$. According to Eq. (2.4.12), making summation about all of the possible values of α yields

$$\begin{aligned} &P\{W_1 = w_1, \cdots, W_n = w_n\} \\ &= \textstyle\sum_d P\{g(X_{D_1}) = w_1, \cdots, g(X_{D_n}) = w_n | D = d\}P\{D = d\} \\ &= \textstyle\sum_d P\{g(X_{D_1}) = w_1, \cdots, g(X_{D_n}) = w_n\}P\{D = d\} \\ &= \textstyle\sum_d (1/2)^n \, P\{D = d\} \\ &= (1/2)^n, \end{aligned}$$

that is, $P\{W_1 = w_1, \cdots, W_n = w_n\} = \prod_i P\{W_i = w_i\}$, where $P\{W_i = w_i\} = 1/2$.

Therefore, the statistic $T = \sum_{i=1}^{n} iW_i$ given by Eq. (2.4.11) is a linear combination of n independent r.v.s with common cdf $Bi(1, 1/2)$. When H_0 is true, the combination is irrelevant to the special distribution of F, so it is distribution-free.

Now considering the case $n = 4$. There are four classes for all possible values of r_i, namely 1,2,3,4. The total number of all possible combination of the signs is $2^4 = 16$. When $T \geq 6$, we can obtain following table.

	Rank				
	1	2	3	4	T-value
Sign	+	+	+	+	10
	−	+	+	+	9
	+	−	+	+	8
	+	+	−	+	7
	−	+	+	+	6
	−	−	+	+	7
	−	+	−	+	6

Since the probability of each combination of signs is 1/16, then we can compute the distribution of T. For example

$$P\{T = 10\} = \frac{1}{16}, \quad P\{T = 6\} = \frac{2}{16}.$$

Let t be the sample value of T, then when $t = 9$, the p-value is

$$p = P\{T \geq 9\} = \frac{1}{8} = 0.125.$$

In addition, we can also compute the limiting distribution of T. By Theorem 2.4.2, we can get

$$E_0T = \frac{n(n+1)}{4}, \quad \text{and} \quad V_0T = \frac{n(n+1)(2n+1)}{24}. \tag{2.4.14}$$

Then applying the Central Limit Theorem, we have

$$\frac{T - E_0T}{V_0T} \xrightarrow{L} N(0,1).$$

Just like the sign test, when n is smaller, the effect of the normal approximation above is not good. Because the distribution of T is symmetrical, we can apply Edgeworth approximation, see Cramér (1946, §12.6 and §17.7). By now the approximated p-value is

$$P\{T \geq t\} = \Phi(s) - \frac{1}{24}\lambda(s^3 - 3s)\phi(s), \tag{2.4.15}$$

where $\Phi(\cdot)$ and $\phi(\cdot)$ are, respectively, the cumulative distribution function and the density function of the standard normal distribution,

$$\lambda = \frac{E_0(T - E_0T)^4}{(V_0T)^2} - 3, \quad \text{and} \quad s = \frac{t - E_0T}{\sqrt{V_0T}}.$$

As discussed in Fellingham and Stoker (1964), Eq. (2.4.15) can be written as

$$P\{T \geq t\} = \Phi(s) + (\frac{3n^2 + 3n - 1}{10n(n+1)(2n+1)})(s^3 - 3s)\phi(s).$$

□

2.4.3 *Wilcoxon Rank Tests*

When utilizing ranks to construct test statistics, another advantage lies in the fact that we can test the relation of medians between two populations. Let $x_1, \cdots, x_m$ be a random sample from $F(x - \theta_x)$, and $y_1, \cdots, y_n$ be a random sample from $F(x - \theta_y)$. Suppose that the parameter space is $\Theta = \{(\theta_x, \theta_y); \theta_x \leq \theta_y\}$, we are interested in testing whether the two medians are equal or not. If let $\Delta = \theta_y - \theta_x$, then $\Theta = \{\Delta; \Delta \geq 0\}$, testing

$$H_0 : \Delta = 0. \tag{2.4.16}$$

Now, it is not necessary to assume the distribution $F(\cdot)$ is symmetrical, although the obtained results are also valid for symmetrical distributions.

Without loss of generality, suppose $x_1 < \cdots < x_m$ and $y_1 < \cdots < y_n$. After mixing and arranging the two groups of samples, we can get

$$z_1 < \cdots < z_{m+n}.$$

By the definition of rank, if $y_i = z_j$, then $r_i = r(y_i) = j$, where $i = 1, \cdots, n$, and the range of j is $1, \cdots, m+n$. Wilcoxon (1945) suggested that one should use the sum of the rank of y as the statistic to test (2.4.16), that is, the **Wilcoxon rank test** is

$$W = \sum_{i=1}^{n} R_i, \tag{2.4.17}$$

where R_i is corresponding to the random variable of r_i. Let $N = m + n$. It is easy to prove that when H_0 is true, we have

$$P(R_i = s) = \frac{(N-1)!}{N!} = \frac{1}{N}, \quad s = 1, \cdots, N;$$

$$P(R_i = s, R_j = t) = \begin{cases} \dfrac{1}{N(N-1)}, & \text{if} \quad s \neq t, \\ 0, & \text{if} \quad s = t. \end{cases}$$

Thus we have, $E_0 R_i = (N+1)/2$, and $V_0 R_i = (N^2-1)/12$; When $i \neq j$, we have $CV_0(R_i, R_j) = -(N+1)/12$. Thus, for Eq. (2.4.17)

$$E_0 W = \frac{1}{2}n(N+1), \quad \text{and} \quad V_0 W = \frac{1}{12}mn(N+1). \tag{2.4.18}$$

Mann and Whitney (1947) introduced a count-type test statistic. Considering the number of that y is larger than x in the mixed samples, namely

$$U = \#(x_j < y_i), \quad i = 1, \cdots, n; \ j = 1, \cdots, m. \tag{2.4.19}$$

Since R_i is the number of samples that is not larger than that of Y_i, we have

$$r_i = \#(x_j < y_i) + \#(y_k < y_i),$$

thus

$$\sum_{i=1}^{n} r_i = \#(x_j < y_i) + \frac{1}{2}n(n+1).$$

Comparing Eq. (2.4.17) with Eq. (2.4.19), we can get

$$U = W - \frac{1}{2}n(n+1).$$

Since this is a shift transformation, we have

$$E_0 U = \frac{1}{2}mn,$$

according to Eq. (2.4.18), we have

$$V_0 U = \frac{1}{12}mn(N+1). \tag{2.4.20}$$

Just like the sign test, it is difficult for us to give the exact distribution of the Wilcoxon rank test and Mann-Whitney count test. But according to Eq. (2.4.19), it is easy to provide a computer program to compute the distribution. For $m = n = 2$, we can compute it as follows:

	Rank					
	1	2	3	4	U	W
Arrangement	y	y	x	x	0	3
	y	x	y	x	1	4
	y	x	x	y	2	5
	x	y	y	x	2	5
	x	y	x	y	3	6
	x	x	y	y	4	7

Thus when H_0 is true, the probability distribution of U is

u	0	1	2	3	4
$P\{U = u\}$	1/6	1/6	2/6	1/6	1/6

Similarly, we can also obtain probability distribution of W. So the above distribution is irrelevant to the special form of the original distribution function F. Therefore it is distribution-free. Moreover, we can obtain the limiting distributions of U and W according to Eqs. (2.4.18) and (2.4.20).

Fix and Hodges (1955) gave the following Edgeworth approximation

$$P\{U \le u\} = \Phi(t) + \frac{m^2 + n^2 + mn + m + n}{20mn(m + n + 1)}(t^3 - 3t)\phi(t),$$

where $t = (u + 0.5 - mn/2)/\sqrt{mn(m+n+1)/12}$, and $\Phi(\cdot)$ and $\phi(\cdot)$ are defined as in Eq. (2.4.15). In virtue of the definition of the statistic U in Eq. (2.4.19), following the idea of Hodges and Lehmann (1962) and Lehmann (1963), we can obtain an estimator of Δ

$$\hat{\Delta} = \text{med}\{y_i - x_j\}. \tag{2.4.21}$$

Statistical properties of the above estimator have been discussed generally in many textbooks. Testing problems of the relationships among several medians will be discussed in Chapter 4.

2.5 U-I Tests

The methods discussed in this section are applicable to some situations with composite null hypotheses. Now we can consider the testing problem which has been mentioned at the beginning of this chapter as given by Eq. (2.1.1), *i.e.*, $H_0 : \theta \in \Theta_0$, where Θ_0 is a proper subset of the parameter space Θ. In some situations, Θ_0 may be regarded as an intersection of some sets. Thus, the testing problem can be formulated as

$$H_0 : \theta \in \bigcap_{i=1}^{k} \Theta_{0i}, \tag{2.5.1}$$

and then the null hypothesis H_0 is the combination of the common part of some subclasses of the null hypotheses, where the subclasses of the null hypotheses are

$$H_{0i} : \theta \in \Theta_{0i}, \quad i = 1, \cdots, k. \tag{2.5.2}$$

The basic idea of the U-I test is, if any one of k subclasses of the null hypotheses mentioned above are rejected, and then we can reject the null hypothesis H_0, where U-I is an abbreviation of the Union-Intersection. Now $T_i(x)$ is a test statistic related to H_{0i}, and when any one of $T_i(x) \geq c$ for some constant c, or any one

$$p_i = \sup_{\theta \in \Theta_{0i}} P_\theta\{T_i(X) \geq T_i(x)\} \tag{2.5.3}$$

is smaller than some constant c', then we can reject H_{0i}, and accordingly reject H_0. Thus, the p-value should be $p_0 = \min\{p_i\}$ for H_0. Now it is quite difficult to adopt the rejecting probability. When $p_0 \leq \alpha/k$, in general, we can reject H_0, where α is a given level probability, say $\alpha = 0.05$. If there are the same distributions among the $T_i(X)$, and the same correlation coefficients between $T_i(X)$ and $T_j(X)$, $i \neq j$, then the **U-I test statistic** may be denoted as

$$T(X) = \max\{T_i(X)\}. \tag{2.5.4}$$

Now the p-value is

$$p = \sup_{\theta \in \Theta_0} P_\theta\{T(X) \geq T(x)\}.$$

When $p \leq \alpha$, we can reject H_0. Similarly, we also can find a t_α, such that

$$\sup_{\theta \in \Theta_0} P_\theta\{T(X) \geq t_\alpha\} = \alpha. \tag{2.5.5}$$

When $T(x) \geq t_\alpha$, we can reject H_0. On the one hand, we will find it is convenient to use the test statistic given in Eq. (2.5.4), but the disadvantage is that sometimes it is difficult to compute the distribution of $T(x)$ under H_0. On the other hand, in most situations, it is convenient to test by using the p-value given in Eq. (2.5.3). However, it also becomes difficult that how to determine the rejecting probability. Whereas, we can still utilize the U-I test to deal with some difficult statistical problems. The second example as follow will illustrate this point.

Example 2.5.1. (Two-sided test for normal mean) Now we will analyze the problem mentioned in Example 2.2.2 with the U-I test. Let $x_1, \cdots, x_n$ be a set of samples from the normal population $N(\mu, \sigma^2)$. Consider the testing problem $H_0 : \mu = \mu_0$. If $\Theta_{01} = \{\mu; \mu \leq \mu_0\}, \Theta_{02} = \{\mu; \mu \geq \mu_0\}$, the testing problem can be expressed as

$$H_0 : \mu \in \Theta_{01} \cap \Theta_{02}, \tag{2.5.6}$$

which is the testing problem composed by the intersection of two one-sided testing problems. From the discussion of Example 2.2.2, the LRTs of $H_{02} : \mu \in \Theta_{02}$ and $H_{01} : \mu \in \Theta_{01}$ all depend on the Student's t-test given by Eq. (2.2.12), *i.e.*, $t = t(x) = \sqrt{n}(\overline{x} - \mu_0)/s$. Thus the two p-values are $p_1 = P_{\mu_0}\{t_1(X) \geq t_1(x)\}$, and $p_2 = P_{\mu_0}\{t_2(X) \leq t_2(x)\}$ respectively, where $t_1(x) = t(x)$, and $t_2(x) = -t(x)$. According to the principle of the U-I test, the p-value corresponding to (2.5.6) is

$$p = \min\{p_1, p_2\}. \tag{2.5.7}$$

Because when H_0 is true, the Student's t-distribution is a symmetrical distribution, from (2.5.4) we can find that the U-I test statistic is

$$t(x) = \max\{t_1(x), t_2(x)\}.$$

Thus the p-value is

$$p = P_{\mu_0}\{|t(X)| \geq |t(x)|\}$$

this is consistent with the expression in Example 2.2.2.

Example 2.5.2. (Testing normal mean and variance simultaneously) Recall that the test about the normal parameters in Examples 2.2.2 and 2.2.3, the mean and variance are discussed respectively. If the mean and variance are discussed simultaneously, the problem will become quite difficult. Now we will analyze the problem.

Let $x_1, \cdots, x_n$ be a set of samples from the normal distribution $N(\mu, \sigma^2)$. When μ_0 and $\sigma_0{}^2$ are known, considering the testing problem

$$H_0 : \mu \leq \mu_0, \sigma^2 \geq \sigma_0{}^2. \tag{2.5.8}$$

Here the parameter space is $\Theta = \{(\mu, \sigma^2); -\infty < \mu < +\infty, \sigma^2 > 0\}$. The problem (2.5.8) is of great significance. Obviously when we design a new teaching method, for example, we expect higher average scores of students with smaller variance. Whereas, dealing with the LR test of that problem, especially computing the p-value, will become very difficult. One can refer to Hsu (1938). Now we will employ the U-I test to analyze that problem. Let

$$\Theta_{01} = \{(\mu, \sigma^2); \mu \leq \mu_0, \sigma^2 > 0\},$$

$$\Theta_{02} = \{(\mu, \sigma^2); -\infty < \mu < +\infty, \sigma^2 \geq \sigma_0{}^2\}.$$

The testing problem (2.5.8) may be denoted as

$$H_0 : \quad \theta \in \Theta_{01} \cap \Theta_{02}, \tag{2.5.9}$$

where $\theta = (\mu, \sigma^2)$. From Example 2.2.2, the LR test statistic for the testing problem $H_{01} : \theta \in \Theta_{01}$ is a Student's t-test statistic, *i.e.*,

$$t(x) = \frac{\sqrt{n}(\overline{x} - \mu_0)}{s},$$

the p-value is $p_1 = P_{\mu_0}\{t(X) \geq t(x)\}$. We can obtain the following LR test about the testing problem $H_{02} : \theta \in \Theta_{02}$

$$\chi^2(x) = \sum_{i=1}^{n}(x_i - \overline{x})^2.$$

When $\chi^2(x) \leq c$ for some constant c, we can reject H_{02}. As the true value of variance is σ^2, $\chi^2(X)/\sigma^2 \sim \chi^2_{n-1}$ and the p-value is

$$p_2 = \sup_{\sigma^2 \geq \sigma_0{}^2} P_{\sigma^2}\{\chi^2(X) \leq \chi^2(x)\} = P_{\sigma_0{}^2}\{\chi^2(X) \leq \chi^2(x)\}.$$

According to the principle of the U-I test, for the testing problem (2.5.9) the p-value is

$$p = \min\{p_1, p_2\}.$$

Generally, when $p \leq \alpha/2$ we can reject H_0.

Example 2.5.3. (Test for multivariate normal mean) This example will serve as a motivation for the topics discussed in the next chapter. We have found that we can construct several test statistics for a single testing problem. How to judge whether a test is good or not? Or how to judge the power of test statistics? From the example below, we can find that it may not be enough just by the p-value.

Let $x_1, \cdots, x_n$ be a set of samples from the p-dimensional normal distribution $N(\boldsymbol{\theta}, I_p)$, where $\boldsymbol{\theta} = (\theta_1, \cdots, \theta_p)'$. Consider the testing problem

$$H_0 : \boldsymbol{\theta} = \mathbf{0}. \tag{2.5.10}$$

It is easy to know that the LR test is

$$\chi^2(\boldsymbol{x}) = n\bar{\boldsymbol{x}}'\bar{\boldsymbol{x}} = n\sum_{j=1}^{p} \bar{x}_j^2, \tag{2.5.11}$$

where $\bar{x}_j = \sum\limits_{i=1}^{n} x_{ij}/n$, $j = 1, \cdots, p$, and $\bar{\boldsymbol{x}} = (\bar{x}_1, \cdots, \bar{x}_p)'$. Under H_0, $\chi^2(\boldsymbol{X}) \sim \chi^2_p$. When α is known, we can calculate χ^2_α, such that

$$P_0(\chi^2(\boldsymbol{X}) \geq \chi^2_\alpha) = \alpha.$$

When $\chi^2(\boldsymbol{x}) \geq \chi^2_\alpha$, we can reject H_0.

We can also employ the U-I test method to test (2.5.10). Let $\Theta_{0j} = \{\boldsymbol{\theta} \in R^p; \theta_j = 0\}$,

$$H_0 : \boldsymbol{\theta} \in \bigcap_{j=1}^{p} \Theta_{0j}.$$

The LR test for $H_{0j} : \boldsymbol{\theta} \in \Theta_{0j}$ is $|\bar{x}_j|$, then the U-I test is

$$U(\boldsymbol{x}) = \max\{|\bar{x}_1|, \cdots, |\bar{x}_p|\}. \tag{2.5.12}$$

Let $y = \max\{\sqrt{n}\bar{x}_1, \cdots, \sqrt{n}\bar{x}_p\}$, when H_0 is true, the density function of y is

$$p[\Phi(y)]^{p-1}\phi(y),$$

where $\Phi(\cdot)$ and $\phi(\cdot)$ are the standard normal cumulative distribution function and density function, respectively. Thus we can get U_α, such that

$$P\{U(\boldsymbol{X}) \geq U_\alpha\} = \alpha.$$

When $U(\boldsymbol{x}) \geq U_\alpha$, we can reject H_0. When $p = 2$, the rejection region of

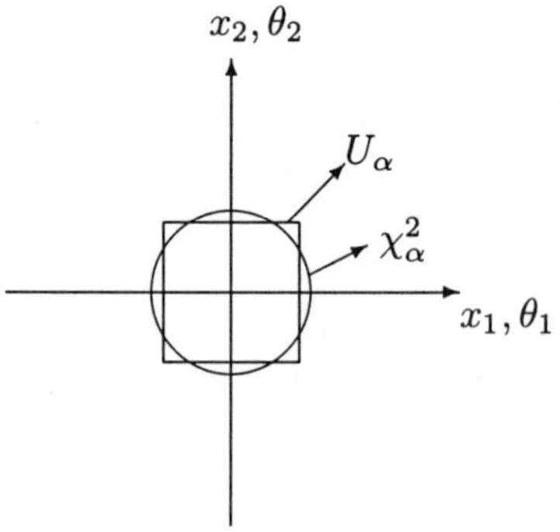

the two test statistics is shown as in the figure. For the LR test, when the sample statistics is outside of the circle shown as χ^2_α, we can reject H_0; for the U-I test, when the sample statistics is outside of the square shown as U_α, we can reject H_0. Thus for the same α, there exist such points that we can reject the LR test but not the U-I test; Similarly, there exist such points that we can reject the U-I test but not the LR test. Which one is better? We will discuss it in the next chapter.

2.6 Empirical Likelihood-Ratio Tests

Let $x_1, \cdots, x_n$ be a set of random samples from $F(x)$, we further considered tests on mean. Let

$$\theta = \int x dF(x). \tag{2.6.1}$$

Suppose we wish to test the null hypothesis

$$H_0 : \theta = \theta_0. \qquad (2.6.2)$$

If H_0 is true, we regard that the true distribution is F_0. As we know, when the form of distribution is already known, we can use the LRT; When the distribution is discrete, we can use the goodness-of-fit test; When the density function of the distribution is symmetrical on mean, we can use the Wilcoxon signed rank test. In this section, we discuss the testing problem (2.6.2) when the priori information of distribution function is completely unknown.

In the last part of Chapter 1, we have discussed that we can estimate the population distribution function with the empirical distribution function and that many optimal properties this estimation procedure has. Now, we will make use of empirical distribution function to consider some testing problems. We use $f(x)$ to denote the density function corresponding to the distribution function $F(x)$. Let

$$p_i = f(x_i) / \sum_{j=1}^{n} f(x_j), \quad i = 1, \cdots, n. \qquad (2.6.3)$$

According to the proof of Theorem 1.4.7, generally, we can assume that the sample values $\{x_1, \cdots, x_n\}$ are different, so, the definition of (2.6.3) is clear, and

$$p_i \geq 0, \ i = 1, \cdots, n; \text{ and } \sum_{i=1}^{n} p_i = 1. \qquad (2.6.4)$$

Here, $p_1, \cdots, p_n$ are called as **weight probabilities** on the basis of $\{x_1, \cdots, x_n\}$. By the formula (1.4.34) in Chapter 1, we can write the empirical likelihood function as follows

$$L(F) = \prod_{i=1}^{n} p_i. \qquad (2.6.5)$$

Let $\mathcal{F}$ denote the set including all of the distribution functions, and F_n denote the empirical distribution function, according to Theorem 1.4.7, we know

$$L(F_n) = \sup_{F \in \mathcal{F}} L(F).$$

By the definition, the weight probabilities corresponding to F_n are $p_1 = \cdots = p_n = 1/n$. At this time, the set of distribution functions satisfying the null hypothesis (2.6.2) and based on the weight probabilities $p_1, \cdots, p_n$ is

$$\mathcal{F}_0 = \{F; \sum_{i=1}^{n} x_i p_i = \theta_0, p_i \geq 0, \sum_{i=1}^{n} p_i = 1\}. \tag{2.6.6}$$

In order to guarantee that $\mathcal{F}_0$ is not an empty set, we need to assume the condition

$$x_{(1)} < \theta_0 < x_{(n)}, \tag{2.6.7}$$

where $x_{(1)} = \min\{x_i\}$ and $x_{(n)} = \max\{x_i\}$. According to the idea of likelihood-ratio test, define

$$\begin{aligned}\Lambda(x; \theta_0) &= \sup_{F \in \mathcal{F}_0} \frac{L(F)}{L(F_n)} \\ &= \sup_{F \in \mathcal{F}_0} \prod_{i=1}^{n} n p_i\end{aligned}$$

as **an empirical likelihood ratio test**, abbreviated to ELR test. Obviously, ELR test also satisfies the three properties of LRT, so, reject the null hypothesis when $\Lambda(x; \theta_0) \leq c$ for some constant c, or equivalently, consider

$$\lambda(x; \theta_0) = -2 \ln \Lambda(x; \theta_0) = \inf_{F \in \mathcal{F}_0} \left[-2 \sum_{i=1}^{n} \ln(n p_i) \right], \tag{2.6.8}$$

reject the null hypothesis when it is bigger than some constant.

Theorem 2.6.1. *If $p_i > 0$ holds for any $i = 1, \cdots, n$, then the test statistic given by Eq. (2.6.8) is*

$$\lambda(x; \theta_0) = 2 \sum_{i=1}^{n} \ln[1 + \beta_0 (x_i - \theta_0)], \tag{2.6.9}$$

where β_0 is the root of $g(\beta) = 0$,

$$g(\beta) = \frac{1}{n} \sum_{i=1}^{n} \frac{x_i - \theta_0}{1 + \beta(x_i - \theta_0)}. \tag{2.6.10}$$

Proof. By Eq. (2.6.8), at this moment, the Lagrangian equation can be written as

$$G(p, \omega, \lambda) = \sum_{i=1}^{n} \ln(n p_i) + \omega(1 - \sum_{i=1}^{n} p_i) + \beta n(\theta_0 - \sum_{i=1}^{n} x_i p_i).$$

Taking partial derivatives with respect to each p_i, and setting them equal to zero, we have

$$\frac{\partial G(p,\omega,\lambda)}{\partial p_i} = \frac{1}{p_i} - \omega - \beta n x_i = 0. \tag{2.6.11}$$

Now, multiplying the two side of the equation above by p_i, we can get

$$\omega p_i + \beta n x_i p_i = 1, \quad i = 1, \cdots, n.$$

Making summation and utilizing the restriction condition, we can get $\omega = n(1-\beta\theta_0)$. Substituting it into Eq. (2.6.11) yields

$$p_i = \frac{1}{n[1+\beta(x_i-\theta_0)]}. \tag{2.6.12}$$

We can get Eq. (2.6.9) by substituting Eq. (2.6.12) into Eq. (2.6.8). Furthermore, we can obtain Eq. (2.6.10) by substituting Eq. (2.6.12) into the restriction condition given by Eq. (2.6.6). It is easy to verify that the Hessian matrix of $\sum \ln(np_i)$ is a negative-definite matrix, thus the desired conclusion holds, and the solution of Eq. (2.6.8) is unique. □

Now, we need to figure out the solution of Eq. (2.6.10). According to the condition (2.6.7) we know that the solution lies in the interval $(-1/(x_{(n)}-\theta_0), -1/(x_{(1)}-\theta_0))$. Since for any β, we have $g'(\beta) < 0$. So, $g(\beta)$ is a strict decreasing function of β. Thus consequently, the solution of $g(\beta) = 0$ in the above interval is unique, and we can use the bisection method to work it out.

Theorem 2.6.2. *When H_0 is true, if the third-order moment of the population F_0 exists, then*

$$\lambda(X;\theta_0) \xrightarrow{L} \chi_1^2.$$

Proof. Since β_0 is the solution of $g(\beta) = 0$, then we have

$$\begin{aligned} 0 = g(\beta_0) &= \frac{1}{n}\sum_{i=1}^{n}(x_i-\theta_0)\left[1-\beta_0(x_i-\theta_0)+\frac{\beta_0^2(x_i-\theta_0)^2}{1+\beta_0(x_i-\theta_0)}\right] \\ &= \bar{x}-\theta_0-\beta_0\left[\frac{1}{n}\sum_{i=1}^{n}(x_i-\theta_0)^2+\frac{\beta_0}{n}\sum_{i=1}^{n}\frac{(x_i-\theta_0)^3}{1+\beta_0(x_i-\theta_0)}\right]. \end{aligned}$$

We can prove that, if H_0 is true, $\max\{\beta_0(x_i-\theta_0)\} = o_p(1)$ (A detailed discussion can be found in Owen (1990)), and since $\bar{x}-\theta_0 = O_p(n^{-1/2})$, then

$$\beta_0 = \frac{\bar{x}-\theta_0}{s^2} + O_p(n^{-1/2}), \tag{2.6.13}$$

where $s^2 = \sum_{i=1}^n (x_i - \theta_0)^2/n$.

Utilizing Taylor series expansion and Eq. (2.6.13) yields

$$\begin{aligned}\lambda(x;\theta_0) &= 2\sum_{i=1}^{n}[\beta_0(x_i-\theta_0) - \tfrac{1}{2}\beta_0^2(x_i-\theta_0)^2 + \tfrac{1}{3}\beta_0^3\epsilon_i] \\ &= n(\bar{x}-\theta_0)^2/s^2 + \tfrac{2}{3}\beta_0^3\sum_{i=1}^{n}\epsilon_i + O_p(n^{-1/2}),\end{aligned}$$

where

$$\beta_0^3\sum_{i=1}^{n}\epsilon_i \le \frac{n(\bar{x}-\theta_0)^2}{s^2}\beta_0\frac{1}{n}\sum_{i=1}^{n}|x_i|^3 + O_p(n^{-1/2}) = O_p(n^{-1/2}).$$

Then, when H_0 is true, we have $n(\hat{x}-\theta_0)^2/s^2 \xrightarrow{L} \chi_1^2$. □

According to the proof, we can see that, empirical likelihood ratio test is similar to t-test. Empirical likelihood-ratio test doesn't assume the concrete form of distribution, but depends on the parameter θ_0 heavily. The idea of empirical likelihood procedure is first proposed by Owen (1988), the discussion about some problems, such as variance and model parameter, can be found in Owen (1988, 1990, 1991), Chuang and Chan (2002), and Wang and Rao (2002a, 2002b). For application of the empirical likelihood procedure in economic models, see Kitamura (2001).

2.7 Problems

2.1. In a 2×2 contingency table, let N_{11} have a hypergeometric distribution and $\hat{\theta}(n_{ij}) = n_{11}n_{22}/n_{12}n_{21}$ be the MLE of odds ratio θ. Then, when $H_0 : \theta = 1$ is true, show that $\ln\hat{\theta}(n_{ij})/\sqrt{\sigma^2}$ approximates to the standard normal distribution $N(0,1)$, where $\sigma^2 = 1/n_{11} + 1/n_{12} + 1/n_{21} + 1/n_{22}$.

2.2. Let $x_1, x_2, \cdots, x_n$ be a set of samples from $N(\mu, \sigma^2)$, where μ and σ^2 are all unknown. Prove that $(n-1)s^2/\sigma^2$ has a χ^2 distribution with $(n-1)$ degrees of freedom, where s^2 is an unbiased estimator of σ^2, that is, $s^2 = \sum_{i=1}^n (x_i - \bar{x})^2/(n-1)$.

2.3. In Example 2.2.3, suppose that the two variances are unknown but equal, verify that the likelihood ratio test statistic is $t(x,y)$; When H_0 is true, verify that $t(X,Y)$ has a Student's t-distribution with $(m+n-2)$ degrees of freedom.

2.4. Suppose that there have k groups of independent samples from the two-parameter exponential distribution:

$$p(x, \alpha_i, \beta_i) = \frac{1}{\beta_i} \exp\left\{-\frac{x-\alpha_i}{\beta_i}\right\}, \quad \alpha_i \le x < +\infty, \quad i = 1, 2, \cdots, k,$$

let n_i denote the sample size of i-th group, $n_i \ge 2, i = 1, 2, \cdots, k$ and $\sum_{i=1}^{k} n_i = n$. Consider the testing problem:

$$H_0 : \alpha_1 = \cdots = \alpha_k; \ \beta_1 = \cdots = \beta_k.$$

Prove that the likelihood ratio statistic for the above test is

$$L = \prod_{i=1}^{k} \frac{d_i^{n_i}}{d^n},$$

where $d_i = \bar{x}_i - x_{(1)i}$, namely the difference of the sample mean of i-th population and the minimal sample observation. Similarly, $d = \bar{x} - x_{(1)}$.

2.5. Let $(x_1, y_1), \cdots, (x_n, y_n)$ be an i.i.d. sample from the bivariate normal distribution $N(\mu_1, \mu_2, \sigma_1, \sigma_2, \rho)$, where ρ is the correlation coefficient. Derive the likelihood ratio test for the testing problem $H_0 : \rho = 0$.

2.6. Consider the following model:

$$x_i = a x_{i-1} + \epsilon_i, \quad i = 1, \cdots, n,$$

where x_i is known. Let $x_0 = 0$, and $\epsilon_1, \cdots, \epsilon_n$ be an i.i.d. with the common distribution $N(0, \sigma^2)$, where a and σ^2 are unknown parameters, and $|a| < 1, \sigma^2 > 0$. Derive the likelihood ratio test for the testing problem $H_0 : a = 0$.

2.7. The following table is about the data of the enrollment of some university in autumn. Let A denote whether is admitted or not, B denote which department, and C denote sex. Consider the following three testing problems:

(a) Testing the independence of A and C. The purpose of the problem is to test whether there exists sex discrimination in the enrollment of the university or not.

(b) Testing the conditional independence of A and C given B. The purpose of the problem is also to examine whether there exists sex discrimination in the enrollment of the university or not.

(c) Give the testing results of the above testing problems and analyze their different meanings.

College (B)	Sex (C)	Admitted (A) Yes	Admitted (A) No
B_1	Male	353	207
	Female	17	8
B_2	Male	120	205
	Female	202	391
B_3	Male	138	279
	Female	131	244
B_4	Male	53	138
	Female	94	299
B_5	Male	22	351
	Female	24	317

2.8. Let $X_1, X_2, \cdots, X_m$ be an i.i.d. sample from F_x, and $Y_1, Y_2, \cdots, Y_N$ be an i.i.d. sample from $G(x)$, where X_i and Y_i are completely independent. Let R denote the rank of X_1 in the mixed samples, derive the distribution of R.

2.9. Let $X_1, X_2, \cdots, X_n$ be an i.i.d. sample from the normal distribution $N(\mu, \sigma^2)$. Considering the testing problem

$$H_0 : \mu \leq 0.$$

Similar to Theorem 2.2.2, calculate the limiting distribution of $-2 \ln \Lambda(x)$. Explain why the result is different from that of Theorem 2.2.2.

2.10. Let $X_1, X_2, \cdots, X_n$ be an i.i.d. sample from F, where F is a symmetrical distribution function. Now let $g(x_1, x_2, \cdots, x_n)$ be the joint distribution of $x_1, x_2, \cdots, x_n$.

(a) Show that $g(X_1, X_2, \cdots, X_n)$ and $g(-X_1, -X_2, \cdots, -X_n)$ have the same distribution.

(b) Show that if $g(X_1, X_2, \cdots, X_n) + g(-X_1, -X_2, \cdots, -X_n) = \mu_0$, the distribution of $g(X_1, X_2, \cdots, X_n)$ is symmetrical about the point $\mu_0/2$.

(c) Apply (b) to the Wilcoxon signed rank statistic, and show that the distribution of T is symmetrical under the null hypothesis and derive the central point of symmetry.

2.11. For any $\alpha \in [0, 1]$, construct an example in which the LRT with testing level α does not exist.

2.12. Consider tests about the independence of contingency table. Let (X, Y) be a random vector having the bivariate normal distribution, namely $(X, Y) \sim N(\mu_1, \mu_2, \sigma_1, \sigma_2, \rho)$, where the correlation coefficient $\rho \in (-1, 1)$. Now partition the whole plane into some squares with side length Δ. Here, the sides must be parallel to the coordinate axes. After partitioning into $I \times J$ cells, similar to the notations in Theorem 2.3.3, let $\chi^2 = \sum_{i=1}^{I} \sum_{j=1}^{J} (nn_{ij} - n_{i+}n_{+j})^2/(nn_{i+}n_{+j})$. Prove that when the null hypothesis is true and $\Delta \to 0$, $\chi^2 \to \rho^2/(1-\rho^2)$.

2.13. Let $X_1, \cdots, X_n$ be random variables with a multinomial distribution $M(n; p_1, \cdots, p_k)$. Consider the testing problem $H_0 : p_1 = p_2$. Verify that the goodness-of-fit test given in Eq. (2.3.3) is

$$\frac{X_1 - X_2}{X_1 + X_2}.$$

When H_0 is true, the limiting distribution of the above statistic is a χ^2 distribution with 1 degree of freedom.

2.14. We recorded the time spent watching television in a day (in hours) by 7 Men and 6 Women

no.	Men	Women
1	2.70	1.67
2	2.03	1.95
3	2.07	2.26
4	2.10	1.57
5	2.31	2.10
6	2.03	2.05
7	1.82	

(a) Assuming a normal model with equal variance, test the hypothesis that mean time watching television is different for Men and Women.

(b) Cross check your results with a non-parametric test.

(c) Comment on the results (a) and (b).

(d) Now, test the hypothesis that Men are watching more television than Women.

2.15. A random sample $\boldsymbol{X} = X_1, X_2, \cdots, X_n$ is selected from the Poisson(θ) distribution, $\theta \in (0, \infty)$. Suppose that the observed sample is $\boldsymbol{x} = (x_1, x_2, \cdots, x_n)$.

(a) Find the maximum likelihood estimator of θ.

(b) Find the generalized likelihood ratio statistic, $\lambda(x)$ for testing

$$H_0 : \theta = 3 \quad \text{vs.} \quad H_1 : \theta \neq 3.$$

(c) Construct the likelihood ratio test, pointing out the problems with its use. In practice these problem may be avoided by approximating the p-value by $P(|\sum_{i=1}^n X_i - 3n| \geq |\sum_{i=1}^n x_i - 3n|)$. Discuss briefly.

(d) Find the p-value, as suggested in (c), for a random sample of size 12 for which the observed values are $x = 1, 0, 3, 1, 1, 4, 1, 1, 3, 2, 3, 2$.

[Use the Central Limit Theorem to approximate the distribution of $\sum_{i=1}^n X_i$.]

2.16. Let $Y_1, \cdots, Y_n$ be independent, identically distributed, with density of the form

$$f(y; \xi) = \exp\left(\xi b(y) - c(\xi) + h(y)\right),$$

with scalar parameter ξ.
Explain in detail how to test $H_0 : \xi = \xi_0$ vs. $H_1 : \xi \neq \xi_0$.
Show that the moment generating function $M(s)$ of $b(Y)$ is $\exp\left(c(\xi + s) - c(\xi)\right)$. Hence show that

$$E(b(Y); \xi) = c'(\xi)$$

$$V(b(Y); \xi) = c''(\xi).$$

Let $S = \sum_{i=1}^n b(Y_i)$, and put $c'(\xi) = \mu$.
Show that S/n is an unbiased estimator of μ, with variance attaining the Cramér-Rao lower bound.

2.17. Let $\boldsymbol{Y}$ (with n components) be distributed according to $N(\boldsymbol{X\beta}, \sigma^2 \boldsymbol{I}_n)$, where $\boldsymbol{X}$ is a known $n \times n$ matrix with rank p $(< n)$; $\boldsymbol{\beta} = (\beta_1, \cdots, \beta_p)'$ and $\sigma > 0$ are parameters. For $0 < \alpha < 1$, try to derive a level-α LRT for the testing problem $H_0 : \beta_1 = 0$ vs. $H_1 : \beta_1 \neq 0$.

Chapter 3

The Comparison of Test Statistics

3.1 Probability of Two Types of Error in Hypothesis Test

As discussed in the last chapter, we can construct different test statistics for the same testing problem. Even in some cases for the same data, using different test statistics may get different conclusions, like Example 2.5.3 of last chapter. Which test statistic is more suitable? We need to analyze this problem carefully and set some criteria.

Review the previous discussion. Let $(\mathcal{X}, \mathcal{A}, \mathcal{P})$ be a statistical space, where $\mathcal{P} = \{P_\theta; \Theta \in \Theta\}$, Θ is a parameter space, and let Θ_0 be a proper subset in Θ. The null hypothesis for testing is:

$$H_0 : \theta \in \Theta_0. \tag{3.1.1}$$

Let $T(X)$ be a test statistic, and $t(x)$ be its sample value, then the p-value is defined as

$$p_t = \sup_{\theta \in \Theta_0} P_\theta\{T(X) \geq t(x)\}.$$

Our criterion is: when p-value is smaller, say $p_t \leq \alpha$, we can reject H_0. Let $S(X)$ be another test statistic, and $S(x)$ be its sample value, p_s is the corresponding p-value. There may exist a sample x such that $p_t \leq \alpha$ and $p_s > \alpha$. According to our criterion, the former should reject H_0 and the latter should not. So which test statistic should be used? Obviously both the two test statistics may make some decision errors, which can be described as

- If H_0 is true, then $T(X)$ is true and $S(X)$ is false;
- If H_0 is not true, then $T(X)$ is false and $S(X)$ is true.

There are two types of error. Since the position is symmetrical for both test statistics, each test statistic may make these two types of error.

How to compute the probability of the error? For the sake of simplicity, after considering it for a long time, Neyman and Pearson introduced the concept of alternative hypothesis (see, Reid, 1982). The parameter space is Θ. That one rejects H_0 means $\theta \in \Theta - \Theta_0$. We call $\Theta_1 = \Theta - \Theta_0$ as an alternative hypothesis. Rewrite the testing problem (3.1.1) as

$$H_0 : \theta \in \Theta_0 \quad \text{vs.} \quad H_1 : \theta \in \Theta_1, \tag{3.1.2}$$

here and thereafter we use vs. to denote the abbreviation of versus. Thus a hypothesis testing problem includes two hypotheses: the null and alternative. As a result, the decision idea of choosing the parameter space had been introduced into the testing problem again. Once we discussed the equivalent method of p-value in Section 2.1, *i.e.* the α-critical value method. Let t_α satisfy

$$\sup_{\theta \in \Theta_0} P_\theta\{T(X) \geq t_\alpha\} = \alpha, \tag{3.1.3}$$

then $p \leq \alpha$ is equivalent to $t(x) \geq t_\alpha$. More generally let A be a set, say $A = (t_\alpha, +\infty)$. A is called a **rejection region** if $t(x) \in A$ rejects H_0. Let

$$\phi(x) = \begin{cases} 1, & \text{if} \quad t(x) \in A, \\ 0, & \text{otherwise.} \end{cases} \tag{3.1.4}$$

Here, $\phi(x)$ is called as **the test function for rejection region** A. Obviously, if A is a measurable set, that is $T^{-1}(A) \in \mathcal{A}$, then $\phi(x)$ is a measurable function whose range is $[0, 1]$. More generally, we define a function as **test function** whose range is $[0, 1]$. Although the definition is very common, we can see in the next section that the best test function has the form defined by the type of (3.1.4) in essence. So, through the test function, we can tie the test statistic and the rejection region together. To be consistent with (3.1.3), it is usual to let

$$\sup_{\theta \in \Theta_0} P_\theta\{T(X) \in A\} = \alpha. \tag{3.1.5}$$

So $p \leq \alpha$ is equivalent to $t(x) \in A$. From Eq. (3.1.5) we can see that the size of rejection region depends on α. For $\theta \in \Theta$,

$$\beta_\phi(\theta) = E_\theta \phi(X) = P_\theta\{T(X) \in A\}$$

is called a **power function** of the test function $\phi(x)$ at θ. Next we will analyze the error probability for test function by the power function.

It is easy to express the two types of errors with the alternative hypothesis: for a certain test function, if $\Theta \in \Theta_0$ and the decision is to reject H_0, we call it the Type I error; If $\theta \in \Theta_1$ and the decision is to accept H_0, we

Table 3.1.1 Two types of errors in hypothesis testing

	Decision	
	Don't reject H_0	Reject H_0
$\theta \in \Theta_0$	Right decision	Type I error
$\theta \in \Theta_1$	Type II error	Right decision

call it the Type II error. Table 3.1.1 gives a summary of possible results of any hypothesis test.

Obviously, the probability of the Type I error is $P_\theta\{T(X) \in A\} = \beta_\phi(\theta)$ for $\theta \in \Theta_0$; the probability of the Type II error is $1 - P_\theta\{T(X) \in A\} = 1 - \beta_\phi(\theta)$ for $\theta \in \Theta_1$. Then both the two probabilities can be represented by the power function of test function. Naturally, we wish to find a test function such that the two probabilities attain their minima simultaneously. However, from Eqs. (3.1.4) and (3.1.5), we can see that this is impossible. Following the thought of Chapter 2, we will protect null hypothesis, that is to say, control the probability of the Type I error.

For the test function $\phi(x)$, if for $\forall\theta \in \Theta_0$, $E_\theta\phi(X) \leq \alpha$, then $\phi(x)$ is called a **level-α test**. If for $\forall\theta \in \Theta_0$, $E_\theta\phi(X) = \alpha$, then $\phi(x)$ is called a **size α test**. Let Φ_α be a set which contains all the tests with the level of α. For ϕ_0 and $\phi_1 \in \Phi_\alpha$, if for $\forall\theta \in \Theta_1$, $1 - \beta_{\phi_0}(\theta) \leq 1 - \beta_{\phi_1}(\theta)$, that is $\beta_{\phi_0}(\theta) \geq \beta_{\phi_1}(\theta)$, then ϕ_0 is said to be **better than** ϕ_1; if for any $\phi \in \Phi_\alpha$, ϕ_0 is better than ϕ, then ϕ_0 is called as a **uniformly most powerful level-α test**, sometimes written as level-α UMP test.

In this way, under controlling the Type I error, we construct a basic criterion of modern statistics of hypothesis testing: **to minimize the Type II error by controlling the Type I error**, that is, for the level α test, when parameter belongs to the alternative hypothesis, the power function is the larger the better. We can see that the basic criterion is based on protecting the null hypothesis.

Example 3.1.1. Let $x_1, \cdots, x_n$ be a set of samples from a normal distribution $N(\theta, 1)$. Let the parameter space be $\Theta = \{\theta_0, \theta_1\}$, that is only two points, here $\theta_0 < \theta_1$. The testing problem is

$$H_0 : \theta = \theta_0 \quad \text{vs.} \quad H_1 : \theta = \theta_1.$$

Because $\bar{x}$ is a sufficient and complete statistic for θ, when H_0 is true, $T(X) = \sqrt{n}(\bar{X} - \theta_0)$ has a $N(0, 1)$ distribution. Then the rejection region is $A = (u_\alpha, +\infty)$. If treating H_0 and H_1 equally, see Fig. 3.1.1, u_α should correspond to the cross of the two distributions. But according to the basic criterion, no matter what the H_1 is, $P_{\theta_0}\{T(X) \in A\} = \alpha$.

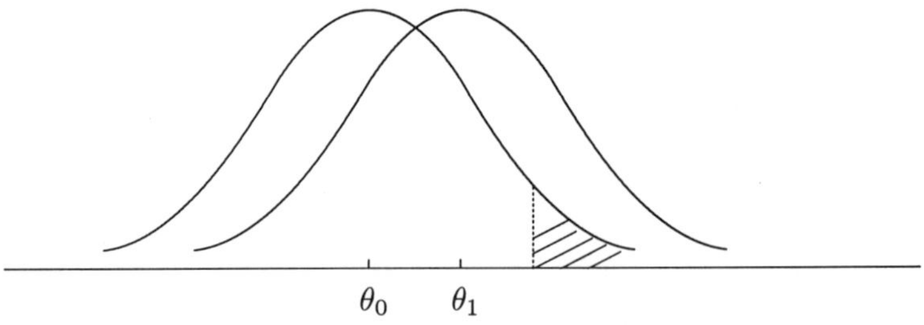

Fig. 3.1.1 Rejection region A

When $n = 1$, the rejection region of $\alpha = 0.05$ is shown as the shadow in Fig. 3.1.1. Even if θ_0 and θ_1 is close to each other, only when $t(x) \in A$ it is possible to reject θ_0 and to accept θ_1.

From the above example, in practice, we should design the testing problem reasonably, that is, reasonably design the null and alternative hypotheses. Furthermore, there is a theorem as follows.

Theorem 3.1.1. *Let $(\mathcal{X}, \mathcal{A}, \mathcal{P})$ be a statistical space, here $\mathcal{P} = \{P_\theta; \theta \in \Theta\}$. If there are only two points in the space and $\Theta_0 = \{\theta_0\}, \Theta_1 = \{\theta_1\}$, then for the testing problem (3.1.2) there exists an α-UMP test.*

Proof. Let Φ_α denote the set including all the α tests. Since $[0, \alpha]$ is a closed set, Φ_α is weakly compact by the Banach-Alaoglu Theorem (see Chapter 3 in Ash (1972)). Let

$$f(\alpha) = \sup_{\phi \in \Phi_\alpha} \beta_\phi(\theta_1).$$

Let $\psi(x) \equiv \alpha$. Since $\psi \in \Phi_\alpha$, $\alpha \leq f(\alpha) \leq 1$. From the definition of supremum, there exists a $\phi_n \in \Phi_\alpha$ such that $\beta_{\phi_n}(\theta_1) \to f(\alpha)$ as $n \to \infty$. From the definition of weak compactness, there exists a subarray of $\{\phi_n\}$, $\{\phi_{n_i}\}$ and $\Phi_0 \in \Phi_\alpha$, such that

$$\beta_{\phi_0}(\theta_1) = \lim_{n_i \to \infty} \beta_{n_i}(\theta_1) = f(\alpha),$$

that is, ϕ_0 is a level-α UMP test, here, $\beta_{n_i}(\theta_1)$ denotes $E_{\theta_1}\phi_{n_i}(X)$. □

In fact, in the above-mentioned theorem ϕ_0 should also be a size of α. If $E_{\theta_0}\phi_0(X) = \alpha' < \alpha$, then from $[0, \alpha'] \subset [0, \alpha]$ we know $f(\alpha') \leq f(\alpha)$, there exists a test $\phi^*(x)$ such that $E_{\theta_0}\phi^*(X) = \alpha$ and $E_{\theta_1}\phi^*(X) \geq E_{\theta_1}\phi_0(X)$. We will provide a constructive proof for this statement in the next section.

3.2 Neyman-Pearson Fundamental Lemma

Almost all the modern statistical testing theory is based on the basic criteria discussed in the last section and the Neyman-Pearson Fundamental Lemma that will been discussed in this section. The Neyman-Pearson Fundamental Lemma is a judgement method, and its starting point and basis is also the likelihood ratio.

Let $(\mathcal{X}, \mathcal{A}, \mathcal{P})$ be a statistical space, where $\mathcal{P} = \{P_\theta; \theta \in \Theta\}$. Suppose that there are only two elements in the parametric space and $\Theta_0 = \{\theta_0\}, \Theta_1 = \{\theta_1\}$. Consider the testing problem

$$H_0 : \theta = \theta_0 \quad \text{vs.} \quad H_1 : \theta = \theta_1. \tag{3.2.1}$$

Here the parametric testing problem can be regarded as a discriminant one on the probability distribution, *i.e.*

$$H_0 : P_\theta = P_{\theta_0} \quad \text{vs.} \quad H_1 : P_\theta = P_{\theta_1}. \tag{3.2.2}$$

Let ν be a σ-finite measure, $\mathcal{P} \ll \nu$. Let $f(x;\theta_i) = dP_{\theta_i}/d\nu, i = 0, 1$. So the parameter ν always exists, e.g., taking $\nu = P_{\theta_0} + P_{\theta_1}$. The likelihood ratio can be written as

$$\lambda(x) = \frac{f(x;\theta_1)}{f(x;\theta_0)}.$$

We can get the following fundamental lemma based on the likelihood ratio.

Theorem 3.2.1. (Neyman-Pearson Fundamental Lemma) *Consider the testing problem (3.2.1). For* $\alpha \in (0,1)$, *there exist* $k > 0$ *and* $r \in [0,1]$, *such that*

$$E_{\theta_0}\phi_0(X) = \alpha, \tag{3.2.3}$$

where

$$\phi_0(x) = \begin{cases} 1, & \text{if} \quad \lambda(x) > k, \\ r, & \text{if} \quad \lambda(x) = k, \\ 0, & \text{if} \quad \lambda(x) < k. \end{cases} \tag{3.2.4}$$

Furthermore, the following results hold:

(a) $\phi_0(x)$ *is a level-*α *UMP test;*

(b) *Any level-*α *UMP test takes the form Eq. (3.2.4) a.e. on* ν, *and satisfies Eq. (3.2.3), except for* $P_{\theta_0}\{\lambda(X) = k\} > 0$ *and* $P_{\theta_1}\{\lambda(X) = k\} = 0$;

(c) *Let* $\beta = E_{\theta_1}\phi_0(X)$, *then* $\alpha < \beta$ *except for* $\lambda(x) = 1$ *a.e.*

Proof. If there exists $k > 0$, satisfying

$$P_{\theta_0}\{\lambda \geq k\} = P_{\theta_0}\{\lambda > k\} = \alpha, \tag{3.2.5}$$

then the existence is proved, and $r = 0$. If

$$P_{\theta_0}\{\lambda \geq k\} > \alpha > P_{\theta_0}\{\lambda > k\},$$

since $P_{\theta_0}\{\lambda = k\} = P_{\theta_0}\{\lambda \geq k\} - P_{\theta_0}\{\lambda > k\} > 0$, let $r = (\alpha - P_{\theta_0}\{\lambda > k\})/P_{\theta_0}\{\lambda = k\}$, then

$$E_{\theta_0}\phi_0(X) = P_{\theta_0}\{\lambda > k\} + rP_{\theta_0}\{\lambda = k\} = \alpha,$$

which completes the proof of Eq. (3.2.3). Let us prove (a). For $\forall \phi \in \Phi_\alpha$, by the definition of test function,

$$\begin{aligned} \lambda(x) > k &\Rightarrow \phi_0(x) - \phi(x) \geq 0, \\ \lambda(x) < k &\Rightarrow \phi_0(x) - \phi(x) \leq 0, \end{aligned}$$

we can get that

$$\begin{aligned} \beta_{\phi_0}(\theta_1) - \beta_\phi(\theta_1) &= \textstyle\int (\phi_0(x) - \phi(x)) f_{\theta_1}(x) d\nu(x) \\ &\geq k \textstyle\int (\phi_0(x) - \phi(x)) f_{\theta_0}(x) d\nu(x) \geq 0. \end{aligned} \tag{3.2.6}$$

Next let us prove (b). Let $\phi^*(x)$ also be a level-α UMP test, *i.e.* $E_{\theta_0}\phi^*(X) \leq E_{\theta_0}\phi_0(X) = \alpha$, $E_{\theta_1}\phi^*(X) \geq E_{\theta_1}\phi_0(X)$. By combining with Eq. (3.2.6), we have

$$\int (\phi_0(x) - \phi^*(x))(f_{\theta_1}(x) - kf_{\theta_0}(x)) d\nu(x) = 0. \tag{3.2.7}$$

Let $C = \{x; \lambda(x) = k\}$. By Eq. (3.2.7), if $\nu(C) = 0$, then $\phi^*(x)$ and $\phi_0(x)$ are equal a.e. on measure ν. If $\nu(C) > 0$, there are two cases. If $P_{\theta_0}\{C\} = 0$, by Eq. (3.2.5), $\phi^*(x)$ and $\phi_0(x)$ are equal a.e.; If $P_{\theta_0}\{C\} > 0$ and $x \in C$, then $\phi^*(x) \in \Phi_\alpha$, $0 \leq r^* \leq r$, where r^* denotes the value of $\phi^*(x)$ under $x \in C$. Thus

$$\begin{aligned} \beta_{\phi^*}(\theta_1) &= P_{\theta_1}\{\lambda > k\} + r^* P_{\theta_1}\{C\} \\ &\leq P_{\theta_1}\{\lambda > k\} + rP_{\theta_1}\{C\} \\ &= \beta_{\phi_0}(\theta_1). \end{aligned}$$

By $\beta_{\phi^*}(\theta) = \beta_{\phi_0}(\theta_1)$, we can get $r^* = r$ or $P_{\theta_1}\{C\} = 0$.

It is easy to prove (c) by the result of (b), and this completes the proof of the theorem. □

Remark 3.2.1. There is a special case of the above theorem (b). Let X be a random sample from a Bernoulli trial (see Example 1.3.4) with $P_\theta\{X=1\}=\theta$ and $P_\theta\{X=0\}=1-\theta$. Consider testing $H_0:\theta=\theta_0=1/3$ vs. $H_1:\theta=\theta_1=1$. Let $\alpha=4/9$. It can be verified that

$$\phi_0(x)=\begin{cases}1, & \text{if} \quad x>0,\\ 1/6, & \text{if} \quad x=0\end{cases}$$

is a level-α UMP test satisfying Eqs. (3.2.3) and (3.2.4). On the other hand,

$$\phi^*(x)=\begin{cases}1, & \text{if} \quad x>0,\\ 1/7, & \text{if} \quad x=0\end{cases}$$

is also a level-α UMP test and different from $\phi_0(x)$. But $P_{\theta_0}\{X=0\}=2/3>0$, and $P_{\theta_1}\{X=0\}=0$. This is a very extreme example. In general, the level-α UMP test satisfying Eqs. (3.2.3) and (3.2.4) is unique a.e.. This is the reason why we only consider the test statistic with the form of rejection region in the last section.

Remark 3.2.2. For a level-α UMP test, it suffices to consider a function of sufficient and complete statistics, and this coincides with Theorem 2.1.1 in Chapter 2. Let $(\mathcal{X},\mathcal{A},\mathcal{P})$ be a statistical space, in which $\mathcal{P}=\{P_\theta;\theta\in\Theta\}, dP_\theta/d\nu=f(x;\theta)$. Let $(\mathcal{T},\mathcal{B},\mathcal{Q})$ be an induced statistical space by the sufficient statistic $T(x)$, where $\mathcal{Q}=\{Q_\theta;\theta\in\Theta\}, Q_\theta\{B\}=P_\theta\{T^{-1}(B)\}$ for $\forall B\in\mathcal{B}$. By Eq. (1.2.31) in Chapter 1, there exists a $\mathcal{B}$ measurable function g and an $\mathcal{A}$ measurable function h, s.t. $f(x;\theta)=g(T(x);\theta)h(x)$. So the likelihood ratio (3.2.2) can be written as

$$\lambda(x)=\frac{f(x;\theta_1)}{f(x;\theta_0)}=\frac{g(T(x);\theta_1)h(x)}{g(T(x);\theta_0)h(x)}, \tag{3.2.8}$$

and this implies that there exists a $\mathcal{B}$ measurable function Λ, s.t. $\lambda(x)=\Lambda(T(x))=\Lambda(T)$, which is a function of T.

Although the Neyman-Pearson Fundamental Lemma seemingly deals with the case which both the null and alternative hypotheses are simple, we can deduce a series of more important results by this lemma in fact. We shall analyze the two examples firstly, then draw some general conclusions.

Example 3.2.1. (Testing binomial parameter) Let $x_1,\cdots,x_n$ be a set of random samples from a Bernoulli trial, where the parameter, success probability, is θ. For $\theta_0\in(0,1)$, consider the problem of testing

$$H_0:\theta\le\theta_0 \quad \text{vs.} \quad H_1:\theta>\theta_0. \tag{3.2.9}$$

By Example 1.3.4 in Chapter 1, we know that $x = x_1 + \cdots + x_n$ is a sufficient statistic of θ, so by Remark 3.2.2, it suffices to consider a sufficient statistic for constructing the LRT.

Choose $\theta_1 > \theta_0$. Consider the problem of testing

$$H_0' : \theta = \theta_0 \quad \text{vs.} \quad H_1' : \theta = \theta_1.$$

Notice that X has a binomial distribution $Bi(n, \theta)$, we can obtain the likelihood ratio as follows:

$$\lambda(x) = \frac{\binom{n}{x} \theta_1^x (1-\theta_1)^{n-x}}{\binom{n}{x} \theta_0^x (1-\theta_0)^{n-x}} = \left(\frac{\theta_1}{\theta_0}\right)^x \left(\frac{1-\theta_1}{1-\theta_0}\right)^{n-x}.$$

By $\theta_1 > \theta_0$, $\lambda(x)$ is a monotone increasing function of x. Corresponding to (3.2.4), the test function can be written as

$$\phi(x) = \begin{cases} 1, & \text{if} \quad x \geq a+1, \\ r, & \text{if} \quad x = a, \\ 0, & \text{if} \quad x < a, \end{cases} \tag{3.2.10}$$

where a and r satisfy

$$\sum_{x=a+1}^{n} \binom{n}{x} \theta_0^x (1-\theta_0)^{n-x} = \alpha_1 \leq \alpha < \sum_{x=a}^{n} \binom{n}{x} \theta_0^x (1-\theta_0)^{n-x},$$

$$r = \frac{\alpha - \alpha_1}{\binom{n}{x} \theta_0^a (1-\theta_0)^a}.$$

Obviously, when $\alpha = \alpha_1$, we have $r = 0$. Because there has no parameter in the test function $\phi(x)$ and we only use the condition $\theta_1 > \theta_0$ in our calculation, the test (3.2.10) is also a level-α UMP test for

$$H_0' : \quad \theta = \theta_0 \quad \text{vs.} \quad H_1 : \theta > \theta_0.$$

Next let us analyze the case of H_0. For $\theta^* < \theta_0$, $\lambda(x)$ is a monotone function of x, then we have $E_{\theta^*}\phi(X) < \alpha$ as discussed in Example 2.2.1 in Chapter 2. Thus

$$\phi(x) \in \Phi_\alpha = \{\phi; E_\theta \phi(X) \leq \alpha, \forall \theta < \theta_0\}.$$

Hence $\phi(x)$ is a level-α UMP test for the problem of testing (3.2.9).

Remark 3.2.3. In fact, the above example has utilized the following result: Consider the problem of testing (3.1.2), and let Θ_{01} be a subset of Θ_0. If $\phi(x)$ is a level-α UMP test for

$$H_0^{'}: \theta \in \Theta_{01} \quad \text{vs.} \quad H_1: \theta \in \Theta_1$$

and $\phi(x) \in \Phi_\alpha$, then $\phi(x)$ is a level-α UMP test for (3.1.2). The reason is that for any $\phi_1(x) \in \Phi_\alpha$, $E_\theta\phi_1(X) \le \alpha$ for $\forall\theta \in \Theta_{01}$, then $E_\theta\phi(X) \ge E_\theta\phi_1(X)$ for $\forall\theta \in \Theta_1$. Notice that the condition $\phi(x) \in \Phi_\alpha$ is very important, meaning that the subset Θ_{01} should be near the vertexes of Θ_1 in Θ_0. In general, a distribution P_{θ_0} is called the **least favorable distribution** if $\phi(x) \in \Phi_\alpha$ in which $\phi(x)$ is a level-α test derived from the Neyman-Pearson Fundamental Lemma when $\theta_0 \in \Theta_0, \theta_1 \in \Theta_1$.

Example 3.2.2. (Testing normal mean) Let $x_1, \cdots, x_n$ be a set of random samples from the normal distribution $N(\theta, \sigma^2)$. For a given θ_0, consider the problem of testing

$$H_0: \quad \theta \le \theta_0 \quad \text{vs.} \quad H_1: \quad \theta > \theta_0. \tag{3.2.11}$$

Assume that σ^2 is known. As $\bar{x}$ is a sufficient and complete statistic of θ, and $\bar{X}$ has the distribution $N(\theta, \sigma^2/n)$, it is equivalent to considering the test (3.2.11) by $\bar{x}$. Select $\theta_1 > \theta_0$ and consider the problem of testing

$$H_0': \quad \theta = \theta_0 \quad \text{vs.} \quad H_1': \quad \theta = \theta_1. \tag{3.2.12}$$

So the likelihood ratio is equal to

$$\lambda(\bar{x}) \sim \exp\left\{\frac{n}{2\sigma^2}[2\bar{x}(\theta_1 - \theta_0) + \theta_0^2 - \theta_1^2]\right\}.$$

Because when $\theta_1 > \theta_0$, $\lambda(\bar{x})$ is the monotone increasing function of $\bar{x}$ corresponding to (3.2.4), the test function is

$$\phi(x) = \begin{cases} 1, & \text{if} \quad \bar{x} \ge c, \\ 0, & \text{if} \quad \bar{x} < c. \end{cases}$$

Since $\sqrt{n}(\bar{X} - \theta_0)/\sigma \sim N(0,1)$ when H_0' is true, we have $c = \sigma u_\alpha/\sqrt{n} + \theta_0$, where u_α satisfies $1 - \Phi(u_\alpha) = \alpha$ and $\Phi(\cdot)$ is the cdf of the standard normal distribution.

Assume that σ^2 is unknown. By Example 1.3.4 and Theorem 1.3.4 in Chapter 1, $(\bar{x}, \hat{\sigma}^2)$ is a sufficient and complete statistic for (θ, σ^2), where $\hat{\sigma}^2 = \sum_{i=1}^n (x_i - \bar{x})^2/(n-1)$. Let $t = (\bar{x} - \theta_0)/\hat{\sigma}$, by Example 1.1.9 in Chapter 1, the statistic t has a noncentral Student's t-distribution with $n-1$ degrees of freedom, $t(n-1, \eta)$, in which $\eta = \theta - \theta_0$. When H_0' is true in (3.2.12), t has a central Student's t-distribution. when H_1' is true, the noncentral parameter is $\eta_1 = \theta_1 - \theta_0 > 0$. So the likelihood ratio is equivalent to

$$\lambda(t) \sim \sum_{s=0}^{\infty}(\frac{\sqrt{2}\eta t}{\sqrt{n-1+t^2}})^s.$$

It is easy to verify that when $\eta > 0$, this is a monotone increasing function of t. Based on the Neyman-Pearson Fundamental Lemma and Remark 3.2.3, the test statistic can be written as

$$\phi(t) = \begin{cases} 1, & \text{if} \quad t \geq t_\alpha, \\ 0, & \text{if} \quad t < t_\alpha, \end{cases}$$

where t_α is the critical value of the Student's t-distribution with $n-1$ degrees of freedom. As $\phi(t) \in \Phi_\alpha$, $\phi(t)$ only depends on $\theta_1 > \theta_0$ and is free of the value of θ_1, by Remark 3.2.3, the statistic t is a level-α UMP test for (3.2.11).

The above two examples utilize the Neyman-Pearson Fundamental Lemma skillfully to construct a level-α UMP test. For these two examples, there are two common characters: (I) the testing problem itself is connected; (II) the likelihood ratio is a monotone function of some variable. Now we analyze these two characters a bit more closely.

The connectivity of the testing problem is essentially concerned with the alternative hypothesis Θ_1. In general, a problem of testing H_0 : $\theta \in \Theta_0$ vs. $H_1 : \theta \in \Theta_1$ is called **one-sided** if there is a convex set C such that $C \supseteq \Theta_1$ and $C \cap \Theta_0 = \phi$. This means that for $\theta_1, \theta_2 \in \Theta_1$, if $\lambda_1\theta_1 + \lambda_2\theta_2 \in \Theta$, then $\lambda_1\theta_1 + \lambda_2\theta_2 \in \Theta_1$, in which λ_1 and λ_2 are nonnegative real values. It can be verified that the above two examples are both one-sided. A test is called **two-sided** if it is not one-sided.

For the second character, we can give a generalized definition as follows. Let $(\mathcal{X}, \mathcal{A}, \mathcal{P})$ be a statistical space, and $f(x;\theta) = dP_\theta/d\nu$ be the density function. For a measurable function $T(x)$ on $(\mathcal{X}, \mathcal{A}, \mathcal{P})$, if for $\theta_1 < \theta_2$, the likelihood ratio $f(x;\theta_2)/f(x;\theta_1)$ is a monotone nondecreasing function on $T(x)$, we say that $f(x;\theta)$ has **monotone likelihood ratio** on $T(x)$. It is easy to find the measurable functions corresponding to the monotone likelihood ratio in the above two examples respectively.

Theorem 3.2.2. *If the density function $f(x;\theta)$ has a monotone likelihood ratio on $T(x)$, then for the one-sided testing problem*

$$H_0 : \theta \leq \theta_0 \quad vs. \quad H_1 : \theta > \theta_0,$$

there exists a level α-UMP test which has the following form

$$\phi(x) = \begin{cases} 1, & \textit{if} \;\; T(x) > c, \\ r, & \textit{if} \;\; T(x) = c, \\ 0, & \textit{if} \;\; T(x) < c, \end{cases} \tag{3.2.13}$$

where c and r satisfy $E_{\theta_0}\phi(X) = \alpha$ and the power function of $\phi(x)$ is a monotone nondecreasing function of θ.

Proof. Take $\theta_1 > \theta_0$ and consider the problem of testing $H_0' : \theta = \theta_0$ vs. $H_1' : \theta = \theta_1$. By the definition of monotone likelihood ratio, it is obvious that the testing form given by Eq. (3.2.13) exactly matches with the condition (3.2.4) in the Neyman-Pearson Fundamental Lemma. Since this test only depends on $\theta_1 > \theta_0$ but is free of the value of θ_1, by Remark 3.2.3, we only need to verify $\phi(x) \in \Phi_\alpha$.

Let $\theta_2 < \theta_0$ and consider the testing problem $H_0'' : \theta = \theta_2$ vs. $H_1'' : \theta = \theta_0$. Similarly, the testing form given by Eq. (3.2.13) exactly matches with the condition 3.2.4 in the Neyman-Pearson Fundamental Lemma, but $E_{\theta_2}\phi(X) = \alpha^*$. From (c) of the Neyman-Pearson Fundamental Lemma, $\alpha^* < \alpha$, and this completes the proof of the first part of the theorem.

By the same method, we can prove the second part, *i.e.* the power function of $\phi(x)$ is a monotone nondecreasing function of θ. To express it more clearly, we give a general proof. Let $g(t)$ be a monotone increasing function of t, we shall prove that $E_\theta g(T(X))$ is a monotone increasing function of θ.

For $\theta_1 < \theta_2$, let

$$\lambda(x) = f(x;\theta_2)/f(x;\theta_1), \;\; A = \{x;\; \lambda(x) > 1\}, \text{ and } B = \{x;\; \lambda(x) < 1\}.$$

From the definition of monotone likelihood ratio and the monotonicity of $g(t)$, for $\forall x_1 \in A$ and $\forall x_2 \in B$,

$$\lambda(x_1) > \lambda(x_2) \Longrightarrow T(x_1) \geq T(x_2) \Longrightarrow g(T(x_1)) \geq g(T(x_2)).$$

By the Integral Mean Value Theorem, there exist $x_1 \in A$ and $x_2 \in B$, s.t.

$$\begin{aligned} E_{\theta_2}g(T(X)) - E_{\theta_1}g(T(X)) &= \int_A g(T(x))[f(x;\theta_2) - f(x;\theta_1)]d\nu(x) \\ &\quad + \int_B g(T(x))[f(x;\theta_2) - f(x;\theta_1)]d\nu(x) \\ &= [g(T(x_1)) - g(T(x_2))] \\ &\quad \times \int_A [f(x;\theta_2) - f(x;\theta_1)]d\nu(x) \\ &\geq 0. \end{aligned}$$

Since $\phi(x)$ is a monotone increasing function on $T(x)$, the theorem is established. □

Example 3.2.3. (One-parameter exponential family and monotone likelihood ratio) Recall that the one-parameter exponential family has the following form of density function

$$f(x;\theta) = c(\theta)\exp\{u(\theta)T(x)\}h(x). \tag{3.2.14}$$

Obviously, if $u(\theta)$ is a monotone increasing function of θ, then $f(x;\theta)$ has a monotone likelihood ratio on $T(x)$. By Theorem 3.2.2, for the one-sided testing problem $H_0 : \theta \leq \theta_0$ vs. $H_1 : \theta > \theta_0$, Equation (3.2.13) gives a level-α UMP test. Thus it is suitable to employ the one-parameter exponential families, such as binomial test mentioned in Example 3.2.1, normal mean test when variance is known mentioned in Example 3.2.2, Poisson and negative binomial, *etc.*

Now we will specially consider the one-sided testing problem of normal variance. Let $x_1, \cdots, x_n$ be a random sample from a normal distribution $N(\mu, \sigma^2)$, where the mean μ **is known**. Consider a one-sided test of variance

$$H_0 : \sigma^2 \leq \sigma_0^2 \quad \text{vs.} \quad \sigma^2 > \sigma_0^2. \tag{3.2.15}$$

Here the joint density function can be written as

$$f(x;\sigma^2) = \left(\frac{1}{\sqrt{2\pi}\sigma}\right)^n \exp\left\{-\frac{1}{2\sigma^2}\sum_{i=1}^{n}(x_i-\mu)^2\right\}.$$

After comparing with Eq. (3.2.14), we have $T(x) = \sum_{i=1}^n (x_i - \mu)^2$. Since $-(2\sigma^2)^{-1}$ is a monotone increasing function of σ^2, so the level-α UMP test has the form as Eq. (3.2.13). By Example 1.1.8 in Chapter 1, when $\sigma^2 = \sigma_0^2$ is known,

$$\chi^2 = \frac{1}{\sigma_0^2}T(X) \sim \chi^2(n), \tag{3.2.16}$$

so $c = \sigma_0^2\chi_\alpha^2(n)$, in which $\chi_\alpha^2(n)$ is the α-critical value of the chi-square distribution with n degrees of freedom. In general, the test statistic defined in (3.2.16) is called a χ^2 **test**.

When the mean μ is unknown, μ is a nuisance parameter. We should take an appropriate variable transformation. Let $\chi^2 = s(x)/\sigma_0^2$, where $s(x) = \sum_{i=1}^n (x_i - \bar{x})^2$. Thus $\chi^2 \sim \sigma^2\chi_{n-1}^2/\sigma_0^2$, where χ_{n-1}^2 represents the chi-square distribution with $n-1$ degrees of freedom. For the Lebesgue measure, the density function of χ^2 is

$$g(t;n-1,\eta) = \frac{1}{2^{(n-1)/2}\Gamma((n-1)/2)}(\eta t)^{n/2-1}e^{-\eta t/2}\eta,$$

in which $\eta = \sigma_0^2/\sigma^2$. Here the testing problem (3.2.15) is equivalent to

$$H_0 : \eta \leq 1 \quad \text{vs.} \quad H_1 : \eta > 1.$$

As $g(t; n-1, \eta)$ also belongs to an exponential family and is independent of the value of μ, a level-α UMP test has the form as (3.2.13), where $c = \sigma_0^2 \chi_\alpha^2(n-1)$.

Example 3.2.4. (One-sided test for uniform distribution parameter) Although the uniform distribution does not belong to a one-parameter exponential family, Theorem 3.2.1 is still available.

Let $x_1, \cdots, x_n$ be a set of random samples from a uniform distribution $U(0, \theta)$. Consider the one-sided testing problem $H_0 : \theta \leq \theta_0$ vs. $H_1 : \theta > \theta_0$. Let $x = (x_1, \cdots, x_n)$. The joint density function is

$$f(x;\theta) = \begin{cases} \dfrac{1}{\theta^n}, & \text{if} \quad 0 \leq x_{\max} \leq \theta, \\ 0, & \text{otherwise,} \end{cases}$$

where $x_{\max} = \max\{x_i\}$. For $\theta_1 < \theta_2$, the likelihood ratio is

$$\lambda(x) = \frac{f(x;\theta_2)}{f(x;\theta_1)} = \begin{cases} \left(\dfrac{\theta_1}{\theta_2}\right)^n, & \text{if} \quad 0 \leq x_{\max} \leq \theta_1, \\ \infty, & \text{if} \quad \theta_1 < x_{\max} \leq \theta_2, \\ 0, & \text{otherwise.} \end{cases}$$

It is a monotone increasing function of $x_{\max}$, *i.e.* $f(x;\theta)$ has a monotone likelihood ratio on $x_{\max}$. By Eq. (3.2.13), a level-α UMP test has the following form

$$\phi(x) = \begin{cases} 1, & \text{if} \quad x_{\max} \geq c, \\ 0, & \text{if} \quad x_{\max} < c, \end{cases}$$

where c satisfies $E_{\theta_0}\phi(X) = \alpha$, *i.e.*

$$\alpha = E_{\theta_0}\phi(X) = n \int_c^{\theta_0} \frac{1}{\theta_0^n} x_{n-1} dx = 1 - \left(\frac{c}{\theta_0}\right)^n.$$

Thus we have $c = \theta_0 \sqrt[n]{1-\alpha}$.

There is another special type of one-sided testing problem. We will expand our analysis based on an example, and then give a general result.

Example 3.2.5. (One-sided test when the alternative hypothesis takes the form of interval) We still consider that $x_1, \cdots, x_n$ is a set of random samples from a normal distribution $N(\theta, \sigma^2)$, where σ^2 is known. For the two given real numbers $\theta_1 < \theta_2$, consider the testing problem

$$H_0 : \theta \leq \theta_1 \text{ or } \theta_2 \leq \theta \quad \text{vs.} \quad H_1 : \theta \in (\theta_1, \theta_2). \tag{3.2.17}$$

It is also a one-sided testing problem. Let $x = (x_1, \cdots, x_n)$, and let $f(x;\theta)$ be the joint density function and

$$f(x;\theta^*) = \frac{1}{2}f(x;\theta_1) + \frac{1}{2}f(x;\theta_2).$$

Obviously it is also a density function. For $\forall\theta_3 \in (\theta_1, \theta_2)$, consider the testing problem

$$H_0' : \theta = \theta^* \quad \text{vs.} \quad H_1' : \theta = \theta_3.$$

Here the likelihood ratio is

$$\lambda(x) = \frac{f(x;\theta_3)}{f(x;\theta^*)}.$$

For computation convenience, we use the reciprocal of $\lambda(x)$. Now $\lambda(x) \geq k$ is equivalent to $1/\lambda(x) \leq 1/k$. We can get

$$1/\lambda(x) \sim \exp\{(\theta_3 - \theta_1)\bar{x}/\sigma^2\} + \exp\{(\theta_3 - \theta_2)\bar{x}/\sigma^2\}.$$

Obviously, it is a convex function of $\bar{x}$. $1/\lambda(x) \leq 1/k$ is equivalent to that there exist c_1 and c_2, s.t. $c_1 \leq \bar{x} \leq c_2$. Based on the Neyman-Pearson Fundamental Lemma, we can construct the test function

$$\phi(x) = \begin{cases} 1, & \text{if} \quad c_1 \leq \bar{x} \leq c_2, \\ 0, & \text{if} \quad \bar{x} < c_1 \text{ or } c_2 < \bar{x}, \end{cases}$$

where c_1 and c_2 satisfy $E_{\theta_1}\phi(X) = E_{\theta_2}\phi(X) = \alpha$. Obviously, the interval $[c_1, c_2]$ is symmetric about $(\theta_1 + \theta_2)/2$, *i.e.* $c_1 + c_2 = \theta_1 + \theta_2$. Thus we have

$$c_1 = \theta_1 + \frac{\sigma}{\sqrt{n}}u_{(1-\alpha)/2}$$

and

$$c_2 = \theta_2 - \frac{\sigma}{\sqrt{n}}u_{(1-\alpha)/2},$$

where $u_{(1-\alpha)/2}$ satisfies $\Phi(u_{(1-\alpha)/2}) = (1-\alpha)/2$ and $\Phi(\cdot)$ is the standard normal distribution function.

Obviously, $\phi(x)$ is independent of the value of θ_3. If $\phi(x) \in \Phi_\alpha$, we can regard $f(x;\theta^*)$ as the least favorable distribution. According to Remark 3.2.3, $\phi(x)$ is a level-α UMP test for the problem of testing (3.2.17). Next we verify $\phi(x) \in \Phi_\alpha$. Without loss of generality, let $(\theta_1, \theta_2) \subset [c_1, c_2]$. As shown in Fig. 3.2.1, the critical region is the interval $[c_1, c_2]$. By the monotonicity of normal distribution on mean, we can easily get that if $\theta < \theta_1$, $E_\theta\phi(X) < E_{\theta_1}\phi(X) = \alpha$; if $\theta > \theta_2$, $E_\theta\phi(X) < E_{\theta_2}\phi(X) = \alpha$. Thus we have $\phi(x) \in \Phi_\alpha$.

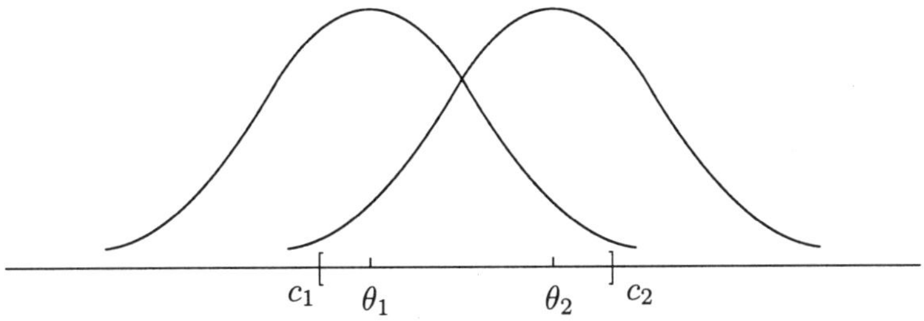

Fig. 3.2.1 Critical region with the alternative hypothesis taking the form of interval

From the above example we can see that the convexity of likelihood ratio plays a key role. Let $(\mathcal{X}, \mathcal{A}, \mathcal{P})$ be a statistical space, $\mathcal{P} = \{P_\theta; \theta \in \Theta\}$, and the density function $f(x;\theta) = dP_\theta/d\nu$. Let $T(x)$ be a measurable function. $f(x;\theta)$ is called to have a **convex likelihood ratio** on $T(x)$, if for any $\theta_1 < \theta_0 < \theta_2$ and $\beta > 0$, $f(x;\theta_1)/f(x;\theta_0) + \beta f(x;\theta_2)/f(x;\theta_0)$ is a convex function of $T(x)$.

Theorem 3.2.3. *If the density function $f(x;\theta)$ has a monotone and convex likelihood ratio on $T(x)$, then for the one-sided testing problem (3.2.17), there exists a level-α UMP test with the following form*

$$\phi(x) = \begin{cases} 1, & \text{if } \ c_1 < T(x) < c_2, \\ r_i, & \text{if } \ T(x) = c_i, \ i = 1, 2, \\ 0, & \text{otherwise}, \end{cases} \tag{3.2.18}$$

where c_i and r_i satisfy $E_{\theta_1}\phi(X) = E_{\theta_2}\phi(X) = \alpha$.

Proof. For the given θ_1 and θ_2, let

$$f^*(x) = \frac{1}{2} f(x;\theta_1) + \frac{1}{2} f(x;\theta_2).$$

Obviously, $f^*(x)$ is also a density function. For $\forall\theta_0 \in (\theta_1, \theta_2)$, consider the testing problem given by (3.2.2)

$$H_0': \quad f(x;\theta) = f^*(x) \quad \text{vs.} \quad H_1': f(x;\theta) = f(x;\theta_0).$$

Then the Neyman-Pearson test function is based on the likelihood ratio

$$\lambda(x) = \frac{f^*(x)}{f(x;\theta_0)}$$

$$1/\lambda(x) \sim \frac{1}{2}\left[\frac{f(x;\theta_1)}{f(x;\theta_0)} + \frac{f(x;\theta_2)}{f(x;\theta_0)}\right].$$

By the definition of convex likelihood ratio, $1/\lambda(x) \le k$ is equivalent to that there exist real numbers c_1 and c_2, such that $c_1 \le T(x) \le c_2$, *i.e.* it has the form as (3.2.18). This test function is independent of the value of θ_0. Next it suffices to verify that $\phi(x) \in \Phi_\alpha$.

For $\forall\theta_3 \le \theta_1$, by the assumption of monotone likelihood ratio and the proof of Theorem 3.2.2, we can get that $E_{\theta_3}\phi(X) \le E_{\theta_1}\phi(X) = \alpha$. Similarly, for $\forall\theta_4 \ge \theta_2$, we have $E_{\theta_4}\phi(X) \le E_{\theta_2}\phi(X) = \alpha$. This completes the proof of the theorem. □

Example 3.2.6. (One-parameter exponential family and convex likelihood ratio) Consider the one-parameter exponential family given by (3.2.14). For $\theta_1 < \theta_0 < \theta_2$ and $\beta > 0$,

$$\begin{aligned}&\frac{f(x;\theta_1)}{f(x;\theta_0)} + \beta\frac{f(x;\theta_2)}{f(x;\theta_0)}\\&= \frac{c(\theta_1)}{c(\theta_0)}\exp\left\{[u(\theta_1) - u(\theta_0)]T(x)\right\} + \beta\frac{c(\theta_2)}{c(\theta_0)}\exp\left\{[u(\theta_2) - u(\theta_0)]T(x)\right\}.\end{aligned}$$

It is a convex function of $T(x)$. Using the result of Example 3.2.3, Theorem 3.2.3 is applicable to the testing problems of many distributions, such as binomial, Poisson, normal with known variance and normal with known mean. Since a noncentral Student's t-distribution also has this property, by Example 3.2.2, it is also applicable for the normal mean test with unknown variance.

It is very lucky to find a level-α UMP test. For most of statistical testing problems, however, the level- α UMP test does not exist. As an example, we can find a level-α UMP test when the dimension of parameter is greater than two. Even if the density belongs to a one-parameter exponential family, the level-α UMP test does not exist for two-sided test, see the following example.

Example 3.2.7. (Two-sided test for normal mean) Let $x_1, \cdots, x_n$ be a set of random samples from a normal distribution $N(\theta, \sigma^2)$ with the the variance σ^2 known. Consider the problem of testing

$$H_0 : \theta = \theta_0 \quad \text{vs.} \quad H_1 : \theta \neq \theta_0, \tag{3.2.19}$$

where θ_0 is a given real number. This is a two-sided test. Let us illustrate that there does not exist a level-α UMP test. Let Φ_α be a set that contains all of the level-α tests for the testing problem (3.2.19). Consider the following one-sided testing problem

$$H_0 : \theta = \theta_0 \quad \text{vs.} \quad H_1' : \theta > \theta_0.$$

By Example 3.2.2, $T(x) = \sqrt{n}(\bar{x} - \theta_0)/\sigma$ is a level-α UMP test. Let $\phi(t)$ be the test function corresponding to $T(x)$. For $\forall \phi_1(x) \in \Phi_\alpha$ and $\forall \theta > \theta_0$, $E_\theta \phi(T(X)) \geq E_\theta \phi_1(X)$. This implies that for any parameter $\theta > \theta_0, \phi(t)$ has the largest power. But it is easy to verify (see the following schematic diagram) that the power function of $\phi(t)$ is monotone decreasing and tends to zero finally when $\theta \to -\infty$. This shows that if the parameter $\theta < \theta_0$,

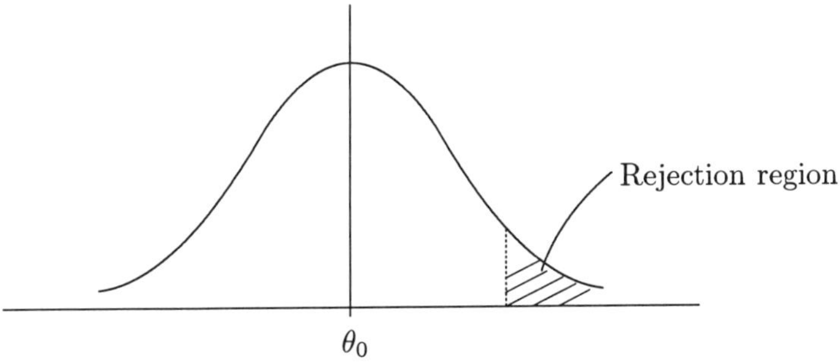

the property owned by the power function of $\phi(t)$ is very bad. Thus for the problem of testing (3.2.19), there does not exist any level-α UMP test. In fact, this example also illustrates that for a distribution with monotone likelihood function and the two-sided testing problem (3.2.19), the level-α UMP test does not exist at all.

3.3 Unbiased Test

The last example in the former section shows that there does not exist any level-α UMP test in many testing problems. In some situations, the cause that leads to such fact lies in that there exists such a level-α test that one can obtain a greater power for some parameters in the alternative hypothesis Θ_1, but a smaller one for other parameters in Θ_1. Of course, such test is not appropriate, which should be deleted out the tests in the comparison of the test functions, and this constructs an appropriate class of test functions. In this appropriate class we will look for the most powerful test, and this will be discussed in this section.

For the testing problem (3.1.4), a test $\phi(x)$ is called a **level-α unbiased test** if

$$E_\theta \phi(X) \leq \alpha, \forall \theta \in \Theta_0; \text{ and } E_\theta \phi(X) \geq \alpha, \forall \theta \in \Theta_1.$$

Let Φ_α^U denote the collection of all level-α unbiased test functions. The test in Example 3.2.6 does not belong to Φ_α^U. A test $\phi_0(x)$ is called a **uniformly most powerful level-α unbiased test**, denoted by level-α UMPU, if $\phi_0(x) \in \Phi_\alpha^U$, for $\forall \phi(x) \in \Phi_\alpha^U$ and $\theta \in \Theta_1$ we have $E_\theta \phi_0(X) \geq E_\theta \phi(X)$.

Based on the discussion about the least favorable distribution, we are more interested in the power functions of the parameters on the boundary between the null hypothesis and the alternative hypothesis. As a result, it is significant to discuss the following set of test statistics. A test function $\phi(x)$ is called a **level-α similar test**, if

$$E_\theta \phi(X) = \alpha, \quad \forall \theta \in \bar{\Theta}_0 \cap \bar{\Theta}_1,$$

where $\bar{\Theta}$ denotes the closure of Θ. Let Φ_α^S denote the collection of all the level-α similar tests. A test $\phi_0(x)$ is called a **uniformly most powerful level-α similar test**, denoted by level-α UMPS test, if $\phi_0(x) \in \Phi_\alpha^S$, and $E_\theta \phi_0(X) \geq E_\theta \phi(X)$ for $\forall \phi(x) \in \Phi_\alpha^S$ and $\theta \in \Theta_1$. In principle, Φ_α^U and Φ_α^S are the subsets of Φ_α, and there is no inclusive relation between the two subsets. However, we have the following theorem.

Theorem 3.3.1. *Let $\phi_0(x)$ be a level-α UMPS test about the testing problem (3.1.3), and $\phi_0(x) \in \Phi_\alpha$. If the power function of each test function in Φ_α^U is continuous with respect to the parameter θ, then $\phi_0(x)$ is a level-α UMPU test.*

Proof. First, under the given conditions, we prove that $\Phi_\alpha^U \subset \Phi_\alpha^S$. Take any $\phi(x) \in \Phi_\alpha^U$, for $\theta \in \bar{\Theta}_0 \cap \bar{\Theta}_1$, there exist $\{\theta_n\} \subset \Theta_0$, such that $\theta_n \to \theta$ when $n \to \infty$. By $E_{\theta_n} \phi(X) \leq \alpha$ and the continuity, we have $E_\theta \phi(X) \leq \alpha$. Similarly we can get $E_\theta \phi(X) \geq \alpha$. Thus we have $E_\theta \phi(X) = \alpha$.

Next it suffices to prove that $\phi_0(x) \in \Phi_\alpha^U$. From the given conditions, it suffices to prove that $E_\theta \phi(X) \geq \alpha$ for $\forall \theta \in \Theta_1$. In fact, let $\psi(x) \equiv \alpha$, then $\psi(x) \in \Phi_\alpha^S$, and we have $E_\theta \phi_0(X) \geq E_\theta \psi(X) = \alpha$. □

Generally the continuity condition of the power function in the above theorem holds, even does so for that of the discrete distributions. So the key of the problem is $\phi_0(x) \in \Phi_\alpha$, and this coincides with Remark 3.2.3. As far as similar tests are concerned, the following theorem is very important.

Theorem 3.3.2. (Generalized Neyman-Pearson Fundamental Lemma) *Let $f_0(x), f_1(x), \cdots, f_m(x)$ be integrable functions in the measure space $(\mathcal{X}, \mathcal{A}, \mu)$. Let $\phi(x)$ be a test function, satisfying*

$$\int \phi(x) f_i(x) d\mu(x) = \alpha_i, \quad i = 1, \cdots, m, \tag{3.3.1}$$

where α_i is a given real-valued number. Let $\phi^*(x)$ satisfy Eq. (3.3.1), and

$$\phi^*(x) = \begin{cases} 1, & \text{if } f_0(x) > \sum_{i=1}^{m} k_i f_i(x), \\ 0, & \text{otherwise.} \end{cases} \tag{3.3.2}$$

Then we have

$$\int \phi^*(x) f_0(x) d\mu(x) \geq \int \phi(x) f_0(x) d\mu(x).$$

Proof. Let $W_1 = \{x;\ f_0(x) > \sum_{i=1}^{m} k_i f_i(x)\}$ and $W_2 = \{x;\ f_0(x) < \sum_{i=1}^{m} k_i f_i(x)\}$. Then we have $\phi^*(x) \geq \phi(x)$, for $x \in W_1$; $\phi^*(x) \leq \phi(x)$, for $x \in W_2$. Thus we have

$$(\phi^*(x) - \phi(x)) \left(f_0(x) - \sum_{i=1}^{m} k_i f_i(x) \right) \geq 0.$$

Taking the integral on both sides, we have

$$\int (\phi^*(x) - \phi(x)) f_0(x) d\mu(x) \geq \sum_{i=1}^{m} k_i \int (\phi^*(x) - \phi(x)) f_i(x) d\mu(x) = 0.$$

□

Theorem 3.3.3. *If the density function $f(x;\theta)$ has a monotone and convex likelihood ratio on $T(x)$, then for $\theta_1 < \theta_2$ there exists a level-α UMPU test for the two-sided testing problem*

$$H_0:\ \theta_1 \leq \theta \leq \theta_2 \quad vs. \quad H_1:\ \theta < \theta_1 \text{ or } \theta > \theta_2, \tag{3.3.3}$$

and it has the following form

$$\phi(x) = \begin{cases} 1, & \text{if } T(x) < c_1 \text{ or } T_2(x) > c_2, \\ r_i, & \text{if } T(x) = c_i, \quad i = 1, 2, \\ 0, & \text{otherwise,} \end{cases} \tag{3.3.4}$$

where c_i and r_i satisfy

$$E_{\theta_1}\phi(X) = E_{\theta_2}\phi(X) = \alpha. \tag{3.3.5}$$

Proof. By Eq. (3.3.5), $\phi(x)$ given by (3.3.4) is a similar test. First, we will prove that it is a level-α UMPS test. Take $\theta_0 < \theta_1$, consider the testing problem

$$H_0':\ \theta < \theta_1 \text{ or } \theta > \theta_2 \quad \text{vs.} \quad H_1':\ \theta = \theta_0.$$

By Theorem 3.3.2, the most powerful test has the following form

$$\phi(x) = \begin{cases} 1, & \text{if } \ f(x;\theta_0) > k_1 f_1(x;\theta_1) + k_2 f_2(x;\theta_2), \\ 0, & \text{otherwise.} \end{cases} \tag{3.3.6}$$

Next we will prove that $k_2 < 0$ by using the method of Proof by Contradiction. Suppose that $k_2 > 0$. When $k_1 > 0$, let

$$\lambda(x) = k_1 f(x;\theta_1)/f(x;\theta_0) + k_2 f(x;\theta_2)/f(x;\theta_0).$$

Since $\theta_0 < \theta_1 < \theta_2$, from the condition of the monotone likelihood ratio, Equation (3.3.6) is equivalent to

$$\phi(x) = \begin{cases} 1, & \text{if } \ T(x) < c, \\ 0, & \text{otherwise.} \end{cases}$$

By Theorem 3.2.2, we know that $E_\theta \phi(T(X))$ is a monotone likelihood function of θ, and this contradicts with Eq. (3.3.5). When $k_1 < 0$, the condition in Eq. (3.3.6) can be rewritten as $f(x;\theta_0) - k_1 f(x;\theta_1) > k_2(x;\theta_2)$, and this also leads to a contradiction. Therefore, we must have $k_2 < 0$. To make sense of (3.3.5), it necessitates $k_1 > 0$. Let

$$\lambda(x) = \frac{f(x;\theta_0)}{f(x;\theta_1)} - k_2 \frac{f(x;\theta_2)}{f(x;\theta_1)}.$$

From the condition of convex likelihood ratio, $\lambda(x) \geq k_1$ is equivalent to that there exist c_1 and c_2 satisfying (3.3.4). Since the test function is independent of the value of θ_0, $\phi(x)$ is a level-α UMPS test.

Finally we prove that $\phi(x) \in \Phi_\alpha$. Note that the problem we are considering is thoroughly contrary to that discussed in Theorem 3.2.3. Let $\beta = 1 - \alpha$ and let $\psi(x)$ have the form of Eq. (3.2.18) satisfying $E_{\theta_1}\psi(X) = E_{\theta_2}\psi(X) = \beta$. Here $\phi(x) = 1 - \psi(x)$. Let $\psi_0(x) \equiv \beta$. By Theorem 3.2.3, $E_\theta \psi(X) \geq E_\theta \psi_0(X) = \beta$ for $\forall \theta \in (\theta_1, \theta_2)$. Thus we have

$$E_\theta \phi(X) = 1 - E_\theta \psi(X) \leq 1 - \beta = \alpha.$$

The theorem is established by using Theorem 3.3.1. □

From Examples 3.2.3 and 3.2.6, we know that the one-parameter exponential family satisfies the conditions in Theorem 3.3.3. So there exists a level-α UMPU test about the two-sided test (3.3.3) for some distributions. Furthermore, we have the following theorem.

Theorem 3.3.4. *If a density function is from the one-parameter exponential family, and then for the two-sided test*

$$H_0: \ \theta = \theta_0 \quad vs. \quad H_1: \ \theta \neq \theta_0 \tag{3.3.7}$$

there exists a level-α UMPU test, with the form (3.3.4), where c_i and r_i satisfy

$$E_{\theta_0}\phi(T(X)) = \alpha, \tag{3.3.8}$$

$$E_{\theta_0}T(X)\phi(T(X)) = \alpha E_{\theta_0}T(X). \tag{3.3.9}$$

Proof. For the given $\Delta_n > 0$, consider the testing problem
$H_0^{(n)} : \theta_0 - \Delta_n \le \theta \le \theta_0 + \Delta_n$ vs. $H_1^{(n)} : \theta < \Theta_0 - \Delta_n$ or $\theta > \theta_0 + \Delta_n$. From Theorem 3.3.3, there exist a level-α UMPU test $\phi_n(x)$, with the form (3.3.4), satisfying (3.3.5). Since Φ_α is weakly compact (cf. the proof of Theorem 3.1.1), there exists a $\phi(x) \in \Phi_\alpha$ such that for $\forall\theta \in \Theta$,

$$E_\theta\phi_n(X) \to E_\theta\phi(X) \text{ when } \Delta_n \to 0.$$

From the construction of $\phi_n(x)$, $\phi(x)$ also has the form (3.3.4). Let the power function be $\beta(\theta) = E_\theta\phi(X)$. From Theorem 1.3.1 in Chapter 1, $\beta(\theta)$ is continuous and derivable, so $\phi(x) \in \Phi_\alpha^U$ satisfies (3.3.8). Since $\beta(\theta_0) = \min_{\theta\in\Theta}\beta(\theta)$, $\beta'(\theta_0) = 0$. From Eq. (3.2.14),

$$\beta'(\theta) = \frac{C'(\theta)}{C(\theta)}E_\theta\phi(T(X)) + E_\theta T(X)\phi(T(X)).$$

From Theorem 1.3.2 in Chapter 1, we have $C'(\theta)/C(\theta) = -E_\theta T(X)$, and thus we have

$$\beta'(\theta) = -E_\theta T(X)E_\theta\phi(T(X)) + E_\theta T(X)\phi(T(X)).$$

Note that $\beta'(\theta_0) = 0$ in conjunction with Eq. (3.3.8) we have Eq. (3.3.9). □

Remark 3.3.1. If the density function $g(t;\theta)$ on $T(x)$ is continuous at $\theta = \theta_0$, and $g(t;\theta_0)$ is symmetrical about θ_0, we have $P_{\theta_0}\{T < \theta_0 - \Delta\} = P_{\theta_0}\{T > \theta_0 + \Delta\}$ for $\Delta > 0$. Now Eq. (3.3.4) can be rewritten as

$$\psi(T) = \begin{cases} 1, & \text{if } \ T < \theta_0 - \Delta \text{ or } T > \theta_0 + \Delta, \\ r, & \text{if } \ T = \theta_0 \pm \Delta, \\ 0, & \text{otherwise,} \end{cases} \tag{3.3.10}$$

where Δ and r satisfy

$$P_{\theta_0}\{T < \theta_0 - \Delta\} + rP_{\theta_0}\{T = \theta_0 - \Delta\} = \frac{\alpha}{2}.$$

This coincides with Eq. (3.3.8). By symmetry and Eq. (3.3.10), we have $E_{\theta_0}(T - \theta_0)\psi(T) = 0$. Thus we have

$$\begin{aligned} E_{\theta_0}T\psi(T) &= E_{\theta_0}(T - \theta_0)\psi(T) + \theta_0 E_{\theta_0}\psi(T) \\ &= \alpha\theta_0. \end{aligned}$$

Note that $E_{\theta_0}T = \theta_0$, then the above equation coincides with Eq. (3.3.9).

Example 3.3.1. (Two-sided test for binomial distribution parameter) Let $X \sim Bi(n;\theta)$. For $\theta \in (0,1)$, consider the two-sided testing problem $H_0 : \theta = \theta_0$ vs. $H_1 : \theta \neq \theta_0$. Binomial distribution belongs to one-parameter exponential family with $T(x) = x$. From Theorem 3.3.4, a level-α UMPU test has the form (3.3.4). Corresponding to Eq. (3.3.8), we have

$$\sum_{x=c_1+1}^{c_2-1} \binom{n}{x} \theta_0^x (1-\theta_0)^{n-x} + \sum_{i=1}^{2}(1-r_i)\binom{n}{c_i}\theta_0^{c_i}(1-\theta_0)^{n-c_i} = 1-\alpha. \tag{3.3.11}$$

Since

$$x\binom{n}{x}\theta_0^x(1-\theta_0)^{n-x} = n\theta_0\binom{n-1}{x-1}\theta_0^{x-1}(1-\theta_0)^{(n-1)-(x-1)},$$

corresponding to Eq. (3.3.9), we can get

$$\sum_{x=c_1+1}^{c_2-1}\binom{n-1}{x-1}\theta_0^{x-1}(1-\theta_0)^{(n-1)-(x-1)} + \sum_{i=1}^{2}(1-r_i)\binom{n-1}{c_i-1}\theta_0^{c_i-1}(1-\theta_0)^{(n-1)-(c_i-1)} = 1-\alpha. \tag{3.3.12}$$

By Eqs. (3.3.11) and (3.3.12), we can get c_i and r_i. However, its computation is rather cumbersome. Since $(X - n\theta_0)/\sqrt{n\theta_0(1-\theta_0)} \xrightarrow{L} N(0,1)$, when H_0 is true, we can use a normal approximation to the binomial distribution if n is large enough.

Example 3.3.2. (Two-sided test for the mean of a normal distribution) Let $x_1, \cdots, x_n$ be a set of samples from $N(\mu, \sigma^2)$. For a given real number μ_0, consider the two-sided testing problem $H_0 : \mu = \mu_0$ vs. $H_1 : \mu \neq \mu_0$.

Case with known variance σ^2. We can apply Theorem 3.3.4 with $T(x) = \bar{x}$. Since $\bar{X} \sim N(\mu, \sigma^2/n)$, which is a symmetrical distribution, by Remark 3.3.1, we have $c_1 = \mu_0 - \Delta, c_2 = \mu_0 + \Delta$. Since $\sqrt{n}(\bar{X} - \mu_0)/\sigma \sim N(0,1)$, when H_0 is true, we can get

$$\Delta = \frac{1}{\sqrt{n}}\sigma\mu_{\frac{\alpha}{2}},$$

where $\mu_{\alpha/2}$ satisfies $1 - \Phi(\mu_{\alpha/2}) = \alpha/2$, and Φ is the standard normal distribution function. Now the rejection region can be written as $\{x; |\bar{x} -$

$\mu_0| \geq \Delta\}$. And then $\sqrt{n}(\bar{x} - \mu_0)/\sigma$ is a level-α UMPU test. This coincides with the likelihood test (cf. Example 2.2.2 in Chapter 2).

Case with unknown variance σ^2. In this case σ^2 is a nuisance parameter. Review the discussion about the one-sided test in Example 3.2.2, let $t = (\bar{x} - \mu_0)/\hat{\sigma}$ with $\hat{\sigma}^2 = \sum_{i=1}^n (x_i - \bar{x})^2/(n-1)$, then t has a noncentral Student's t-distribution $t(n-1, \eta)$ with $n-1$ degrees of freedom and noncentrality parameter $\eta = \mu - \mu_0$. Thus the testing problem is equivalent to $H_0 : \eta = 0$ vs. $H_1 : \eta \neq 0$. Since the noncentral Student's t-distribution has a monotone and convex likelihood ratio, Theorem 3.3.4 is applicable. When H_0 is true, the t has a central Student's t-distribution, which is a symmetrical distribution. According to Remark 3.3.1, we have $c_1 = -\Delta, c_2 = \Delta$, where $\Delta = t_{\alpha/2}$ is an $\alpha/2$ critical value of the Student's t-distribution with $n-1$ degrees of freedom. Then the t test is a level-α UMPU test.

Example 3.3.3. (Two-sided test for the variance of a normal distribution) Let $x_1, \cdots, x_n$ be a set of samples from $N(\mu, \sigma^2)$. For the given $\sigma_0^2 > 0$, consider the two-sided testing problem $H_0 : \sigma^2 = \sigma_0^2$ vs. $H_1 : \sigma^2 \neq \sigma_0^2$.

Case with known mean μ. From Theorem 1.3.4 in Chapter 1, $\sum_{i=1}^n (x_i - \mu)^2$ is a complete and sufficient statistic for σ^2. Let $t(x) = \sum_{i=1}^n (x_i - \mu)^2/\sigma_0^2$. Then $T(X) \sim \sigma_0^2 \chi_n^2/\sigma^2$, and the density function with respect to the Lebesgue measure is

$$g(t; n, \eta) = \frac{1}{2^{\frac{n}{2}}\Gamma(\frac{n}{2})}(\eta t)^{\frac{n}{2}-1} e^{-\frac{\eta t}{2}} \eta,$$

where $\eta = \sigma_0^2/\sigma^2$, and $\eta > 0$. Now the testing problem is equivalent to $H_0 : \eta = 1$ vs. $H_1 : \eta \neq 1$. This belongs to the exponential family. From Theorem 3.3.4, there exists a level-α UMPU test

$$\phi(t) = \begin{cases} 1, & \text{if} \quad t \leq c_1 \text{ or } t \geq c_2, \\ 0, & \text{otherwise.} \end{cases}$$

When H_0 is true, $T(X)$ has a chi-square distribution with n degrees of freedom, corresponding to Eqs. (3.3.8) and (3.3.9), we have

$$\int_{c_1}^{c_2} g(t; n, 1) dt = 1 - \alpha, \tag{3.3.13}$$

and

$$\int_{c_1}^{c_2} t g(t; n, 1) dt = (1 - \alpha) ET = n(1 - \alpha). \tag{3.3.14}$$

Note that $tg(t;n,1) = ng(t;n+2,1)$, then Eq. (3.3.14) can be written

$$\int_{c_1}^{c_2} g(t;n+2,1)dt = 1-\alpha. \tag{3.3.15}$$

Solving Eqs. (3.3.13) and (3.3.15), we can get c_1 and c_2. However, its computation is rather complicated. Though the χ^2 distribution is not symmetrical, we can still make use of the idea of Remark 3.3.1, let c_1 and c_2 satisfy

$$\int_0^{c_1} g(t;n,1)dt = \int_{c_2}^{+\infty} g(t;n,1)dt = \frac{\alpha}{2}.$$

Case with unknown mean μ. Let $T(x) = \sum_{i=1}^n (x_i - \bar{x})^2/\sigma_0^2$ be a test statistic. We can get a level-α UMPU test by a similar method mentioned above, except that the degrees of freedom are $n-1$. In other words, $g(t;n,\eta)$ is substituted by $g(t;n-1,\eta)$.

We can see that the above test statistic coincides with the likelihood ratio statistic.

Example 3.3.4. (Two-sided test for two normal populations) Let $x_1, \cdots, x_n$ be a set of samples from $N(\mu_1, \sigma_1^2)$ and $y_1, \cdots, y_m$ from $N(\mu_2, \sigma_2^2)$. Consider the two-sided test for the mean and the variance, respectively.

Comparing the means. Consider the two-sided testing test $H_0 : \mu_1 = \mu_2$ vs. $H_1 : \mu_1 \neq \mu_2$. When the two variances σ_1^2 and σ_2^2 are known, or unknown but equal, recalling the discussion in Example 2.2.3 in Chapter 2, we can formulate the problem as the case of Example 3.3.2, so we can get a level-α UMPU test. When the variances are unknown and unequal, this is the Behrens-Fisher problem.

Comparing the variances. Consider the two-sided testing problem $H_0 : \sigma_1^2 = \sigma_2^2$ vs. $H_1 : \sigma_1^2 \neq \sigma_2^2$. Now we will discuss the case that both μ_1 and μ_2 are unknown. Let $T(x,y) = S_1^2/S_2^2$, where

$$S_1^2 = \frac{1}{n-1}\sum_{i=1}^n (x_i - \bar{x})^2, \text{ and } S_2^2 = \frac{1}{m-1}\sum_{i=1}^m (y_i - \bar{y})^2.$$

This is a function of sufficient statistic. Since S_1^2/σ_1^2 and S_2^2/σ_2^2 have χ^2 distributions with $n-1$ degrees of freedom and $m-1$ degrees of freedom, respectively, $T(X,Y) \sim \sigma_2^2 F(n-1,m-1)/\sigma_1^2$ by Example 1.1.10 in Chapter 1, and the density function with respect to the Lebesgue measure is

$$h(t;\eta) = c\eta(\eta t)^{(n-3)/2}\left(1 + \frac{n-1}{m-1}\eta t\right)^{-(n+m)/2+1}, \quad t > 0,$$

where $\eta = {\sigma_1}^2/{\sigma_2}^2$, and c is a constant. Thus the testing problem is equivalent to $H_0 : \eta = 1$ vs. $H_1 : \eta \neq 1$. Since $h(t; \eta)$ belongs to the exponential family, there exists a level-α UMPU test with rejection region $\{t;\ t \leq c_1 \text{ or } t \geq c_2\}$ by Theorem 3.3.4, where c_1 and c_2 satisfy

$$\int_{c_1}^{c_2} h(t; 1)dt = 1 - \alpha$$

and

$$\int_{c_1}^{c_2} \frac{t-1}{(n-1)t + (m-1)} h(t; 1)dt = 0.$$

For convenience, let $F(c_1) = \alpha/2$, and $1 - F(c_2) = \alpha/2$, where F denotes an F-distribution function with $n-1$ and $m-1$ degrees of freedom.

Refer to Example 2.2.3 in Chapter 2, we see that it coincides with the likelihood ratio test.

3.4 Comparing Power Functions

For many testing problems, there does not exist a level-α UMPU test. More seriously, it is very difficult to define a class of test function in which we can find out the most powerful test. Therefore, for the testing problems, we often need to make a concrete analysis of concrete problems, study characteristics of the change of power functions, and compare the performances of test function based on the above analysis. Now, we explore some techniques in virtue of a concrete testing problem.

3.4.1 *Test Procedures*

Let $\boldsymbol{x} = (x_1, x_2)'$ be a sample from a two-dimensional normal distribution $N(\boldsymbol{\mu}, \Lambda)$, where $\boldsymbol{\mu} = (\mu_1, \mu_2)'$, and the covariance matrix Λ is known. Here, we take only one sample just for convenience of discussion below. However, we can regard $\boldsymbol{x}$ as the sample mean when the covariance matrix is known. Now consider the following one-sided testing problem

$$H_0 : \boldsymbol{\mu} = 0 \quad \text{vs.} \quad H_1 : \boldsymbol{\mu} > 0, \tag{3.4.1}$$

where $\boldsymbol{\mu} > 0$ denotes $\mu_i \geq 0$ for $i = 1, 2$ with at least one nonzero. Without loss of generality, suppose that the covariance matrix Λ satisfies $VX_1 = VX_2 = 1$, and $CV(X_1, X_2) = \rho$. Let

$$A = \begin{pmatrix} \frac{1}{\sqrt{1-\rho^2}} & -\frac{\rho}{\sqrt{1-\rho^2}} \\ 0 & 1 \end{pmatrix},$$

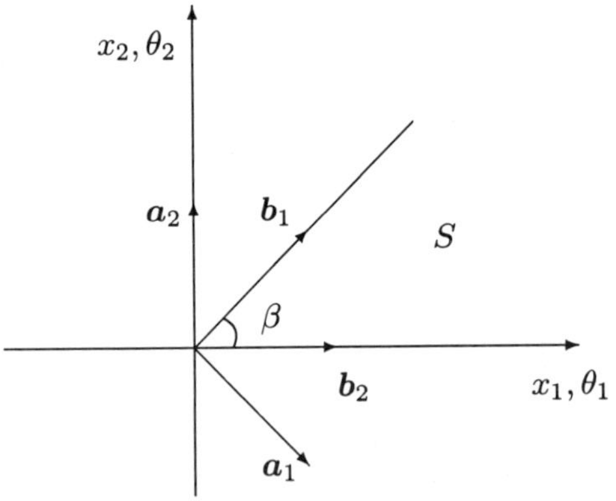

Fig. 3.4.1 Region corresponding to the alternative hypothesis when $\rho < 0$

then we have $A\Lambda A' = I$. Let $\boldsymbol{y} = A\boldsymbol{x}$ and $\boldsymbol{\theta} = A\boldsymbol{\mu}$, then $\boldsymbol{Y} \sim N(\boldsymbol{\theta}, I)$. Since $\boldsymbol{\mu} > \mathbf{0}$ is equivalent to $A^{-1}\boldsymbol{\theta} > \mathbf{0}$, then the testing problem (3.4.1) is equivalent to $H_0 : \boldsymbol{\theta} = \mathbf{0}$ vs. $H_1 : \boldsymbol{\theta} \in S - \{\mathbf{0}\}$ with

$$S = \{\boldsymbol{\theta} \in R^2; \sqrt{1-\rho^2}\theta_1 + \rho\theta_2 \geq 0, \theta_2 \geq 0\}. \tag{3.4.2}$$

Let $\boldsymbol{a}_1 = (\sqrt{1-\rho^2}, \rho)'$, and $\boldsymbol{a}_2 = (0,1)'$. When $\rho < 0$, the region S is illustrated in Fig. 3.4.1, where the angle included in the region is $\beta = \arccos(-\rho)$. Furthermore, let $\boldsymbol{b}_1 = (-\rho, \sqrt{1-\rho^2})'$, and $\boldsymbol{b}_2 = (0,1)'$. Then Eq. (3.4.2) can be rewritten as

$$S = \{\boldsymbol{\theta} \in R^2;\ \boldsymbol{\theta} = \gamma_1\boldsymbol{b}_1 + \gamma_2\boldsymbol{b}_2 \text{ for } \gamma_i \geq 0 \text{ and } i = 1,2\}. \tag{3.4.3}$$

We call $\boldsymbol{b}_i$ in S as an edge vector. It can be seen that S is a set of nonnegative linear combinations of edge vectors, which will be called a **polyhedral convex cone**. Now we need to test whether the mean $\boldsymbol{\theta}$ is at origin or in the region S. For $\boldsymbol{y}, \boldsymbol{z} \in R^2$, define the following inner product and norm respectively,

$$(\boldsymbol{y}, \boldsymbol{z}) = \boldsymbol{y}'\boldsymbol{z}, \text{ and } ||\boldsymbol{y}|| = (\boldsymbol{y}, \boldsymbol{y})^{1/2}. \tag{3.4.4}$$

Then Eq. (3.4.2) can be rewritten as $S = \{\boldsymbol{\theta} \in R^2; (\boldsymbol{a}_1, \boldsymbol{\theta}) \geq \mathbf{0}, (\boldsymbol{a}_2, \boldsymbol{\theta}) \geq \mathbf{0}\}$.

The χ^2 test. This is a likelihood ratio test for the two-sided testing problem $H_0 : \boldsymbol{\mu} = \mathbf{0}$ vs. $H_1 : \boldsymbol{\mu} \neq \mathbf{0}$ or equivalently, $H_0 : \boldsymbol{\theta} = \mathbf{0}$ vs. $H_1 : \boldsymbol{\theta} \neq \mathbf{0}$. It is easy to get the following likelihood-ratio test statistic

$$\chi^2 = \boldsymbol{y}'\boldsymbol{y} = \boldsymbol{x}'\Lambda^{-1}\boldsymbol{x}. \tag{3.4.5}$$

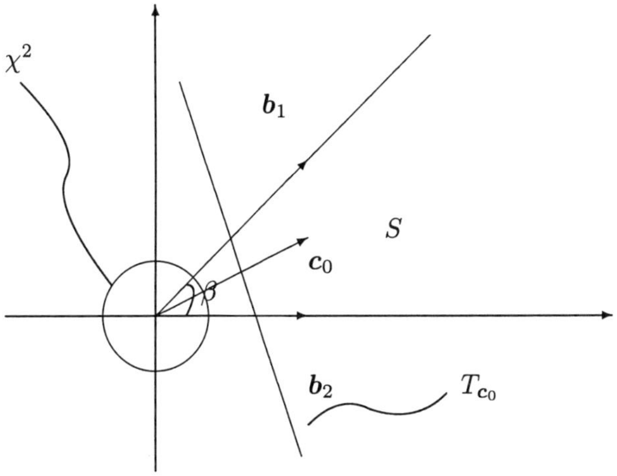

Fig. 3.4.2 Rejection region of χ^2 test and T_{c_0} test

Since $\chi^2 = ||\boldsymbol{y}||^2$, Equation (3.4.5) characterizes the distance between the sample point and the origin. When H_0 is true, χ^2 has a chi-square distribution with 2 degrees of freedom. The rejection region is $\{\boldsymbol{y} \in R^2; \chi^2 \geq {\chi^2}_\alpha(2)\}$.

The best linear test. Take any $\boldsymbol{c} \in S$ satisfying $||\boldsymbol{c}|| = 1$. Define a **linear test statistics** as $T_{\boldsymbol{c}} = \boldsymbol{c}'\boldsymbol{y}$. Then, when H_0 is true, $T_{\boldsymbol{c}} \sim N(0,1)$. Let $\phi_{\boldsymbol{c}}(\boldsymbol{y})$ be a test function corresponding to the rejection region $\{\boldsymbol{y}; T_{\boldsymbol{c}} \geq u_\alpha\}$, where u_α satisfies $1 - \Phi(u_\alpha) = \alpha$. According to discussion in Section 3.2, $\phi_{\boldsymbol{c}}(\boldsymbol{y})$ is a level-α UMP test for the following one-sided testing problem

$$H_0 : \boldsymbol{\theta} = \boldsymbol{0} \quad \text{vs.} \quad H_c : \boldsymbol{\theta} \in \eta \boldsymbol{c} \ (\eta > 0). \tag{3.4.6}$$

Now we discuss the power function of $\phi_{\boldsymbol{c}}(\boldsymbol{y})$. For $\boldsymbol{\theta} \in S - \{\boldsymbol{0}\}$, since $E_{\boldsymbol{\theta}} T_{\boldsymbol{c}} = \boldsymbol{c}'\boldsymbol{\theta}$, the power function is

$$\beta(\boldsymbol{c}, \boldsymbol{\theta}) = E_{\boldsymbol{\theta}} \phi_{\boldsymbol{c}}(\boldsymbol{Y}) = 1 - \Phi(u_\alpha - \boldsymbol{c}'\boldsymbol{\theta}). \tag{3.4.7}$$

Thus, the power function depends on the inner product $\boldsymbol{c}'\boldsymbol{\theta} = (\boldsymbol{c}, \boldsymbol{\theta}) = ||\boldsymbol{\theta}|| \cos A(\boldsymbol{c}, \boldsymbol{\theta})$, where $A(\boldsymbol{c}, \boldsymbol{\theta})$ denotes the angle between vector $\boldsymbol{c}$ and $\boldsymbol{\theta}$. When the norm $||\boldsymbol{\theta}||$ is a constant, the power function (3.4.7) is a monotone decreasing function of the angle. Since

$$\min_{\boldsymbol{\theta} \in S} \beta(\boldsymbol{c}, \boldsymbol{\theta}) = \min\{\beta(\boldsymbol{c}, \boldsymbol{b}_1), \beta(\boldsymbol{c}, \boldsymbol{b}_2)\}, \tag{3.4.8}$$

the power attains its minimum at one of the two edge vectors for any $T_{\boldsymbol{c}}$. When $||\boldsymbol{\theta}||$ is fixed, let $T_{\boldsymbol{c}_0} = \boldsymbol{c}_0{}'\boldsymbol{y}$ be the linear test statistic such that the

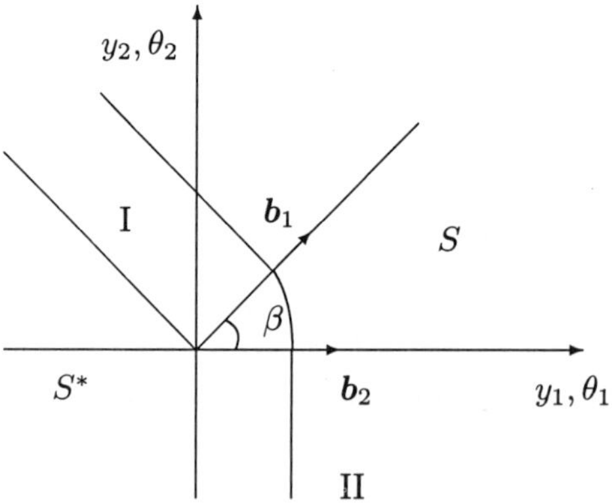

Fig. 3.4.3 Rejection region of $\bar{\chi}^2$ test

minimal power function attains the maximum, where $\boldsymbol{c}_0$ satisfies

$$\min_{\boldsymbol{\theta}\in S} \beta(\boldsymbol{c}_0, \boldsymbol{\theta}) = \max_{\boldsymbol{c}\in S} \min_{\boldsymbol{\theta}\in S} \beta(\boldsymbol{c}, \boldsymbol{\theta}).$$

By Eq. (3.4.8), $\boldsymbol{c}_0$ satisfies

$$\beta(\boldsymbol{c}_0, \boldsymbol{b}_1) = \beta(\boldsymbol{c}_0, \boldsymbol{b}_2) \Leftrightarrow (\boldsymbol{c}_0, \boldsymbol{b}_1) = (\boldsymbol{c}_0, \boldsymbol{b}_2), ||\boldsymbol{c}_0|| = 1. \tag{3.4.9}$$

Solving Eq. (3.4.9), we can get $\boldsymbol{c}_0 = (\sqrt{(1-\rho)/2}, \sqrt{(1+\rho)/2})'$. $T_{\boldsymbol{c}_0}$ will be called a **best linear test** for the testing problem (3.4.1) or (3.4.2).

Likelihood-ratio test. Recall the definition of likelihood-ratio test, we need to obtain the MLE of parameters under H_0 and $H_0 \cup H_1$ respectively. Obviously, the MLE of parameter under H_0 is $\mathbf{0}$. Let $\hat{\boldsymbol{\theta}}$ be the MLE of parameter under $H_0 \cup H_1$, then $\hat{\boldsymbol{\theta}}$ satisfies

$$||\boldsymbol{y} - \hat{\boldsymbol{\theta}}||^2 = \min_{\boldsymbol{\theta}\in S} ||\boldsymbol{y} - \boldsymbol{\theta}||^2.$$

Based on Fig. 3.4.3 we can see that the sample space is divided into four parts. Here, S^* will be called as the dual cone of S, satisfying $S^* = \{\boldsymbol{y} \in R^2; (\boldsymbol{y}, \boldsymbol{z}) \le 0, \forall \boldsymbol{z} \in S\}$. It can be verified that $\hat{\boldsymbol{\theta}} = \boldsymbol{y}$ when $\boldsymbol{y} \in S$; $\hat{\boldsymbol{\theta}} = \mathbf{0}$ when $\boldsymbol{y} \in S^*$; $\hat{\boldsymbol{\theta}}$ is the projection of $\boldsymbol{y}$ on the half line containing $\boldsymbol{b}_1$ when $\boldsymbol{y} \in$ I; $\hat{\boldsymbol{\theta}} = (y_1, 0)'$ when $\boldsymbol{y} \in$ II. Then the likelihood-ratio statistic is

$$\bar{\chi}^2 = ||\boldsymbol{y} - \hat{\boldsymbol{\theta}}||^2 - ||\boldsymbol{y}||^2 = ||\hat{\boldsymbol{\theta}}||^2. \tag{3.4.10}$$

Thus we can obtain the rejection region $G = \mathrm{I} \cup \mathrm{II} \cup S \cup S^*$ as shown in Fig. 3.4.3, that is,

$$G = \begin{cases} y_1^2 + y_2^2 \ge c_\alpha^2, & \text{if } \sqrt{1-\rho^2}y_1 + \rho y_2 \ge 0 \text{ and } y_2 \ge 0, \\ -\rho y_1 + \sqrt{1-\rho^2}y_2 \ge c_\alpha, & \text{if } -\rho y_1 + \sqrt{1-\rho^2}y_2 \ge 0 \text{ and} \\ & \qquad \sqrt{1-\rho^2}y_1 + \rho y_2 \le 0, \\ y_1 \ge c_\alpha, & \text{if } y_1 \ge 0 \text{ and } y_2 \le 0, \end{cases}$$

where c_α satisfies $P_\theta\{\boldsymbol{Y} \in G\} = \alpha$. Now we need to obtain the distribution of $\bar{\chi}^2$ under H_0. Since the space R^2 is divided into four parts, by Eq. (3.4.10), $\bar{\chi}^2 \sim \chi_2^2$, when $\bar{\chi}^2 \in S$; $\bar{\chi}^2 \sim \chi_1^2$, when $\bar{\chi}^2 \in \mathrm{I}\bigcup\mathrm{II}$; $\bar{\chi}^2 = 0$, when $\bar{\chi}^2 \in S^*$. Here, χ_i^2 denotes a chi-square distribution with i degrees of freedom. Furthermore, when H_0 is true, $P_0\{\boldsymbol{Y} \in S\} = \beta/2\pi = \arccos(-\rho)/2\pi$, $P_0\{\boldsymbol{Y} \in \mathrm{I}\cup\mathrm{II}\} = 1/2$, and $P_0\{\boldsymbol{Y} \in S^*\} = 1/2 - P_0\{\boldsymbol{Y} \in S\}$. Then we have the following theorem.

Theorem 3.4.1. *When H_0 is true and $c > 0$, we have*

$$P_0\{\bar{\chi}^2 \geq c^2\} = \frac{1}{2}P_0\{\chi_1^2 \geq c^2\} + \frac{1}{2\pi}\arccos(-\rho)P\{\chi_2^2 \geq c^2\};$$

and

$$P_0\{\bar{\chi}^2 = 0\} = \frac{1}{2} - \frac{1}{2\pi}\arccos(-\rho).$$

For fixed α, we can get the critical value c_α of rejection region G. It can be seen that the distribution in Theorem 3.4.1 is a weighted average of χ^2 distribution, where the weights are the probabilities of sample points falling into the corresponding regions. The above distribution is usually called a $\bar{\chi}^2$ distribution, and the test statistic denoted by Eq. (3.4.10) is called a **$\bar{\chi}^2$ test** .

3.4.2 *Power Functions and Their Comparison*

Power function of test statistics. For the testing problem (3.4.1) or equivalently (3.4.2), we have already constructed three test statistics. Now, it is necessary to compare the power functions of the three test statistics. For any $\boldsymbol{\theta} \in S$, we compute the three power functions respectively.

For the χ^2 test given by Eq. (3.4.5), the power function depends on the noncentral χ^2 distribution with 2 degrees of freedom:

$$P_{\text{square}}(\boldsymbol{\theta}) = P\{\chi_2^2(\Delta^2) \geq \chi_\alpha^2\}, \tag{3.4.11}$$

where $\Delta^2 = \boldsymbol{\theta}'\boldsymbol{\theta} = \boldsymbol{\mu}'\Lambda^{-1}\boldsymbol{\mu}$ is the noncentrality parameter. For the best linear test $T_{c_0} = \boldsymbol{c}_0'\boldsymbol{y}$, with $\boldsymbol{c}_0$ satisfying Eq. (3.4.9), the power function is

$$P_{\text{linear}}(\boldsymbol{\theta}) = 1 - \Phi(u_\alpha - \boldsymbol{c}_0'\boldsymbol{\theta}) = 1 - \Phi(u_\alpha - ||\boldsymbol{\theta}||\cos A(\boldsymbol{c}_0, \boldsymbol{\theta})). \tag{3.4.12}$$

For the $\bar{\chi}^2$ test given by Eq. (3.4.10), the computation of the power function is very complicated. Let $S_\Delta = \{\boldsymbol{\theta} \in R^2; ||\boldsymbol{\theta}||^2 = \Delta^2, \boldsymbol{\theta} \in S\}$. For $\boldsymbol{\theta} \in S_\Delta$, take the following polar coordinate transformation

$$y_1 = r\cos\psi, \qquad y_2 = r\sin\psi;$$
$$\theta_1 = \Delta\cos\phi, \qquad \theta_2 = \Delta\sin\phi,$$

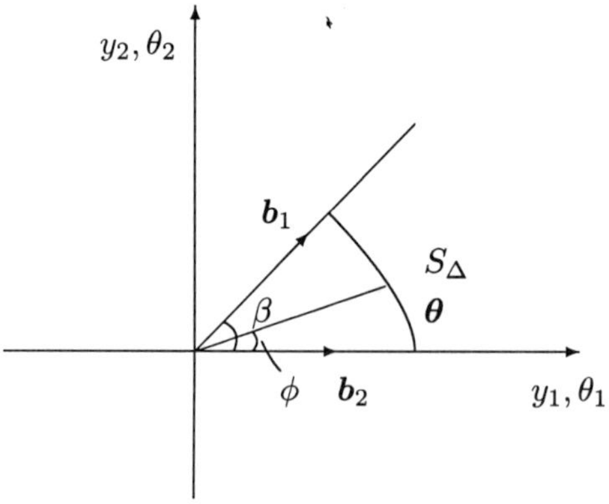

Fig. 3.4.4 The choice of parameter

where $0 \leq r$, $\Delta < \infty$, $0 \leq \psi \leq 2\pi$, and $0 \leq \phi \leq \beta$.

As Fig. 3.4.4 shown, when Δ is fixed, the power function of $\bar{\chi}^2$ test is a function of ϕ with $\phi = \arg(\boldsymbol{\theta})$. Then we have

$$P_{\text{likelihood}}(\boldsymbol{\theta}) = P_{\boldsymbol{\theta}}\{\bar{\chi}^2 \geq c_\alpha^2\} = P_1 + P_2,$$

$$P_1 = \Phi(c_\alpha - \Delta\cos\phi)\Phi(\Delta\sin\phi) + \Phi(\Delta\sin(\beta - \phi))\Phi(\Delta\cos(\beta - \phi) - c_\alpha),$$

and

$$P_2 = \frac{1}{2\pi}e^{-\Delta^2/2}\int_{c_\alpha}^{\infty} re^{-r^2/2}\int_0^\beta e^{-\Delta r\sin\psi}d\psi dr.$$

Now, we compare the power functions and study their properties.

Power function comparison. Obviously, the size of power function is related to the position (the length Δ and the angle ϕ) of parameter $\boldsymbol{\theta} \in S$, see Fig. 3.4.4. By the definition of S_Δ, we first fix the length $\Delta^2 = \boldsymbol{\theta}'\boldsymbol{\theta} = \boldsymbol{\mu}'\Lambda^{-1}\boldsymbol{\mu}$. Sometimes Δ is called a Mahalanobis distance (see Anderson, 1984, p. 206). Thus, the power function is a function of the angle $\phi = \arg(\boldsymbol{\theta})$. As Fig. 3.4.2 shown, for the χ^2 test, we get the same power function irrespective of the change of the angle. For the test $T_{\boldsymbol{c}_0}$, we can obtain the biggest power at $\phi = \beta/2$, but the smallest power at $\phi = 0$ or $\phi = \beta$. For the $\bar{\chi}^2$ test, by the symmetry of the rejection region, the property of its power function is similar to that of $T_{\boldsymbol{c}_0}$ test. So, we choose the following parameters to compare. First, take $\beta = \pi/6$, $\pi/4, \pi/3, \pi/2$, then take $\Delta = 1, 2, 3, 4$ and $\phi = 0, \beta/2$. Through computation, we can obtain Tables 3.4.1-3.4.3.

Table 3.4.1 Comparison of power functions when $\beta = \pi/4$.

	ϕ	$\Delta = 1$	$\Delta = 2$	$\Delta = 3$	$\Delta = 4$
ρ_{square}	0	0.13278	0.41555	0.77078	0.95671
	$\frac{\beta}{2}$	0.13278	0.41555	0.77078	0.95671
ρ_{linear}	0	0.23545	0.58038	0.87007	0.97985
	$\frac{\beta}{2}$	0.25950	0.63874	0.91231	0.99074
$\rho_{\text{likelihood}}$	0	0.23400	0.59094	0.88562	0.98555
	$\frac{\beta}{2}$	0.24938	0.61660	0.89721	0.98574

Table 3.4.2 Comparison of power functions when $\beta = \pi/2$.

	ϕ	$\Delta = 1$	$\Delta = 2$	$\Delta = 3$	$\Delta = 4$
ρ_{square}	0	0.13278	0.41555	0.77078	0.95671
	$\frac{\beta}{2}$	0.13278	0.41555	0.77078	0.95671
ρ_{linear}	0	0.17418	0.40878	0.68311	0.88170
	$\frac{\beta}{2}$	0.25950	0.63874	0.91231	0.99074
$\rho_{\text{likelihood}}$	0	0.19470	0.52853	0.85103	0.97850
	$\frac{\beta}{2}$	0.22716	0.57424	0.87392	0.98285

Table 3.4.3 Comparison of power functions when $\beta = 3\pi/4$.

	ϕ	$\Delta = 1$	$\Delta = 2$	$\Delta = 3$	$\Delta = 4$
ρ_{square}	0	0.13278	0.41555	0.77078	0.95671
	$\frac{\beta}{2}$	0.13278	0.41555	0.77078	0.95671
ρ_{linear}	0	0.10344	0.18956	0.30965	0.45455
	$\frac{\beta}{2}$	0.25950	0.63874	0.91231	0.99074
$\rho_{\text{likelihood}}$	0	0.16374	0.48012	0.82131	0.97149
	$\frac{\beta}{2}$	0.20089	0.52892	0.84725	0.97686

Note that $\beta = \arccos(-\rho)$, then the ρ's are $-\sqrt{3}/2$, $-\sqrt{2}/2$, $-1/2$ and 0 when $\beta = \pi/6, \pi/4, \pi/3$ and $\pi/2$, respectively. As Tables 3.4.1-3.4.3 shown, the best linear test T_{c_0} has bigger power when the angle of the parameter $\boldsymbol{\theta}$ lies in the middle part of the alternative hypothesis S, *i.e.* nearby $\beta/2$; but it has smaller power when ϕ is near 0 or β, especially for large β. On the contrary, the likelihood-ratio test $\bar{\chi}^2$ is relatively stable, but the computation of its statistic and distribution is rather difficult. Consequently, we suggest that one should use the best linear test when β is smaller; otherwise, one should utilize the Likelihood-ratio test. It can be seen from this example that we should compare several testing methods for some testing problems, and choose the test statistic with bigger power and easier computation. For more detailed discussion for the problem, see Shi (1988c).

3.5 Comparing Robustness

From the discussion in the previous chapters, we can see that most excellent tests are based on likelihood ratio, consequently depend on known distribution functions. Sometimes, the information about distribution functions may not be very reliable, thus we hope that the test statistic does not depend upon distribution functions heavily. Consider the following problem.

Recall the problem discussed in Subsection 2.4.2. Let $F(x)$ be a continuous distribution function, which is symmetrical about the point 0. For an unknown parameter θ, let $x_1, \cdots, x_n$ be a set of samples drawn from $F(x-\theta)$, consider the one-sided testing problem

$$H_0 : \theta = 0 \quad \text{vs.} \quad H_1 : \theta > 0. \tag{3.5.1}$$

Since the normal distribution is symmetrical about the mean, the t test is a level-α UMP test for the one-sided test given by (3.5.1) according to Example 3.2.2. Now we should consider whether the t test still has some optimal properties for general symmetrical distributions or not. Just like the estimation problem (see Example 1.4.8 in Chapter 1), we will call this class of problems as **robustness** study, which focuses on the limiting properties of power functions.

3.5.1 *Efficiency of Tests*

Dependency of power on sample size. Consider the normal distribution, when the variance σ^2 is known, for the testing problem (3.5.1) a level-α UMP test statistic is $Z_n = \sqrt{n}\bar{X}/\sigma$, with the rejection region $Z_n \geq u_\alpha$, where u_α satisfies $1 - \Phi(u_\alpha) = \alpha$. For a fixed parameter $\theta_1 > 0$, the power function is

$$P_{\theta_1}\{Z_n \geq u_\alpha\} = 1 - \Phi(u_\alpha - \sqrt{n}\theta_1/\sigma). \tag{3.5.2}$$

Obviously $P_{\theta_1}\{Z_n \geq u_\alpha\} \to 1$ when $n \to \infty$. Therefore, if we want to study the limiting properties of the power functions, our study will be restricted to an arbitrary small neighborhood where the parameter belongs to the null hypothesis. The power function tends to 1 with rate $\sqrt{n}$, so we can establish an alternative hypothesis depending on sample size:

$$H_0 : \theta = 0 \quad \text{vs.} \quad H_{1n} : \theta_n = \theta/\sqrt{n}, \tag{3.5.3}$$

where $\theta > 0$ is given. So, we have $\theta_n \to 0$ when $n \to \infty$. By Eq. (3.5.2), the power function can be written as

$$P_{\theta_n}\{Z_n \geq u_\alpha\} = 1 - \Phi(u_\alpha - c\theta). \tag{3.5.4}$$

This is a constant. According to the Central Limit Theorem, we can consider that almost all power functions of test statistics for the testing problem (3.5.1) can be the written as the form (3.5.4). Since c depends on both the mean and the variance of the test statistic Z_n when the parameter is θ_n, we can obtain different c for different test statistics. If c is independent of the parameter θ, we can use c to explore the properties of the power function.

Pitman efficiency. This is a kind of efficiency criterion focusing on the mean (Noether, 1955). Let Z_n be the test statistic for the testing problem (3.5.1), with rejection region $Z_n \geq k_n$, and satisfy the following regularity conditions.

(1) $\lim_{n\to\infty} P_0\{Z_n \geq k_n\} = \alpha$;
(2) There exist two sequences $\{\mu_n(\theta)\}$ and $\{\sigma_n(\theta)\}$, such that

$$(Z_n - \mu_n(0))/\sigma_n(0) \xrightarrow{L} N(0,1)$$

uniformly in a small neighborhood of $\theta = 0$;
(3) $\mu_n(\theta)$ is derivable in the above neighborhood, and let $\mu_n'(0)$ denote the derivative at $\theta = 0$;
(4) For the θ_n given in Eq. (3.5.3), $\sigma_n(\theta_n)/\sigma_n(0) \to 1$, and $\mu_n'(\theta_n)/\mu_n'(0) \to 1$;
(5) There exists a $c \geq 0$ such that $\lim_{n\to\infty} \mu_n'(0)/(\sqrt{n}\sigma_n(0)) = c$.

Here, c is called the test **efficiency** or **Pitman efficiency**. The above Condition (2) requires the uniform convergence. To verify this condition, we need some stronger form of the Central Limit Theorem, such as Berry-Esseen Theorem (Chung, 1974; Serfling, 1980). The following theorem demonstrates the relationship between the efficiency and the power function (3.5.4).

Theorem 3.5.1. *Let Z_n be a test statistic for the testing problem (3.5.1) and satisfy the regularity conditions, then for the alternative hypothesis in (3.5.3) we have the following asymptotic power*

$$\lim_{n\to\infty} P_{\theta_n}\{Z_n \geq k_n\} = 1 - \Phi(u_\alpha - c\theta), \tag{3.5.5}$$

where c is the test efficiency, and u_α satisfies

$$\lim_{n\to\infty} P_0\{Z_n \geq k_n\} = 1 - \Phi(u_\alpha) = \alpha. \tag{3.5.6}$$

Proof. For $\forall\varepsilon > 0$ and $\theta > 0$, by the regularity conditions there exists an N large enough such that, when $n > N$,

$$\left| P_{\theta_n}\{Z_n \geq k_n\} - 1 + \Phi\left(\frac{k_n - \mu_n(\theta_n)}{\sigma_n(\theta_n)}\right) \right| < \varepsilon.$$

Expanding $\mu_n(\theta_n)$ at the point 0 yields

$$\mu_n(\theta_n) = \mu_n(0) + \theta\mu_n'(\theta_n^*),$$

where $\theta_n^* = \theta^*/\sqrt{n}$, $0 < \theta^* < \theta$. Thus the variable in Φ can be written as

$$\frac{k_n - \mu_n(0)}{\sigma_n(0)}\left(\frac{\sigma_n(0)}{\sigma_n(\theta_n)}\right) - \frac{\theta}{\sqrt{n}}\frac{\mu_n'(\theta_n^*)}{\sigma_n(0)}\left(\frac{\sigma_n(0)}{\sigma_n(\theta_n)}\right).$$

By the regularity conditions, we can obtain Eqs. (3.5.5) and (3.5.6). □

Example 3.5.1. (Efficiency of the Student's t-test) The test statistic is $t = \sqrt{n}\bar{x}/s$, where $s^2 = \sum_{i=1}^n (x_i - \bar{x})^2$ is the deviation square sum of the sample. Now consider the continuous symmetrical distribution. Since $s^2/n \to \sigma_f^2$, where

$$\sigma_f^2 = \int x^2 f(x) d\mu(x),$$

we have $\mu_n(\theta) = \sqrt{n}\theta/\sigma_f$, and $\mu_n'(0) = \sqrt{n}\sigma_f$. Furthermore, $\sigma_n(0) = 1$, by the regularity conditions, the efficiency of the t test is $c_t = 1/\sigma_f$.

Example 3.5.2. (Efficiency of the sign test) Recall the sign test $G = \sum_{i=1}^n g(x_i)$ given by Eq. (2.4.3), where $g(x) = 1$ for $x > 0$; otherwise, $g(x) = 0$. From the discussion in Section 2.4, $\mu_n(\theta) = E_\theta G = n(1 - F(-\theta))$ for $\theta > 0$, thus we have $\mu_n'(0) = nf(0)$. Furthermore, since $\sigma_n^2 = n/4$, we can obtain the efficiency of G by the regularity conditions as $c_G = 2f(0)$.

Example 3.5.3. (Efficiency of the Wilcoxon signed rank test) The test statistic is $T = \sum_{i=1}^n r_i g(x_i)$, where r_i denotes the corresponding rank of $|x_i|$. At first, we prove that

$$T = \#\left(\frac{x_i + x_j}{2} > 0, i \leq j\right), \tag{3.5.7}$$

i.e. the T test is exactly the number that the sum of every two samples is positive. Suppose that there are positive samples, say $x_{i_1}, \cdots, x_{i_p}$. First we establish a closed interval with the center at the origin and length x_{i_1}, then the rank of x_{i_1} equals to the number of the samples in the interval including x_{i_1}. Obviously, all sums of every sample in the interval with x_{i_1} are positive, thus the rank of x_{i_1} equals to the number of these summations.

Similarly, we can calculate the rank of $x_{i_2}, \cdots, x_{x_p}$, and obtain Eq. (3.5.7). For $\theta > 0$, we can also get

$$T(\theta) = \#\left(\frac{x_i + x_j}{2} > \theta, i \le j\right).$$

Let $T_{ij} = g(x_i + x_j)$, where g is defined as given in Example 3.5.2, then $T = \sum\sum_{i \le j} T_{ij}$. Thus, for $\theta > 0$

$$E_\theta T = \sum_{i \le j}\sum E_\theta T_{ij} = nE_\theta T_{11} + \frac{1}{2}n(n-1)ET_{12}, \qquad (3.5.8)$$

where

$$E_\theta T_{11} = P_\theta\{X_1 > 0\} = 1 - F(-\theta),$$

and

$$\begin{aligned} E_\theta T_{12} = P_\theta\{X_1 + X_2 > 0\} &= \int [1 - F(-x - \theta)] f(x - \theta) d\mu(x) \\ &= \int [1 - F(-y - 2\theta)] f(y) d\mu(y). \end{aligned}$$

The regularity conditions imply that we can differentiate under integral sign. According to the symmetry and Eq. (3.5.8), we can get

$$\mu_n'(0) = nf(0) + n(n-1)\int f^2(x)d\mu(x).$$

By Eq. (2.4.14) in Chapter 2, $\sigma_0^2 = n(n+1)(2n+1)/24$. Thus the efficiency of the Wilcoxon signed rank test is

$$c_T = \sqrt{12}\int f^2(x)d\mu(x).$$

According to Eq. (3.5.5), we can obtain the asymptotic power functions of the above three test statistics under the alternative hypothesis given by Eq. (3.5.3).

3.5.2 *Asymptotic Relative Efficiency*

To compare the asymptotic power functions conveniently, we give the following definition. Let $Z_n^{(1)}$ and $Z_n^{(2)}$ be the two test statistics for the testing problem (3.5.1), and satisfy the regularity conditions. Then,

$$e(Z_n^{(1)}, Z_n^{(2)}) = c_1^2/c_2^2 \qquad (3.5.9)$$

will be called as the **asymptotic relative efficiency** of $Z_n^{(1)}$ with respect to $Z_n^{(2)}$, where c_i is the efficiency of $Z_n^{(i)}$. Obviously $e(Z_n^{(1)}, Z_n^{(2)}) > 1$ implies the power of $Z_n^{(1)}$ is superior to that of $Z_n^{(2)}$. Sometimes, Eq. (3.5.9) is called as the **Pitman efficiency**. According to the above three examples, we can get

$$e(G,t) = 4\sigma_f^2 f^2(0),$$

$$e(G,t) = f^2(0) \Big/ \left[\sqrt{3} \int f^2(x) d\mu(x)\right]^2,$$

and

$$e(T,t) = 12\sigma_f^2 \left[\int f^2(x) d\mu(x)\right]^2. \qquad (3.5.10)$$

Thus, we can obtain the Pitman efficiency once the density function is given.

Example 3.5.4. (Efficiency comparison between the Wilcoxn signed rank test and the Student's *t*-test) This example was first studied by Tukey (1960), and was discussed in detail by Hettmansperger (1984). Now we utilize the latter's results. Continue to consider the example discussed in Example 1.4.8 of Chapter 1, which is a mixture normal model. Suppose that the distribution function is

$$F_\alpha(x) = (1-\alpha)\Phi(x) + \alpha\Phi\left(\frac{x}{3}\right),$$

where $\Phi(\cdot)$ denotes the standard normal distribution function, and $\alpha \in [0,1]$. It is easy to get the density function as follows

$$f_\alpha(x) = (1-\alpha)\phi(x) + \frac{\alpha}{3}\phi\left(\frac{x}{3}\right),$$

where $\phi(\cdot)$ is the density function of $\Phi(\cdot)$. We can get that

$$\sigma_f^2 = \int x^2 f_\alpha(x) d\mu(x) = 1 + 8\alpha,$$

$$\int f_\alpha^2(x) d\mu(x) = \frac{(1-\alpha)^2}{2\sqrt{\pi}} + \frac{\alpha^2}{6\sqrt{\pi}} + \frac{\alpha(1-\alpha)}{\sqrt{5\pi}}.$$

According to Eq. (3.5.10), we can get

$$e(T,t) = \frac{3(1+8\alpha)}{\pi}\left[(1-\alpha)^2 + \frac{\alpha^2}{3} + \frac{2\alpha(1-\alpha)}{\sqrt{5}}\right]^2.$$

Table 3.5.1 provides some calculating results.

When $\alpha = 0$, it is the standard normal distribution, and the t test is level-α UMP, but the difference of efficiency between the T test and the

Table 3.5.1 Relative efficiency of the T-test to the t-test.

α	0	.01	.03	.05	.10	.15	.50
$e(T,t)$	0.955	1.009	1.108	1.196	1.373	1.497	2.661

t test is not obvious. When the distribution is polluted, such as more than 3% contaminated, the T test is more effective than the t test. When the contaminated percentage attains 50%, the efficiency of the T test is as much 2.6 times as that of the Student's t-test. Thus, as far as the fluctuations of distribution are concerned, the T test is more robust than the Student's t-test, that is to say, the t test depends on the assumption of normal distribution heavily. Thus, in practical application, the Student's t-test will show good performance only when the information about the norma distribution is very certain.

Asymptotic relative efficiency and sample size. Corresponding to the robustness of the estimation, we can explain the Pitman efficiency as follows: it is the ratio of required sample sizes for the test statistics to attain the same power.

For the two test statistics $Z_n^{(i)}, i = 1, 2$ satisfying the regularity conditions, let θc in Eq. (3.5.4) be $\theta_1 c_1$ and $\theta_2 c_2$ respectively. On the one hand, in order to achieve the same power, we have

$$\theta_1 c_1 = \theta_2 c_2, \tag{3.5.11}$$

namely $\theta_2/\theta_1 = c_1/c_2$. On the other hand, they should attain to the same power under the same parameter, thus, according to the alternative hypothesis in (3.5.3), we have $\theta_1/\sqrt{n_{1j}} = \theta_2/\sqrt{n_{2j}}$, namely $\theta_2/\theta_1 = \sqrt{n_{2j}}/\sqrt{n_{1j}}$. Consequently, we have

$$\lim_{j\to\infty} \frac{n_{2j}}{n_{1j}} = \frac{c_1^2}{c_2^2}. \tag{3.5.12}$$

Using Eq. (3.5.9), we can also obtain the Pitman efficiency, $e(Z_n^{(1)}, Z_n^{(2)}) = n_{2j}/n_{1j}$. Obviously, $e(Z_n^{(1)}, Z_n^{(2)}) > 1$ is equivalent to $n_{2j} > n_{1j}$. That is to say, in order to achieve the same asymptotic power, $Z_n^{(2)}$ needs more samples than $Z_n^{(1)}$.

Next, we will verify (3.5.12) through comparing the Wilcoxon signed rank T test with the sign G test. For $\theta > 0$, the power of the level-α T test should be

$$P_\theta\{T \geq u_\alpha\} \approx 1 - \Phi\left(\frac{u_\alpha - E_\theta T}{\sqrt{V_\theta T}}\right).$$

By the proof of Theorem 3.5.1 in conjunction with the calculation of Example 3.5.3, we can obtain the following approximate power function

$$1 - \Phi\left(u_\alpha - \theta\sqrt{n}\sqrt{12}\int f^2(x)d\nu(x)\right).$$

For a given β satisfying $\alpha < \beta < 1$, let u_β satisfy $1 - \Phi(u_\beta) = \beta$. Thus

$$u_\beta \approx u_\alpha - \theta\sqrt{n}\sqrt{12}\int f^2(x)d\nu(x).$$

Furthermore we can get the solution n as follows

$$n_T \approx \frac{(u_\alpha - u_\beta)^2}{\theta^2 12(\int f^2(x)d\nu(x))^2}.$$

For the G test, we can apply a similar method in conjunction with the Example 3.5.2 to get

$$n_G \approx \frac{(u_\alpha - u_\beta)^2}{\theta^2 4 f^2(0)}.$$

Thus we have

$$e(G,T) = \frac{n_T}{n_G} \approx \frac{f^2(0)}{3(\int f^2(x)d\mu(x))^2}.$$

This coincides with Eq. (3.5.10), consequently proves Eq. (3.5.12). It can be seen that it is simpler to use the regularity conditions to compute the Pitman efficiency directly; However, it is more intuitional to compare based on sample sizes.

According to the discussion of this section, for the testing problem, it is difficult to tell which test method is absolutely the best one. As a result, we should make a concrete analysis of concrete problems according to the practical situation.

3.6 Interval Estimation Based on Test

In Section 1.4, we have discussed the problem of parameter estimation. We know that an estimate is a function of sample. We can obtain a concrete estimator of parameter once we obtained sample values, and we call this method as **point estimation**. In fact, since the sample is a random variable, different samples will result in different point estimation. As we can imagine intuitively, to give an interval estimate of parameter is more reasonable. That is to say, for given sample values, we need to construct an interval within which the true parameter will lie with high probability.

Let $(\mathcal{X}, \mathcal{A}, \mathcal{P})$ be a statistical space, where $\mathcal{P} = \{P_\theta; \theta \in \Theta\}$ and the parameter space $\Theta \subseteq R$. Let $\hat{\theta}_L(x)$ and $\hat{\theta}_U(x)$ be two measurable functions of the sample, and satisfy $\hat{\theta}_L(x) \le \hat{\theta}_U(x)$ for all $x \in \mathcal{X}$. For the given sample x, if the interval $[\hat{\theta}_L(x), \hat{\theta}_U(x)] \subseteq \Theta$, then $[\hat{\theta}_L(x), \hat{\theta}_U(x)]$ is called an **interval estimation** of the parameter θ. We will estimate the probability within which the true parameter lies in based on testing method. We will start our study from an example.

Example 3.6.1. (Interval estimation of normal mean) Let $x_1, \cdots, x_n$ be set of samples from a normal distribution $N(\mu, \sigma^2)$, according to the discussion in Section 1.4, we know that $\bar{x} = \sum_{i=n}^n x_i/n$ and $s^2 = \sum_{i=n}^n (x_i - \bar{x})^2/(n-1)$ are the minimum variance unbiased estimates of μ and σ^2 respectively. This is a kind of point estimation. For given μ_0, and for testing problem $H_0 : \mu = \mu_0$ vs. $H_1 : \mu \ne \mu_0$, from Example 3.3.2, we know that $t = \sqrt{n}(\bar{x} - \mu_0)/s$ is a level-α UMPU test with rejection region

$$W_{\mu_0} = \{x; |\bar{x} - \mu_0| \ge \frac{s}{\sqrt{n}} t_{\alpha/2}\},$$

where $t_{\alpha/2}$ denotes $\alpha/2$-critical value of the Student's t-distribution with $n-1$ degrees of freedom. Since what we need to consider is the interval within which the parameter lies in, let

$$\begin{aligned} W_{\mu_0}^c &= R - W_{\mu_0} \\ &= \left\{x;\ \mu_0 - \frac{s}{\sqrt{n}} t_{\alpha/2} \le \bar{x} \le \mu_0 + \frac{s}{\sqrt{n}} t_{\alpha/2}\right\} \qquad (3.6.1) \\ &= \left\{x;\ \bar{x} - \frac{s}{\sqrt{n}} t_{\alpha/2} \le \mu_0 \le \bar{x} + \frac{s}{\sqrt{n}} t_{\alpha/2}\right\} \end{aligned}$$

it is easy to obtain that $P_{\mu_0}\{W_{\mu_0}^c\} = 1 - \alpha$. By the arbitrariness of μ_0, we also have $P_\mu\{W_\mu^c\} = 1 - \alpha$. Let $\hat{\theta}_L(x) = \bar{x} - s t_{\alpha/2}/\sqrt{n}$, and $\hat{\theta}_U(x) = \bar{x} + s t_{\alpha/2}/\sqrt{n}$, then we have

$$P_\mu\{\hat{\theta}_L(X) \le \mu \le \hat{\theta}_U(X)\} = 1 - \alpha. \qquad (3.6.2)$$

Thus, we have constructed an interval and known the probability of which the true parameter falls in the interval.

Next, consider the one-sided testing problem $H_0 : \mu \le \mu_0$ vs. $H_1 : \mu > \mu_0$. From Example 3.2.2, the t test is a level-α UMP test. The acceptance region corresponding to Eq. (3.6.1) is

$$\begin{aligned} W_{\mu_0}^c &= \left\{ \bar{x} \le \mu_0 + \frac{s}{\sqrt{n}} t_\alpha \right\} \\ &= \left\{ \mu_0 \ge \bar{x} - \frac{s}{\sqrt{n}} t_\alpha \right\}. \end{aligned} \tag{3.6.3}$$

Let $\hat{\theta}_L(x) = \bar{x} - st_\alpha/\sqrt{n}$, then we can obtain a one-sided interval estimation $[\hat{\theta}_L(x), \infty)$, and for any μ we have

$$P_\mu(\mu \ge \hat{\theta}_L(X)) = 1 - \alpha. \tag{3.6.4}$$

Similarly, corresponding to the one-sided testing problem $H_0 : \mu \ge \mu_0$ vs. $H_1 : \mu < \mu_0$, we can obtain a one-sided interval estimation $(-\infty, \hat{\theta}_U(x))$ with $\hat{\theta}_U(x) = \bar{x} + st_\alpha/\sqrt{n}$, and for any μ we have

$$P_\mu\{\mu \le \hat{\theta}_U(X)\} = 1 - \alpha. \tag{3.6.5}$$

From the above example, we can see that interval estimation has close relation to hypothesis test. And this relation motivated Jerzy Neyman to propose the theory of interval estimation based on a long-time thinking after he had proposed the fundamental lemma. In fact, the research of interval estimation promotes further understanding about hypothesis test.

3.6.1 *Interval Estimation and Acceptance Region*

In testing problems, we pay our attention to the rejection region. However, when we consider the interval estimation, as Example 3.6.1 shown, we will pay our attention to the acceptance region. We only consider the situation in which the acceptance region takes the form of interval.

For $\theta_0 \in \Theta$, consider the problem of testing $H_0 : \theta = \theta_0$. Let W be the rejection region corresponding to test $\phi(x)$ with test statistic $t(x)$. Let $A = W^c$ be the acceptance region of $\phi(x)$. In general, the size of acceptance region depends on θ_0, and $\theta_0 \in A$, thus we can denote acceptance region as $A(\theta_0)$, and regard it as an interval with center θ_0. We have

$$P_{\theta_0}\{T(X) \in W\} = \alpha \quad \Longleftrightarrow \quad P_{\theta_0}\{T(X) \in A(\theta_0)\} = 1 - \alpha. \tag{3.6.6}$$

On the other hand, we can transform the acceptance region into an interval with the statistic $t(x)$ as center, as Eqs. (3.6.1) and (3.6.3) in Example 3.6.1, the interval will be called an **interval estimator** of parameter, and denoted as $B(t(x))$. Then we have

$$t(x) \in A(\theta_0) \quad \Longleftrightarrow \quad \theta_0 \in B(t(x)). \tag{3.6.7}$$

Thus, we can understand a testing problem as follows: accepting the null hypothesis when the sample values fall within the acceptance region is equivalent to accepting the null hypothesis when the parameter falls within interval estimation.

As the same with the testing problem, in principle we can construct many interval estimators. How to evaluate an interval estimator? Intuitively, a good interval estimator should satisfy the following two conditions:

(1) **Reliability.** The probability which the true parameter falls in the interval should be large.
(2) **Accuracy.** The width of interval should be short. Similar to two types of errors in testing problems, the above two conditions can not be satisfied simultaneously. In general, the greater the degree of confidence the wider the confidence interval must be. From Eq. (3.6.6), based on the principle of hypothesis test, the principle of interval estimation is: **Under the guarantee of reliability, the higher the accuracy, namely the smaller the width of interval, the better the interval estimation.**

For a given $\alpha \in (0,1)$, say $\alpha = 0.05$, an interval estimator of parameter $[\hat{\theta}_L(x), \hat{\theta}_U(x)]$ is called a **confidence interval** with **confidence level** $1-\alpha$, if

$$\inf_{\theta\in\Theta} P_\theta\{\theta \in [\hat{\theta}_L(x), \hat{\theta}_U(x)]\} \geq 1-\alpha. \tag{3.6.8}$$

In addition, $\hat{\theta}_L(x)$ and $\hat{\theta}_U(x)$ are called a **lower confidence bound** and an **upper confidence bound** respectively.

Based on the above discussion, we can obtain the following theorem.

Theorem 3.6.1. *Let $T(X)$ be the test statistic for $H_0 : \theta = \theta_0$, and let $A(\theta_0)$ denote a level-α acceptance region. For a sample value $t(x)$, let*

$$B(t(x)) = \{\theta; t(x) \in A(\theta)\}.$$

Then $B(T(X))$ is a $1-\alpha$ confidence interval. On the contrary, let $B(T(X))$ be a $1-\alpha$ confidence interval. For $\theta_0 \in \Theta$, let

$$A(\theta_0) = \{t(x); \theta_0 \in B(t(x))\}. \tag{3.6.9}$$

Then $A(\theta_0)$ is a level-α acceptance region for the problem of testing $H_0 : \theta = \theta_0$.

Proof. If $A(\theta_0)$ is the acceptance region of a level-α test, then from Eq. (3.6.6),

$$P_{\theta_0}\{T(X) \notin A(\theta_0)\} \le \alpha \quad \Longleftrightarrow \quad P_{\theta_0}\{T(X) \in A(\theta_0)\} \ge 1 - \alpha.$$

Furthermore from Eq. (3.6.7), $P_{\theta_0}\{\theta_0 \in B(T(X))\} \ge 1 - \alpha$. Since $\theta_0 \in \Theta$ is arbitrary, then we have

$$P_\theta\{\theta \in B(T(X))\} \ge 1 - \alpha, \quad \forall \theta \in \Theta.$$

Conversely, if $B(T(X))$ is a $1 - \alpha$ confidence interval, then for any $\theta \in \Theta$, we have $P_\theta\{\theta \notin B(T(X))\} \le \alpha$. Then from Eqs. (3.6.7) and (3.6.9), we have

$$P_{\theta_0}\{T(X) \notin A(\theta_0)\} = P_{\theta_0}\{\theta_0 \notin B(T(X))\} \le \alpha.$$

That is, $A(\theta_0)$ is the acceptance region of a level-α test. □

In the above discussion, we only considered the situation of simple null hypothesis. However, for a complex null hypothesis such as a one-sided test in Example 3.6.1, the above result still holds by taking θ_0 as the parameter corresponding to the least favorable distribution. Now we need to discuss whether a good interval estimator can correspond to a good test statistic?

3.6.2 *Uniformly Most Accurate Confidence Interval*

Since what we are discussing is about interval estimation, without loss of generality, let the parameter space also be an interval, namely $\Theta = [a, b]$, where a and b can be finite or infinite, and the closed interval can be open on one or both sides. If $\alpha > 0$, then the $1-\alpha$ confidence interval $[\hat{\theta}_L(X), \hat{\theta}_U(X)]$ may have three cases: $a < \hat{\theta}_L(X)$ and $\hat{\theta}_U(X) = b$; $a = \hat{\theta}_L(X)$ and $\hat{\theta}_U(X) < b$; $a < \hat{\theta}_L(X)$ and $\hat{\theta}_U(X) < b$. We call the first two cases as **one-sided confidence intervals**, the third as **two-sided confidence interval**.

At first, we discuss the cases of one-sided confidence interval, mainly discuss the first case. Let $\hat{\theta}_L(X)$ be a $1 - \alpha$ lower confidence bound and $a < \hat{\theta}_L(X)$. If θ is the true value, naturally we hope that $\theta' < \hat{\theta}_L(X)$ if $\theta' < \theta$, or as a concession we have a larger $P_\theta\{\theta' < \hat{\theta}_L(X)\}$, namely a smaller $P_\theta\{\theta' \ge \hat{\theta}_L(X)\}$. In general, for any $1 - \alpha$ lower confidence bound $\hat{\theta}_L^*(X)$, if $\theta' < \theta$ we have

$$P_\theta\{\theta' \ge \hat{\theta}_L(X)\} \le P_\theta\{\theta' \ge \hat{\theta}_L^*(X)\},$$

then $\hat{\theta}_L(X)$ is called a $1-\alpha$ **uniformly most accurate lower confidence bound**, denoted by $(1 - \alpha)$-**UMA lower confidence bound.**

Now, from the above discussion, we naturally associate the one-sided test, and have the following theorem.

Theorem 3.6.2. *Let $A(\theta_0)$ be the acceptance region of a level-α UMP test for the one-sided testing problem $H_0: \theta \leq \theta_0$ vs. $H_1: \theta > \theta_0$. For $t(x) \in A(\theta_0)$, if $B(T(X))$ can be equivalently written as*

$$B(T(X)) = \{x;\ \theta_0 \geq \hat{\theta}_L(X)\}, \tag{3.6.10}$$

as Eq. (3.6.4) shown, then $\hat{\theta}_L(X)$ is a $(1-\alpha)$-UMA lower confidence bound.

Proof. From Theorem 3.6.1, we know that $B(T(X))$ is a $1-\alpha$ confidence interval. Let $\hat{\theta}^*_L(X)$ be any $1-\alpha$ lower confidence bound. Define a test, with acceptance region

$$A^*(\theta_0) = \{x;\ \theta_0 \geq \hat{\theta}^*_L(x)\}.$$

From Theorem 3.6.1, the test is of level α. By the definition of UMP test, for $\theta > \theta_0$ we have

$$P_\theta\{T(x) \in A(\theta_0)\} \leq P_\theta\{x \in A^*(\theta_0)\},$$

which is equivalent to $P_\theta\{\theta_0 \geq \hat{\theta}_L(x)\} \leq P_\theta\{\theta_0 \geq \hat{\theta}^*_L(x)\}$, that is $\hat{\theta}(x)$ is a $(1-\alpha)$-UMA lower confidence bound. □

We can apply the same method to discuss the second case, and define a $1-\alpha$ **uniformly most accurate upper confidence bound**, denoted as $(1-\alpha)$-UMA upper confidence bound. In the same way, we can also give a similar result as Theorem 3.6.2.

Theorem 3.6.3. *Let $A(\theta_0)$ be the acceptance region of level-α UMP test for the one-sided testing problem $H_0: \theta \leq \theta_0$ vs. $\theta > \theta_0$. For $t(x) \in A(\theta_0)$, if $B(T(x))$ can be equivalently written as*

$$B(T(x)) = \{x;\ \theta_0 \leq \theta_0(x)\}, \tag{3.6.11}$$

as Eq. (3.6.5) shown, then $\theta_0(x)$ is a $(1-\alpha)$-UMA upper confidence bound.

Example 3.6.2. (One-sided interval estimation for normal variance) For the given normal distribution $N(\mu, \sigma^2)$, we want to construct a $(1-\alpha)$-UMA lower confidence bound for σ^2. For this purpose, we take a sample $x_1, \cdots, x_n$. For a given $\sigma_0^2 > 0$, consider the one-sided testing problem

$$H_0: \sigma^2 \leq \sigma_0^2 \quad \text{vs.} \quad H_1: \sigma^2 > \sigma_0^2. \tag{3.6.12}$$

When the mean μ is known, from Example 3.2.3, we know that the acceptance region of a level-α UMP test is

$$A(\sigma_0^2) = \{x;\ T(x) \le \sigma_0^2 \chi_\alpha^2(n)\}, \tag{3.6.13}$$

where $T(x) = \sum_{i=1}^n (x_i - \mu)^2$ and $\chi_\alpha^2(n)$ denotes α-critical value of the χ^2 distribution with n degrees of freedom. Since Eq. (3.6.13) can be rewritten as $B(T(x)) = \{x; \sigma_0^2 \ge T(x)/\chi_\alpha^2(n)\}$, by Theorem 3.6.2, we can obtain a $(1-\alpha)$-UMA lower confidence bound for σ^2 as follows

$$\hat{\theta}_L(x) = T(x)/\chi_\alpha^2(n).$$

Similarly, we can obtain a $(1-\alpha)$-UMA upper confidence bound for σ^2 as $\hat{\theta}_U(x) = T(x)/\chi_{1-\alpha}^2(n)$.

When the mean μ is unknown, from Example 3.2.3, we know that the $(1-\alpha)$-UMA lower confidence bound for σ^2 is $\hat{\theta}_L(x) = s(x)/\chi_\alpha^2(n-1)$, where $s(x) = \sum_{i=1}^n (x_i - \bar{x})^2$.

In the discussion about testing problems, generally, a level-α UMP test dose not exist for two-sided tests. Likewise, generally speaking, a $(1-\alpha)$-UMA confidence interval does not exist for a two-sided confidence interval, either. This needs a further analysis.

3.6.3 *Uniformly Most Accurate Unbiased Confidence Interval*

Now, we further discuss the definition of the lower confidence bound. If $\hat{\theta}_L(x)$ is a $1-\alpha$ lower confidence bound, from Eq. (3.6.8), we have

$$P_\theta\{\theta \ge \hat{\theta}_L(x)\} \ge 1-\alpha \text{ for } \forall \theta \in \Theta. \tag{3.6.14}$$

Further, as we have discussed that, if the true parameter value is θ, then when $\theta' < \theta$, we hope that $P_\theta\{\theta' \ge \hat{\theta}_L(x)\}$ is smaller. Thus, we introduce the definition below: $\hat{\theta}_L(x)$ is called a $1-\alpha$ **unbiased lower confidence bound**, if $\hat{\theta}_L(x)$ satisfies Eq. (3.6.14) and

$$P_\theta\{\theta' \ge \hat{\theta}_L(x)\} \le 1-\alpha \text{ for } \forall \theta' < \theta. \tag{3.6.15}$$

Further, $\hat{\theta}_L(x)$ is called a $1-\alpha$ uniformly most accurate unbiased lower bound, denoted as $(1-\alpha)$-UMAU lower confidence bound, if for any $1-\alpha$ unbiased confidence bound $\hat{\theta}_L^*(x)$, we have

$$P_\theta\{\theta' \ge \hat{\theta}_L(x)\} \le P_\theta\{\theta' \ge \hat{\theta}_L^*(x)\} \text{ for } \forall \theta' < \Theta.$$

Similarly, we can define a $1-\alpha$ uniformly most accurate unbiased upper bound.

The interval estimator $[\hat{\theta}_L(x), \hat{\theta}_U(x)]$ is called a $1-\alpha$ unbiased confidence interval, if it satisfies Eq. (3.6.8) and

$$P_\theta\{\theta' \in [\hat{\theta}_L(x), \hat{\theta}_U(x)]\} \le 1-\alpha \text{ for } \forall \theta' \neq \theta.$$

Further, it is called a $1-\alpha$ uniformly most accurate unbiased confidence interval, denoted as $(1-\alpha)$-UMAU confidence interval, if for any $1-\alpha$ unbiased confidence interval $[\hat{\theta}_L^*(x), \hat{\theta}_U^*(x)]$ and $\theta' \neq \theta$ we have

$$P_\theta\{\theta' \in [\hat{\theta}_L(x), \hat{\theta}_U(x)]\} \le P_\theta\{\theta' \in [\hat{\theta}_L^*(x), \hat{\theta}_U^*(x)]\}.$$

Similar to Theorem 3.6.2, we can obtain the following theorem.

Theorem 3.6.4. *Let $A(\theta_0)$ be the acceptance region of a level-α UMPU test for the two-sided testing problem $H_0 : \theta = \theta_0$ vs. $H_1 : \theta \neq \theta_0$. For $t(x) \in A(\theta_0)$, if $B(T(x))$ can be equivalently written as*

$$B(T(x)) = \{x; \hat{\theta}_L(x) \le \theta_0 \le \hat{\theta}_U(x)\}, \tag{3.6.16}$$

as Eq. (3.6.2) shown, then $[\hat{\theta}_L(x), \hat{\theta}_U(x)]$ is a $(1-\alpha)$-UMAU confidence interval.

Proof. From Theorem 3.6.1, we know that $[\hat{\theta}_L(x), \hat{\theta}_U(x)]$ is a $1-\alpha$ confidence interval. Furthermore, for $\theta \neq \theta_0$, we have

$$P_{\theta_0}\{\theta \in [\hat{\theta}_L(x), \hat{\theta}_U(x)]\} = P_{\theta_0}\{T(x) \in A(\theta_0)\} \le 1-\alpha.$$

Hence it is a $1-\alpha$ unbiased confidence interval. For any $1-\alpha$ confidence interval $[\hat{\theta}_L^*(x), \hat{\theta}_U^*(x)]$, define a test function with acceptance region

$$A^*(\theta_0) = \{x;\ \hat{\theta}_L^*(x) \le \theta_0 \le \hat{\theta}_U^*(x)\}.$$

From Theorem 3.6.1, this is of level α. For $\theta \neq \theta_0$, from the definition of unbiased confidence interval, we have

$$P_\theta\{X \in A^*(\theta_0)\} = P_\theta\{\hat{\theta}_L(X) \le \theta \le \hat{\theta}_U(X)\} \le 1-\alpha.$$

Thus the test is a level-α unbiased test. According the assumption of α-UMPU test, for $\forall \theta \neq \theta_0$, we have

$$\begin{aligned} &P_\theta\{T(X) \in A(\theta_0)\} \le P_\theta\{X \in A^*(\theta_0)\} \\ &\qquad \Rightarrow P_\theta\{\hat{\theta}_L(X) \le \theta_0 \le \hat{\theta}_U(X)\} \le P_\theta\{\hat{\theta}_L^*(X) \le \theta_0 \le \hat{\theta}_U^*(X)\}. \end{aligned}$$

Thus $[\hat{\theta}_L(x), \hat{\theta}_U(x)]$ is a $(1-\alpha)$-UMAU confidence interval. □

In the above three theorems, the requirement for $B(T(x))$ in Eqs. (3.6.10), (3.6.11) and (3.6.16) is generally applicable. If $\hat{\theta}_L(x)$ and $\hat{\theta}_U(x)$ are monotone increasing functions of x, then their inverse functions exist, and we have

$$\theta \in \{\theta; \hat{\theta}_L(x) \leq \theta \leq \hat{\theta}_U(x)\} \quad \Longleftrightarrow \quad x \in \{x; \hat{\theta}_U^{-1}(\theta) \leq x \leq \hat{\theta}_L^{-1}(\theta)\}.$$

Thus, the requirement is applicable for the distributions with monotone likelihood ratio. Specially, it is applicable for one-parameter exponential family.

Example 3.6.3. (Interval estimator of binomial parameter) Based on the above analysis, we have related confidence estimate with hypothesis test very well. However, there is a problem that needs special attention, namely the situation when the measure $\nu\{x; T(x) = c\} \neq 0$, which mainly involves the discrete distribution. Now, we will discuss this problem.

Let x be a sample from the binomial distribution $Bi(n, \theta)$, we want to construct a $(1-\alpha)$-UMAU confidence interval for θ. Choose arbitrarily a $\theta_0 \in (0, 1)$, and consider the two-sided testing problem

$$H_0 : \theta = \theta_0 \quad \text{vs.} \quad H_1 : \theta \neq \theta_0.$$

Example 3.3.1 has given a level-α UMPU test. In Example 3.3.1, we can see that the computation of rejection region is rather troublesome. Now we will discuss interval estimation of parameter.

Suppose the sample size n is large. Let $\hat{\theta} = x/n$. From Example 1.4.7 in Chapter 1, under the hypothesis H_0, as $n \to \infty$, we have

$$\frac{\sqrt{n}(\hat{\theta} - \theta_0)}{\sqrt{\theta_0(1-\theta_0)}} \xrightarrow{L} N(0, 1).$$

Hence, when n is large enough, from Example 3.3.1, the acceptance region of level-α UMPU test is approximately

$$A(\theta_0) \approx \{x; \sqrt{\theta_0(1-\theta_0)}c/\sqrt{n} - \theta_0 \leq \theta_0 \leq \sqrt{\theta_0(1-\theta_0)}c/\sqrt{n} + \theta_0\},$$

where $c = u_{\alpha/2}$ satisfies $1 - \Phi(u_{\alpha/2}) = \alpha/2$. From $\hat{\theta} - \theta \leq \sqrt{\theta(1-\theta)}c/\sqrt{n}$, we have

$$(1 + c^2/n)\theta^2 - (2\hat{\theta} + c^2/n)\theta + \hat{\theta}^2 \leq 0.$$

When $\hat{\theta} \in (0, 1)$ is fixed, the above quadratic function is convex, and its corresponding equation has two real roots, hence we can obtain Eq. (3.6.16) with

$$\begin{aligned} \hat{\theta}_L(x) &= (x + c^2/2 - c\sqrt{x(n-x)/n + c^2/4})/(n + c^2), \\ \hat{\theta}_U(x) &= (x + c^2/2 + c\sqrt{x(n-x)/n + c^2/4})/(n + c^2). \end{aligned} \qquad (3.6.17)$$

Table 3.6.1 Approximate requirement for sample size when the length of the interval is different.

d	0.01	0.02	0.03	0.04	0.05
n	9604	2401	1069	625	400

We still have another simple approximation method, however the corresponding error may be larger. From Example 1.4.7 in Chapter 1, we know that maximum likelihood estimate is consistent, then the acceptance region may be approximately by

$$A(\theta_0) = \left\{x;\ \sqrt{\hat{\theta}(1-\hat{\theta})}c/\sqrt{n} - \theta_0 \le \hat{\theta} \le \sqrt{\hat{\theta}(1-\hat{\theta})}c/\sqrt{n} + \theta_0\right\}.$$

Thus it is easy to obtain Eq. (3.6.16) with

$$\begin{aligned} \hat{\theta}_L(x) &= \hat{\theta} - \sqrt{\hat{\theta}(1-\hat{\theta})}c/\sqrt{n}, \\ \hat{\theta}_U(x) &= \hat{\theta} + \sqrt{\hat{\theta}(1-\hat{\theta})}c/\sqrt{n}. \end{aligned} \tag{3.6.18}$$

The above analysis is based on the assumption of large sample size, then a key problem is how many samples are required at least.

Requirements for sample size n. Sample size is related to the accuracy and reliability of confidence estimate. As we have discussed before, the accuracy of estimate is related to the length of interval, and the reliability is just confidence level (or coefficient), now suppose the length of interval is $2d$, namely, the interval is $(\hat{\theta} - d, \hat{\theta} + d)$ with $d > 0$; the confidence level is 95%.

We can obtain that the length of the interval given by Eq. (3.6.17) is $2c\sqrt{x(n-x)/n + c^2/4}/(n+c^2)$ with $c = 1.96$. Since $0 \le x \le n$, we have $x(n-x)/n \le n/4$. According to the requirement for the length of the interval, $c/\sqrt{n+c^2} = 2d$, we may require that $n \ge c^2/(2d)^2 - c^2$.

For Eq. (3.6.18), since $\hat{\theta}(1-\hat{\theta}) \le 0.25$, we can obtain $n \ge c^2/(2d)^2$. It can be seen that the two approximation methods have not much difference in the requirements for sample size. When $\alpha = 0.05$, the difference is $(1.96)^2 \approx 4$. Table 3.6.1 gives the requirements for sample size for different d when $\alpha = 0.05$.

Thus it can be seen that, the interval estimation has quite much requirement for sample size, and that when the length of interval becomes longer, the requirement for sample size decreases rapidly. When we arrange a drug trial, or design a poll, it often involves a Bernoulli trial or binomial distribution. Hence, we can refer to Table 3.6.1 when we arrange the sample size of experiments.

Accurate interval estimation. For a given $x = 0, 1, \cdots, n$, a binomial distribution function is

$$F(\theta|x) = \sum_{y=0}^{x} \binom{n}{y} \theta^y (1-\theta)^{n-y}.$$

It is easy to verify that it is a monotone decreasing function of θ. Hence, when $x > 0$, if let

$$\hat{\theta}_L(x) = \sup\{\theta; F(\theta|x-1) \geq \alpha_1\},$$

and

$$\hat{\theta}_U(x) = \inf\{\theta; F(\theta|x) \leq \alpha_2\},$$

then $[\hat{\theta}_L(x), \hat{\theta}_U(x)]$ forms a $1-\alpha$ confidence interval, where α_1 and α_2 are positive and satisfy $\alpha_1+\alpha_2 = \alpha$. Thus, we can consider the lower confidence bound and the upper confidence bound respectively. From Example 3.2.1 and Theorem 3.6.2, we know that the lower confidence bound $\hat{\theta}_L(x)$ based on a level-α_1 UMP test is the solution of the following equation

$$\sum_{y=0}^{x-1} \binom{n}{y} \theta^y (1-\theta)^{n-y} = 1 - \alpha_1.$$

Since the left-hand side of the above equation is a binomial expansion, based on the relation between the binomial expansion and the β function, $\hat{\theta}_L(x)$ equivalently is the solution of the following equation

$$\frac{\Gamma(n+1)}{\Gamma(k)\Gamma(n-k+1)} \int_0^{\theta} z^{k-1}(1-z)^{n-k} dz = \alpha.$$

The left-hand side of the above equation is a β distribution function with k and $n-k+1$ degrees of freedom. Based on the relation between the β-distribution and the F-distribution, if we let $F(x|s,t)$ denote the cdf of F-distribution with s and t degrees of freedom, we have

$$F = \frac{\beta}{1-\beta} \frac{n-k+1}{k} \sim F(2k, 2(n-k+1)),$$

then $\hat{\theta}_L(x)$ is also the solution of

$$F\left(h(\theta)|2k, 2(n-k+1)\right) = \alpha_1,$$

where $h(\theta) = \theta(n-k+1)/[k(1-\theta)]$. Solving the above equation yields

$$\hat{\theta}_L(x) = k/[k+(n-k+1)F_{\alpha_1}(2(n-k+1), 2k)],$$

Table 3.6.2 Approximate requirement for sample size when the length of the interval is different.

d	0.01	0.02	0.03	0.04	0.05
n	9701	2448	1098	623	401

Table 3.6.3 Requirements for sample size when the reliability is lower.

d	0.01	0.02	0.03	0.04	0.05
n	1235	332	157	94	63

where $F_\alpha(s,t)$ denotes the upper α-quantile of F-distribution with s and t degrees of freedom. Similarly, we can obtain,

$$\hat{\theta}_U = (k+1)F_{\alpha_2}(2(k+1),2(n-k))/[(n-k)+(k+1)F_{\alpha_2}(2(k+1),2(n-k))].$$

Obviously, when $k=0$ or $k=n$, $\hat{\theta}_L=0$ or $\hat{\theta}_U=n$, then it turns to be a one-sided interval estimation. We can verify that when n is given, the maximum value of the length of the interval $\hat{\theta}_U-\hat{\theta}_L$ can be attained at $k=[n/2]$, where $[x]$ denotes the maximum integer not exceeding x. Similar to Table 3.6.1, Table 3.6.2 provides the accurate requirements for sample size when $\alpha=0.05$ and $\hat{\theta}_U-\hat{\theta}_L=2d$ for different d.

Comparing with Table 3.6.1, we can see that the normal approximation is quite good. On the other hand, the requirement for sample size in Table 3.6.2 is quite large. When the sample size is smaller, we have to sacrifice the accuracy and the reliability of interval estimation. Table 3.6.3 provides the requirements for sample size when $\alpha=0.50$.

Thus it can be seen that we must make a comprehensive consideration of all factors according to concrete problems in application.

3.6.4 *Uniformly Most Accuracy and Interval Length*

Recall that the principle of interval estimation, under the guarantee of reliability, the higher the accuracy, namely the smaller the interval length, the better the interval estimate. Then, can a $(1-\alpha)$-UMAU confidence interval guarantee that it is the shortest length among all the $1-\alpha$ unbiased interval estimates?

Let $[\hat{\theta}_L(x),\hat{\theta}_U(x)]$ be a $1-\alpha$ confidence interval, $E_\theta[\hat{\theta}_U(X)-\hat{\theta}_L(X)]$ is defined as the **interval length** at θ. For the given θ, let

$$A(\theta)=\{x;\ \hat{\theta}_L(x)\le\theta\le\hat{\theta}_U(x)\}, \tag{3.6.19}$$

which is a set of x. We have the following theorem.

Theorem 3.6.5. *If* $[\hat{\theta}_L(x), \hat{\theta}_U(x)]$ *is a* $(1-\alpha)$*-UMAU confidence interval, then for any* $1-\alpha$ *unbiased confidence interval* $[\hat{\theta}_L^*(x), \hat{\theta}_U^*(x)]$, *we have*

$$E_{\theta_0}[\hat{\theta}_U(X) - \hat{\theta}_L(X)] \le E_{\theta_0}[\hat{\theta}_U^*(X) - \hat{\theta}_L^*(X)]$$

for $\forall\theta_0 \in \Theta$.

Proof. Since the parameter space considered is (a, b), let $d\nu(\theta)$ be the Lebesgue measure in this space. For a given $\theta \in \Theta$, let $A^*(\theta)$ have the form (3.6.19) corresponding to $[\hat{\theta}_L^*(x), \hat{\theta}_U^*(x)]$. Then we have

$$\begin{aligned}
E_{\theta_0}[\hat{\theta}_U(x) - \hat{\theta}_L(x)] &= \int_{\mathcal{X}} [\hat{\theta}_U(x) - \hat{\theta}_L(x)] dP_{\theta_0}(x) \\
&= \int_{\mathcal{X}} \int_{\hat{\theta}_L(x)}^{\hat{\theta}_U(x)} d\theta dP_{\theta_0}(x) \\
&= \int_{\theta \neq \theta_0} \int_{A(\theta)} dP_{\theta_0} d\theta \\
&= \int_{\theta \neq \theta_0} P_{\theta_0}\{A(\theta)\} d\theta \\
&\le \int_{\theta \neq \theta_0} P_{\theta_0}\{A^*(\theta)\} d\theta \\
&= E_{\theta_0}[\hat{\theta}_U^*(x) - \hat{\theta}_L^*(x)].
\end{aligned}$$

In the above equations, $\theta \neq \theta_0$ implies $\Theta - \{\theta_0\}$, and the inequality follows from the definition of $(1-\alpha)$-UMAU confidence interval. □

The above problem was first explored by Pratt (1961), Madansky (1962) considered the situation of $(1-\alpha)$-UMAU confidence interval later. Maatta and Casella (1987) studied the situation based on test statistics.

3.7 Problems

3.1. Let $x_1, \cdots, x_n$ be a set of samples from $F(x-\theta)$, consider the one-sided testing problem

$$H_0 : \theta = 0 \quad \text{vs.} \quad H_1 : \theta > 0,$$

where $F(x)$ is an absolute continuous distribution function and its density function is symmetrical about the origin. Taking the mixture normal distribution $F(x) = (1-\alpha)\Phi_x + \alpha\Phi_x$ as an example, discuss the efficiency of the Wilcoxon signed rank test and the Student's t-test and their robustness.

3.2. Let $x = (x_1, \cdots, x_n)$ be a sample from a normal population $N(\mu_1, \sigma^2)$, and $y = (y_1, \cdots, y_n)$ be a sample from a normal population $N(\mu_2, \sigma^2)$, where the x and y are independent and (μ_1, μ_2, σ^2) are parameters. Consider the following two problems:

(a) Find the maximum likelihood estimates of the parameters under the restrictions of $\mu_1 \leq \mu_2$ and $\mu_1 = \mu_2$ respectively. Furthermore, find the likelihood ratio test for the problem of testing $H_0 : \mu_1 = \mu_2$ vs. $H_1 : \mu_1 \leq \mu_2$.

(b) Find the maximum likelihood estimate of the parameter without any restriction. Furthermore, construct a 95% confidence interval for the parameter $\mu_2 - \mu_1$.

3.3. Let $x_1, \cdots, x_n$ be a set of samples from a log-normal population $LN(\mu, \sigma^2)$, that is, $\ln(X_i)$ is i.i.d. from $N(\mu, \sigma^2)$, $i = 1, \cdots, n$, where (μ, σ^2) are parameters. Consider the following two problems:

(a) Find the maximum likelihood estimates of the parameters in the parameter space $\{(\mu, \sigma^2) : \mu \in R, \sigma^2 > 0\}$ and Construct a 95% confidence interval for μ.

(b) Find the maximum likelihood estimates of the parameters under the restriction of $\mu \leq 0$. Furthermore, construct a likelihood ratio test for the problem of testing $H_0 : \mu = 0$ vs. $H_1 : \mu \leq 0$.

3.4. Let $x_1, \cdots, x_n$ be a set of samples from a population with density function $p(x, \mu, \sigma)$, where $\mu \in R, \sigma > 0$. Let $x = (x_1, \cdots, x_n)$. Suppose that $(T_1(x), T_2(x))$ are sufficient statistics for the parameters (μ, σ) with joint density function

$$f(t_1, t_2, \mu, \sigma) = \sqrt{\frac{n}{2\pi}} \exp\left[-\frac{n(t_2 - \mu)^2}{2}\right] \cdot \sqrt{\frac{n}{2\pi\sigma}} \exp\left[-\frac{n(t_1 - t_2)^2}{2\sigma}\right].$$

Consider the following three problems:

(a) Let $\Psi_1 = \{\psi(T_2(x))\}$, where $\psi(T_2(x))$ is a test function generated by $T_2(x)$. For the problem of testing $H_0 : \mu \leq 0$ vs. $H_1 : \mu > 0$, find a level $\alpha = 0.05$ UMP test $\psi^*(T_2(x))$ among the class of test functions Ψ_1.

(b) Let $\Psi_2 = \{\psi((T_1(x), T_2(x)))\}$, where $\psi((T_1(x), T_2(x)))$ is a test function generated by $T_1(x), T_2(x)$. For the problem of testing $H_0 : \mu \leq 0$ vs. $H_1 : \mu > 0$, prove that $\psi^*(T_2(x))$ is also a level-$\alpha = 0.05$ UMP test among the class of test function Ψ_2.

(c) Construct a 95% uniformly most accurate lower confidence bound for the parameter μ.

3.5. Let $x = (x_1, \cdots, x_n)$ be a set of samples from an exponential distribution with density function

$$p(x, \theta) = \begin{cases} \exp\{-(x-\theta)\}, & \text{if } x \geq \theta; \\ 0, & \text{otherwise.} \end{cases}$$

Consider the problem of testing $H_0 : \theta = 0$ vs. $H_1 : \theta > 0$, construct a level-α UMP test.

3.6. Let $x = (x_1, \cdots, x_n)$ and $y = (y_1, \cdots, y_n)$ be independent samples from normal distributions $N(\mu_1, \sigma_1^2)$ and $N(\mu_2, \sigma_2^2)$, respectively. For the problem of testing

$$H_0 : \frac{\sigma_1^2}{\mu_1} = \frac{\sigma_2^2}{\mu_2} \quad \text{vs.} \quad H_1 : \text{ not } H_0,$$

construct a level-α UMP test.

3.7. Let $x = (x_1, \cdots, x_n)$ be a set of samples from a normal population $N(\mu, \sigma^2)$, consider the problem of testing

$$H_0 : \mu_1 = \mu_2, \sigma^2 = \sigma_0^2 \quad \text{vs.} \quad H_1 : \text{ not } H_0,$$

construct a level-α UMPU test.

3.8. Assume that $x_i = \theta x_{i-1} + \varepsilon_i$, $i = 1, \cdots, n$, where $x_0 = 0$ and $\{\epsilon_i\}$ are i.i.d. from a normal distribution $N(0, \sigma^2)$.

(a) Prove that the density of $x = (x_1, \cdots, x_n)$ has the following form

$$f(x, \theta) = \left(\frac{1}{\sqrt{2\pi\sigma^2}}\right)^n \exp\left\{-\frac{1}{2\sigma^2}\sum_{i=1}^{n}(x_i - \theta x_{i-1})^2\right\}.$$

(b) For the problem of testing $H_0 : \theta = 0$ vs. $H_1 : \theta \neq 0$, figure out the likelihood ratio statistic.

3.9. Research shows that some hypertension drugs such as Propranolol can relieve the symptoms of stage fright (*Time Magazine*, July 5, 1982, p. 58). To test the effect of Propranolol on the performance anxiety, each of 29 professional and student musicians gave two solo recitals before an audience of critics and faculty members. Ninety minutes before one recital, they were given Propranolol; on the other occasion, a placebo. Stage fright were measured (70 is a normal resting rate). The beat-per-minute (BPM) data about eight of the performers are reported in the following table:

Treatment	1	2	3	4	5	6	7	8
Propranolol	85	107	69	122	106	121	137	87
Placebo	126	140	95	148	142	172	133	143

Let X denote the difference of the BPM between the placebo treatment and the Propranolol treatment, and θ the median of the distribution of r.v. X. For $\alpha \approx 0.05$, construct a Wilcoxon signed rank test for the problem of testing $H_0 : \theta = 0$ vs. $H_1 : \theta > 0$. Furthermore, construct a point estimate of θ and a confidence interval with confidence level about 95%.

3.10. Suppose that $X_1, \cdots, X_n \overset{\text{i.i.d.}}{\sim} U(0, \theta_1)$ and $Y_1, \cdots, Y_m \overset{\text{i.i.d.}}{\sim} U(0, \theta_2)$, where $X = (X_1, \cdots, X_n)$ and $Y = (Y_1, \cdots, Y_m)$ are mutually independent. Prove that there exists a level-α UMP test for the problem of testing $H_0 : \theta_1 \leq \theta_2$ vs. $H_1 : \theta_1 > \theta_2$.

3.11. Suppose that $X_1, \cdots, X_n \overset{\text{i.i.d.}}{\sim} N(\mu_1, \sigma_1^2)$ and $Y_1, \cdots, Y_m \overset{\text{i.i.d.}}{\sim} N(\mu_2, \sigma_2^2)$. Consider the problem of testing $H_0 : \sigma_2^2 \leq \sigma_1^2$ vs. $H_1 : \sigma_2^2 > \sigma_1^2$.

(a) Suppose that both μ_1 and μ_2 are known, there exists a UMP test for the above testing problem.

(b) Suppose that both μ_1 and μ_2 are unknown, there has no UMP test for the above testing problem.

(c) If either μ_1 or μ_2 is known, but NOT both, how about the above testing problem?

3.12. Let $p(x; \mu) = c(\mu) \exp\{\mu x\} h(x)$ be the density function of $T(x)$, where $c(\mu) > 0$. Suppose that a level-α UMP test $\phi(T)$ for the problem of testing

$$H_0 : \mu \leq \mu_0 \quad \text{vs.} \quad \mu > \mu_0$$

has the following form

$$\phi(T) = \begin{cases} 1, & \text{if } \ T > C, \\ r, & \text{if } \ T = C, \\ 0, & \text{if } \ T < C, \end{cases}$$

where the constants r $(0 \leq r \leq 1)$ and c are determined by $E_{\mu_0} \phi(T(X)) = \alpha$. If the power function is $g(\mu) = E_{\mu_0} \phi(T(X))$, prove that $g'(\mu_0) > 0$.

3.13. For a multinomial distribution, consider the problem of testing $H_0 : p_1 = p_{10}, \cdots, p_k = p_{k0}$.

(a) Exemplify that the χ^2 goodness-of-fit under H_0 is not always unbiased.

(b) The χ^2 goodness-of-fit test under H_0 is unbiased if and only if $p_{10} = \cdots = p_{k0} = 1/k$.

3.14. Suppose X follows from the one-parameter exponential family $C(\theta)\exp(\theta T(x))$. Consider the problem of testing $H_0 : \theta = \theta_0$ vs. $H_1 : \theta \neq \theta_0$, where a level-$\alpha$ UMPU test has the form

$$\phi_\alpha(T) = \begin{cases} 1, & \text{if } T < c_{1\alpha} \text{ or } T > c_{2\alpha}, \\ r_{i\alpha}, & \text{if } T = c_{i\alpha},\ i = 1, 2, \\ 0, & \text{if } c_{1\alpha} < T < c_{2\alpha}. \end{cases}$$

Prove that the acceptance region $(c_{1\alpha}, c_{2\alpha})$ of ϕ_α shrinks with the increasing of α, *i.e.*, we have $c_{1\alpha'} \geq c_{1\alpha}$ when $\alpha' > \alpha$, and $r_{1\alpha'} \geq r_{1\alpha}$ if $c_{1\alpha'} = c_{1\alpha}$; at the meantime, $c_{2\alpha'} \leq c_{2\alpha}$, and $r_{2\alpha'} \geq r_{2\alpha}$ if $c_{2\alpha'} = c_{2\alpha}$.

3.15. The merits of a new technique of a production line are under investigation to see if this technique gives a lower number of defective items in general than the old technique. Ten days of production are obtained by both techniques with the following results:

Day	1	2	3	4	5	6	7	8	9	10
New	172	165	206	184	174	142	190	169	161	200
Old	201	179	159	192	177	170	182	179	169	210

Using level $\alpha = 0.05$,

(a) Analyze the data using the sign test, mentioning any assumptions used. [Hints: Write p_+=probability that the new technique gives a lower number of defective items and set up the null hypothesis $H_0 : P_+ = 0.5$ and alternative $H_1 : p_+ < 0.5$. Notice that this is a one-sided test.]

(b) Analyze the data using the Wilcoxon signed rank test.

3.16. The waiting time, X between tremors recorded at a seismological station is believed to follow an exponential distribution with parameter θ. That is, the probability density function of X is,

$$f(x) = \frac{1}{\theta}\exp\left(-\frac{x}{\theta}\right), \quad x > 0.$$

A single observation, x is available and it is required to test the null hypothesis $H_0 : \theta = 10$ vs. $H_1 : \theta \neq 10$. The following test is proposed:

$$\text{Reject } H_0 \text{ if } x < 8 \text{ or } x > 12.$$

(a) What is the significance level of this test?

(b) Calculate the probabilities of Type II errors when $\theta = 2$ and $\theta = 12$.

(c) Calculate the powers of the test when $\theta = 2$ and $\theta = 12$ (give your answers to three decimal places).

3.17. Tracheobronchial clearance was studied on seven pairs of twins by measuring the retention percentage for an inhaled radioactive spray. One member of each pair lived in a rural area and the other in an urban area. The retention rates after one hour were

Pair	1	2	3	4	5	6	7
Rural (x_i)	10.1	51.8	33.5	32.8	69.0	38.9	54.6
Urban (y_i)	28.1	36.2	40.7	38.8	71.0	47.0	57.0

(a) Test the null hypothesis of no difference in median retention rate using the sign test. You should obtain the exact p-value.

(b) Test the null hypothesis of a higher median retention rate for rural dwellers using the Wilcoxon signed rank test. You should obtain the exact p-value.

(c) You are given that the 28 averages of differences $(d_i + d_j)/2$, where $d_i = x_i - y_i$, are

−18.00	−13.05	−12.60	−12.00	−10.20	−10.00	−8.10
−7.65	−7.20	−7.05	−6.60	−6.00	−5.25	−5.05
−4.80	−4.60	−4.20	−4.00	−2.40	−2.20	−2.00
−1.20	3.75	4.20	4.80	6.60	6.80	15.60

Calculate a 95% Wilcoxon confidence interval for the population median difference (rural−urban).

3.18. (a) Let $X_{(1)}, \cdots, X_{(n)}$ be an ordered, random sample from a continuous distribution function F.
Derive the distribution of $R = F(X_{(n)}) - F(X_{(1)})$.
Hence show that, for small ϵ, δ, the sample size n required to ensure that

$$P(R \geq 1 - \epsilon) \geq 1 - \delta$$

is approximately θ/ϵ, where θ is the unique positive solution to $1 + \theta - \delta e^{\theta} = 0$.

(b) Describe in detail the Wilcoxon signed rank test, used to test whether a continuous distribution is symmetric about a point θ_0.

3.19. Let $X_1, \cdots, X_n$ be a random sample from a continuous distribution that is symmetric about the unknown median $\theta, -\infty < \theta < \infty$. Explain carefully how to test $H_0 : \theta = 0$ vs. $H_1 : \theta > 0$ using the Wilcoxon signed rank test.
Show that the null mean and variance of the Wilcoxon signed rank statistic are $n(n+1)/4$ and $n(n+1)(2n+1)/24$ respectively. State, without proof, the asymptotic null distribution of this statistic.
Each of the $n(n+1)/2$ averages $(X_i + X_j)/2, i \le j = 1, \cdots, n$, is called a **Walsh average**. It is proposed to test H_0 vs. H_1, using as statistic the total number of Walsh averages greater than 0. Show that this test is equivalent to the Wilcoxon signed rank test.

3.20. Suppose that a product lifetime X is distributed according to the Weibull distribution with the pdf

$$m\eta^{-m}x^{m-1}\exp\left\{-\left(\frac{x}{\eta}\right)\right\}^m \quad \text{for } x > 0,$$

where $m > 0$ is known and $\eta > 0$ is a parameter. Let $L(> 0)$ denote the lower specification limits, then $R = P(X > L)$ represents the reliability of the product. Suppose that n products are randomly taken from a large number of products. The n products are inspected independently for failure, and each failure time is recorded. The inspection is terminated until $r(\ge 1)$ failures are observed in the sample of n. For $0 < \alpha < 1$, construct a $(1-\alpha)$-UMA lower confidence bound for the reliability R.

3.21. Suppose that $X_1, \cdots, X_n$ are i.i.d. r.v.s with the common uniform distribution $U(0, \theta)$, where the parameter $\theta > 0$.

(a) For a given $\theta_0 > 0$, find a level-α UMP test for the testing problem $H_0 : \theta \le \theta_0$ vs. $H_1 : \theta > \theta_0$.

(b) For $0 < \alpha < 1$, construct a $(1-\alpha)$-UMA upper confidence bound for θ.

3.22. Suppose that $X_1, \cdots, X_n$ are i.i.d. r.v.s with the common normal distribution $N(\mu, \sigma^2)$, where $\sigma > 0$ is known and μ is a parameter. Let $p = P(X_1 < L) + P(X_1 > U)$, where U and L are known and satisfy $U > L$. Furthermore, suppose that $p_0 \in (0, 1/2)$ is known and satisfies $(U - L)/\sigma = -2Z_{p_0/2}$, where Z_q denotes the lower q-quantile of the standard normal distribution, *i.e.* Z_q satisfies $q = \int_{-\infty}^{Z_q} \exp\{-t^2/2\}/\sqrt{2\pi}dt$. For $0 < \alpha < 1$, find a level-α UMP unbiasd test for the testing problem $H_0 : p \le p_0$ vs. $H_1 : p > p_0$.

3.23. Let $X_1, X_2, \cdots, X_n$ be i.i.d. with the cdf

$$\theta k^{\theta} x^{-(\theta+1)} \quad \text{for } x > k > 0 \text{ and } \theta > 0,$$

where k is known and θ is a parameter.

(a) Find the cdf of $2\theta \ln(x_1/k)$.

(b) For $0 < \alpha < 1$, find a α-UMP test for the testing problem $H_0 : \theta \leq \theta_0$ vs. $H_1 : \theta > \theta_0$ (where $\theta_0 > 0$ is known).

(c) For $0 < \alpha < 1$, find a $(1-\alpha)$-UMA lower confidence bound for θ.

Chapter 4

Parametric Tests for Simple Models

Up to now, we have discussed the following problem: assuming that an r.v. has a certain distribution, then constructing a test statistic based on a sample from the distribution and using it to test the parameters in the distribution. however, in many applications, one problem usually involves many r.v.'s which are related to each other. For researching convenience, we construct a statistical model or stochastic model based on some assuming conditions so that we can use mathematical formulations to express the relationships among the r.v.'s clearly. In this chapter, we will discuss the parametric tests for simple statistical model. We mainly discuss the testing problem for one factor model. Although it is the simplest statistical model, the corresponding method is not difficult to be generalized.

4.1 One-Factor Model: Analysis of Variance

Assume that there are k independent populations, $D_1, \cdots, D_k$, and draw n_i samples from the population D_i, $i = 1, \cdots, k$. If we only consider observation error, we can establish the following model:

$$y_{ij} = \theta_i + \epsilon_{ij}, \quad i = 1, \cdots, k;\ j = 1, \cdots, n_i, \tag{4.1.1}$$

where y_{ij} denotes the observation, θ_i denotes the unknown parameter called mean effects, and ϵ_{ij} denotes the observation error. In general, assume that the ϵ_{ij}'s are mutually independent and $\epsilon_{ij} \sim N(0, \sigma_i^2)$. So we can regard $y_{i1}, \cdots, y_{in_i}$ as a set of samples from $N(\theta_i, \sigma_i^2)$. Consider the testing problem

$$H_0: \theta_1 = \cdots = \theta_k \quad \text{vs.} \quad H_1: \text{not } H_0, \tag{4.1.2}$$

that is, H_1 denotes that there exist $i \neq i'$ such that $\theta_i \neq \theta_{i'}$. It can be seen that, although the k populations are originally independent, and consequently the models given by Eq. (4.1.1) are also independent, these models are linked together by the testing problem (4.1.2). We can see later that the problem analysis would finally result in the comparison among variances in essence. So this kind of problem is called **analysis of variance**, denoted as ANOVA. Since there is only one factor in Eq. (4.1.1), this simplest situation is called **one-way ANOVA**.

Sometimes Eq. (4.1.1) can be written as

$$y_{ij} = \mu + \nu_i + \epsilon_{ij}, \quad j = 1, \cdots, n_i; \; i = 1, \cdots, k, \tag{4.1.3}$$

where μ is an unknown common parameter shared by all the k populations. Equation (4.1.3) is called a **saturated model**. Without loss of generality, we can assume $\sum_{i=1}^{k} \nu_i = 0$, this is because model (4.1.3) is equivalent to

$$y_{ij} = (\mu' + \bar{\nu}') + (\nu_i' - \bar{\nu}') + \epsilon_{ij},$$

indeed, it suffices to set $\mu = \mu' + \bar{\nu}'$ and $\nu_i = \nu_i' - \bar{\nu}'$, where $\bar{\nu}' = \sum_{i=1}^{k} \bar{\nu}_i'/k$. Here the corresponding testing problem for the saturated model (4.1.3) is $H_0: \nu_1 = \cdots = \nu_k = 0$, where $\sum_{i=1}^{k} \nu_i = 0$ and μ is a nuisance parameter. In this section, we mainly discuss the model given by Eq. (4.1.1). Discussion about general saturated models will be delayed to later sections.

The ANOVA evolved as a method of analyzing agricultural experiments. There are k varieties, and let population D_i denote the i-th varies. We can divide a block land almost without difference on fertility into n equal small blocks, variety D_i is planted in n_i-th block, such that $n_1 + \cdots + n_k = n$. Finally we decide whether there is a statistical difference among the varieties based on the yield data. Now this idea has been applied in many experimental data analysis or trials data analysis.

Example 4.1.1. (Establishment of model) The experimental data in this example are collected in the Laboratory of the First Department of Internal Medicine of Kyushu University in Japan (Shi *et al.*, 1988; Shi, 1991). Among several aspects of the behaviors of rats, the behavior of food intake of two kinds of Zucker rats (*i.e.* Obese and Lean) is considered. Furthermore, assume that obese is caused by some hidden gene (Zucker and Zucker, 1961).

Choose four rats in each kind of Zucker rats, denoted as O_i and L_i, $i = 1, \cdots, 4$. Amount of food intake was measured by the number of pelleted chows consumed by these Zucker rats during a day period (8 A.M. - 8 P.M.) and a night period (8 P.M. - 8 A.M.) at four different ages, namely 12, 23, 30 and 43 weeks, corresponding to human ages 18, 30, 40 and 60.

Table 4.1.1 Amount of food intake of the obese Zucker rats.

	Age in weeks	12	23	30	43
O_1	A	216	231	264	265
	B	450	428	442	377
	A+B	666	659	706	642
O_2	A	290	345	351	284
	B	308	289	327	314
	A+B	598	634	678	598
O_3	A	180	276	289	373
	B	474	434	381	321
	A+B	654	710	670	694
O_4	A	154	175	226	248
	B	467	480	419	386
	A+B	621	655	645	634

Table 4.1.2 Amount of food intake of the lean Zucker rats.

	Age in weeks	12	23	30	43
L_1	A	157	153	178	133
	B	316	314	316	330
	A+B	473	467	494	463
L_2	A	146	175	193	154
	B	333	282	304	290
	A+B	479	457	497	444
L_3	A	132	189	204	168
	B	331	336	324	271
	A+B	463	525	528	439
L_4	A	181	163	188	156
	B	275	305	281	247
	A+B	456	468	469	403

Tables 4.1.1 and 4.1.2 report the amounts consumed by these rats during the two time periods (denoted by A and B), respectively.

For different research purposes, we can construct different models to analyze the data set. According to different research goals, the original experimental aim is to study whether there occurs an essential difference on food intake behavior of two kinds of rats with the age increasing, and we will discuss two problems in later sections. Now we only research whether the amount of food intake of two kinds of rats changes with the age increasing. So there are 4 populations corresponding to ages, drawing a sample y_{ij}

from each population, the population mean is θ_i, that is, $EY_{ij} = \theta_i, j = 1, \cdots, 4, i = 1, \cdots, 4$. We need to test the following testing problem

$$H_0: \ \theta_1 = \theta_2 = \theta_3 = \theta_4, \tag{4.1.4}$$

and the alternative hypothesis is that H_0 is not true.

Consider an **LR test** for the testing problem (4.1.2). Assume that k populations have the same variance, that is, $\sigma_1^2 = \cdots = \sigma_k^2 = \sigma^2$. When the variances are unequal, that is a version of the Behrens-Fisher problem.

Now the likelihood function is

$$L(\boldsymbol{y}; \theta_1, \cdots, \theta_k) = \left(\frac{1}{2\pi\sigma^2}\right)^{n/2} \exp\left\{-\frac{1}{2\sigma^2}\sum_{i=1}^{k}\sum_{j=1}^{n_i}(y_{ij} - \theta_i)^2\right\},$$

where $n = n_1 + \cdots + n_k$ and σ^2 is a nuisance parameter, the parameter space of the mean is R^k. It is easy to obtain the MLEs of θ_i and σ^2 as follows

$$\begin{aligned} \hat{\theta}_i &= \bar{y}_i = \sum_{j=1}^{n_i} y_{ij}/n_i, \quad i = 1, \cdots, k, \\ \hat{\sigma}^2 &= \sum_{i=1}^{k}\sum_{j=1}^{n_i}(y_{ij} - \bar{y}_i)^2/n = Q_1^2/n. \end{aligned} \tag{4.1.5}$$

When H_0 is true, let $\theta_1 = \cdots = \theta_k = \theta$, then the MLEs of θ and σ^2 are

$$\begin{aligned} \hat{\theta} &= \bar{y} = \sum_{i=1}^{k} n_i\bar{y}_i/n, \\ \hat{\sigma}_0^2 &= \sum_{i=1}^{k}\sum_{j=1}^{n_i}(y_{ij} - \bar{y})^2/n = Q^2/n. \end{aligned} \tag{4.1.6}$$

Based on the discussion in Section 2.2, the LR test is

$$\Lambda(\boldsymbol{y}) = \frac{L(\boldsymbol{y}; \hat{\theta}, \cdots, \hat{\theta})}{L(\boldsymbol{y}; \hat{\theta}_1, \cdots, \hat{\theta}_k)} \propto \frac{\hat{\sigma}^2}{\hat{\sigma}_0^2}. \tag{4.1.7}$$

Let $Q_2^2 = \sum_{i=1}^{k} n_i(\bar{y}_i - \bar{y})^2$. By $\sum_{i=1}^{k}\sum_{j=1}^{n_i}(y_{ij} - \bar{y}_i)(\bar{y}_i - \bar{y}) = 0$, it is easy to obtain that

$$Q^2 = Q_1^2 + Q_2^2. \tag{4.1.8}$$

Now let us analyze the intuitive meaning of Eq. (4.1.8). Let $\boldsymbol{y}$ denote a vector of all observations, that is,

$$\boldsymbol{y} = (y_{11}, \cdots, y_{1n_1}, \cdots, y_{k1}, \cdots, y_{kn_k})'$$

is an n-dimensional vector. Now $\Theta = R^k$, and $\Theta_0 = \{\boldsymbol{\theta} \in \boldsymbol{R}^k;\ \theta_1 = \cdots = \theta_k\}$ is a linear subspace in Θ. Figure 4.1.1 shows that the projection of $\boldsymbol{y}$

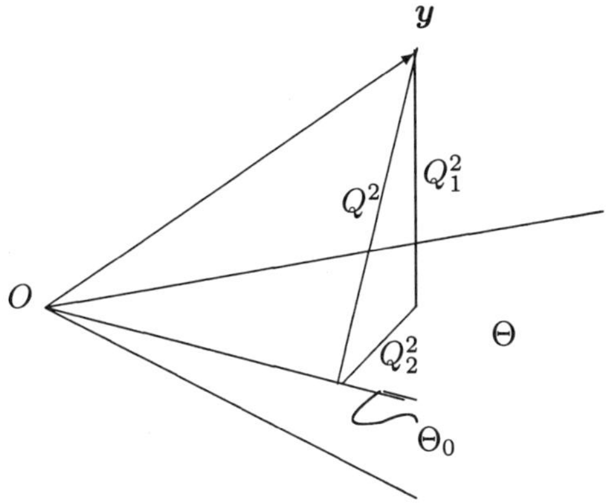

Fig. 4.1.1 Geometrical explanation of the sum of squares

onto Θ_0 and the projection of $\boldsymbol{y}$ onto Θ then onto Θ_0 form a right triangle. By the Pythagorean theorem, we can obtain Eq. (4.1.8).

Considering from the point of view of data analysis, we know that Q represents the total sum of squares between all the y_{ij}'s and the overall mean $\bar{y}$ (SST), which describes the diversity of all data; Q_1 represents the sum of squares due to differences within groups (SSW), which describes the diversity of each group; Q_2 represents the sum of squares due to differences between groups (SSB), which describes the diversity between groups. Therefore, Equation (4.1.8) can be written as

$$\text{SST=SSW+SSB}. \tag{4.1.9}$$

It is easy to verify that for $\theta \in \Theta$, we have $E_\theta Q_1^2 = (n-k)\sigma^2$, and $E_\theta Q_2^2 = (k-1)\sigma^2 + \sum_{i=1}^k n_i(\theta_i - \bar{\theta})^2$, where $\bar{\theta} = \sum_{i=1}^k n_i\theta_i/n$. Hence, for any parameter $\theta \in \Theta$, $Q_1^2/(n-k)$ is an unbiased estimator of σ^2, albeit only if $\theta \in \Theta_0$, then $Q_2^2/(k-1)$ is an unbiased estimator for σ^2, otherwise its estimator may be larger. Let

$$F(y) = \frac{Q_2^2/(k-1)}{Q_1^2/(n-k)}. \tag{4.1.10}$$

Comparing with (4.1.7), we reject H_0 if the LR test is small enough or equivalently, and reject H_0 if $F(y)$ is large enough. The statistic given by Eq. (4.1.10) is called an F-statistic, the corresponding test is called an F-test. If consider the residual errors within group as random errors and

Table 4.1.3 ANOVA table of two kinds of Zucker rats.

$i =$	1	2	3	4	
$\bar{y}_i^O =$	634.75	664.5	674.75	642	$\bar{y}^O = 654.00$
$\bar{y}_i^L =$	467.75	479.25	578.75	437.25	$\bar{y}^L = 490.75$

Table 4.1.4 ANOVA table of the obese Zucker rats.

Source of variation	Sum of squares	Degrees of freedom	F-value	F_α	Significance
Age in weeks	$Q_2^2 = 4221.5$	3	1.34	$F_{0.05} = 3.49$	not reject
Error	$Q_1^2 = 12606.5$	12		$F_{0.01} = 5.95$	H_0
Total	$Q^2 = 16828$	15			

errors between groups caused by factor levels, then the intuitive sense of F-test is very obvious.

By the direct sum decomposition Eq. (4.1.8), Q_1^2 and Q_2^2 are mutually independent. As there are k linear relationships $\sum_{j=1}^{n_i}(y_{ij} - \bar{y}_i) = 0$ in Q_1^2, the rank of Q_1^2 is $n-k$. As there is only one linear relationship $\sum_{i=1}^{k} n_i(\bar{y}_i - \bar{y}) = 0$ in Q_2^2, the rank of Q_2^2 is $k-1$. Since Q_1^2/σ^2 and Q_2^2/σ^2 have the χ^2 distributions with degrees of freedom $n-k$ and $k-1$ respectively by Theorem 2.2.2, we can obtain the following theorem.

Theorem 4.1.1. *When H_0 is true, the LR test statistic $F(Y)$ given by (4.1.10) has an F-distribution with $k-1$ and $n-k$ degrees of freedom. The level-α LR test rejects H_0 if $F(y) \geq F_\alpha$, where F_α is the upper α quantile of the F-distribution with $k-1$ and $n-k$ degrees of freedom.*

Example 4.1.2. (ANOVA table) Analyze the data given in Example 4.1.1. Now we are interested in the problem whether the amount food intake of two kinds of Zucker rats will change with the age increasing or not. So, we only need to study the total amount of food intake per day, namely, $A + B$. We use y_{ij}^O and y_{ij}^L to express the amount of food intake of the Lean and Obese rats respectively, where i denotes the age in weeks, j denotes different Zucker rats. We can obtain the following table.

Then, we can obtain the ANOVA tables of two kinds of Zucker rats (see Tables 4.1.4 and 4.1.5). We can see from these tables that for obese Zucker rats under given weeks of age there is no significant difference in the amount of food intake, since $1.34 < 3.49$, namely, $F(y) < F_{0.05}$; there is significant difference for lean Zucker rats, but not extremely significant difference, since $3.49 < 4.51 < 5.95$, namely, $F_{0.05} < F(y) < F_{0.01}$. In

Table 4.1.5 ANOVA table of the lean Zucker rats.

Source of variation	Sum of squares	Degrees of freedom	F-value	F_α	Significance
Age in weeks	$Q_2^2 = 7674.7$	3	4.51	$F_{0.05} = 3.49$	under size 0.05
Error	$Q_1^2 = 6809.3$	12		$F_{0.01} = 5.95$	reject H_0
Total	$Q^2 = 14484$	15			

application of ANOVA, we usually call $F(y) > F_{0.05}$ having a significant difference, and $F(y) > F_{0.01}$ having an extremely significant difference. We will apply some other methods later to analyze this example again.

4.2 One-Factor Model: Multiple Comparisons

In this section, based on the idea of the U-I test (see Section 2.5), we will further discuss the testing problem (4.1.2). For one-factor model (4.1.1), consider the **paired-comparison** problem to investigate whether there is difference between any two levels, θ_i and θ_j, or not. The testing problem can be expressed as

$$H_0^{ij} : \theta_i = \theta_j \quad \text{vs.} \quad H_1^{ij} : \theta_i \neq \theta_j. \tag{4.2.1}$$

Let $\boldsymbol{\theta} = (\theta_1 \cdots \theta_k)'$ and $\Theta_{ij} = \{\boldsymbol{\theta} \in \boldsymbol{R}^k;\ \theta_i = \theta_j\}$, then (4.2.1) can be rewritten as

$$H_0^{ij} : \boldsymbol{\theta} \in \Theta_{ij} \quad \text{vs.} \quad H_1^{ij} : \boldsymbol{\theta} \notin \Theta_{ij}.$$

Here, we can construct $k(k-1)/2$ testing problems. Then, the testing problem given by (4.1.2) can be written as

$$H_0 : \boldsymbol{\theta} \in \bigcap_{i<j} \Theta_{ij} \quad \text{vs.} \quad H_1 : \boldsymbol{\theta} \notin \bigcap_{i<j} \Theta_{ij}. \tag{4.2.2}$$

Usually this kind of testing problem is called **multiple comparisons**.

Generally, let $\boldsymbol{a} = (a_1, \cdots, a_k)'$ satisfying $\sum_{i=1}^k a_i = 0$. Consider the problem of testing

$$H_0^{\boldsymbol{a}} : \boldsymbol{a}'\boldsymbol{\theta} = 0 \quad \text{vs.} \quad H_1^{\boldsymbol{a}} : \boldsymbol{a}'\boldsymbol{\theta} \neq 0. \tag{4.2.3}$$

It can be seen that (4.2.3) implies (4.2.1) as long as $\boldsymbol{a}$ is chosen appropriately. However, (4.2.3) has a wider applicability, especially it covers an important testing problem with the following form:

$$\frac{1}{m}(\theta_1 + \cdots + \theta_m) = \frac{1}{k-m}(\theta_{m+1} \cdots + \theta_k).$$

Recall Example 4.1.2, we may be interested in the problem of $(\theta_1+\theta_2)/2 = (\theta_3+\theta_4)/2$, that is, whether the amount of food intake of Zucker rats can be divided into two stages obviously or not. Sometimes, the problem of this type is called a **linear combination problem**. Let

$$A = \left\{ \boldsymbol{a} \in \boldsymbol{R}^k;\ \sum_{i=1}^{k} a_i = 0 \right\}.$$

Then the testing problem given by (4.1.2) can be rewritten as

$$H_0 : \boldsymbol{a}'\boldsymbol{\theta} = 0 \quad \forall \boldsymbol{a} \in A \quad \text{vs.} \quad H_1 : \boldsymbol{a}'\boldsymbol{\theta} \neq 0 \quad \exists \boldsymbol{a} \in A. \tag{4.2.4}$$

Usually this type of testing problem is called **generalized multiple comparison**. It can be seen that (4.1.2) can be expressed as (4.2.2) and (4.2.4) respectively. Here, an interesting problem is whether or not we obtain different statistics from different research viewpoints. Now we analyze this problem in detail.

4.2.1 *Multiple Comparisons*

Since the testing problem (4.2.1) is to compare the means of two normal populations, recalling the discussion in Example 2.2.3, the test statistic is

$$t_{ij}(\boldsymbol{y}) = c_{ij}\frac{\bar{y}_i - \bar{y}_j}{\hat{\sigma}}, \tag{4.2.5}$$

where $c_{ij} = \sqrt{n_i n_j/(n_i - n_j)}$, and $\hat{\sigma}^2 = Q_1^2/(n-k)$ is an unbiased estimator for σ^2 (see Eqs. (4.1.5) and (4.1.10)). Since we assume that the variances of k populations are equal, therefore, as an estimator of σ^2, $\hat{\sigma}^2$ is better than that based on two populations samples. If H_0^{ij} is true, then $t_{ij}(\boldsymbol{Y})$ has a Student's t-distribution with $n-k$ degrees of freedom. Hence, if

$$|t_{ij}(\boldsymbol{y})| \geq t_{\alpha/2}(n-k),$$

then reject H_0^{ij}. At the same time, we can obtain a $1-\alpha$ confidence interval for $\theta_i - \theta_j$, $B_{ij}(\alpha)$,

$$\left[\bar{y}_i - \bar{y}_j - (\hat{\sigma}/c_{ij})t_{\alpha/2}(n-k),\ \bar{y}_i - \bar{y}_j + (\hat{\sigma}/c_{ij})t_{\alpha/2}(n-k)\right]. \tag{4.2.6}$$

Now consider the problem of testing (4.2.2). From (4.2.5), the p-value for the problem of testing (4.2.1) can be written as

$$p_{ij} = P\Big\{|t_{ij}(\boldsymbol{Y})| \geq |t_{ij}(\boldsymbol{y})|\Big\}. \tag{4.2.7}$$

For different values of i and j, we have $r = k(k-1)/2$ different cases. So the p-value for the problem of testing (4.2.2) is $p_0 = \min\{p_{ij}\}$. We have discussed that it is difficult to decide the rejection probability of p_0, *i.e.*, to control the Type I error probability. We usually use the method mentioned in Section 2.5: If $p_0 \leq \alpha/r$ then reject H_0. Sometimes this method is called **least significant difference (LSD)**. When r or k is large enough, the LSD will become very small, this is unfavorable for the rejection of H_0. However, our testing goal is to reject H_0. For this reason, many authors have discussed this problem. Some important results can be found in Miller (1981) and Hsu (1996).

For discussing this problem in a more intuitive way, consider the interval estimation corresponding to the problem of testing (4.2.2). By the symmetry of $\theta_i - \theta_j$, we only need to choose a good interval estimate with the form

$$[\bar{y}_i - \bar{y}_j - d, \bar{y}_i - \bar{y}_j + d]\,. \tag{4.2.8}$$

By the definition, the d in confidence interval with confident level $1-\alpha$ as (4.2.8) should satisfy that under H_0

$$P\left\{\bigcap_{i<j}\left[\bar{Y}_i - \bar{Y}_j - d, \bar{Y}_i - \bar{Y}_j + d\right]\right\} \geq 1-\alpha. \tag{4.2.9}$$

According to the idea of the U-I test, substituting $t_{\alpha/2r}(n-k)$ for $t_{\alpha/2}(n-k)$ in (4.2.6), we can obtain the interval $B_{ij}(\alpha/r)$.

Theorem 4.2.1. *The following Interval estimate*

$$\theta_i - \theta_j \in B_{ij}(\alpha/r), \quad i \neq j$$

is a confidence interval with confidence level $1-\alpha$.

Proof. If suffices to verify (4.2.9). By the Bonferroni inequality

$$\begin{aligned}
P\{\bigcap_{i<j} B_{ij}(\alpha/r)\} &= 1 - P\{\bigcup_{i<j} B^c_{ij}(\alpha/r)\} \\
&\geq 1 - \sum_{i<j} P\{B^c_{ij}(\alpha/r)\} \\
&= 1 - \sum_{i<j} P\{c_{ij}|\bar{Y}_i - \bar{Y}_j|/\hat{\sigma} \geq t_{\alpha/2r}(n-k)\} \\
&= 1 - \sum_{i<j} \alpha/r \\
&= 1 - \alpha.
\end{aligned}$$

□

It can be seen from the proof of the theorem's proof that the confidence level of the confidence interval $B_{ij}(\alpha/r)$ is larger than $1-\alpha$, which yields a lager confidence interval, thus with a lower accuracy. Let us see the following example.

Example 4.2.1. (Interval estimation) Continue to analyze the data in Example 4.1.1. From Table 4.1.3 in Example 4.1.2, an unbiased estimator for σ^2 about the lean Zucker rats is

$$\hat{\sigma}^2 = Q_1^2/(n-k) = 6809.3/12 = 567.54$$

In this case $r = 4(4-1)/2 = 6$. When $\alpha = 0.05$, $\alpha/12 = 0.004$, then $t_{0.004}(12) \approx 3.1$ and $1/c_{ij} \approx 0.7$. So we can obtain $(\hat{\sigma}/c_{ij})t_{\alpha/2r}(n-k) \approx 49.5$. The interval estimator for the difference is $\theta_i - \theta_j \in \bar{y}_i - \bar{y}_j \pm 49.5$, such as $\theta_1 - \theta_2 \in -11.50 \pm 49.5$. It can be seen that the interval length is relatively long.

Tukey's method. Let us construct a confidence interval with the confidence level exactly $1-\alpha$. Assume that **the sample sizes are equal**, *i.e.* $n_1 = \cdots = n_k = n_0$, where $n_0 = n/k$. Let

$$E_{ij} = \left[\bar{y}_i - \bar{y}_j - \hat{\sigma}\sqrt{2/n_0}q_\alpha,\ \bar{y}_i - \bar{y}_j + \hat{\sigma}\sqrt{2/n_0}q_\alpha\right], \tag{4.2.10}$$

where q_α satisfies

$$P\left\{\frac{|\bar{y}_i - \theta_i - (\bar{y}_j - \theta_j)|}{\hat{\sigma}\sqrt{2/n_0}} \le q_\alpha, \quad \forall i > j\right\} = 1 - \alpha.$$

Then q_α is also the solution of the following equation

$$k\int_0^\infty \int_{-\infty}^{+\infty} [\Phi(z) - \Phi(z - \sqrt{2}q_\alpha s)]^{k-1} d\Phi(z) g(s) ds = 1 - \alpha,$$

where $\Phi(\cdot)$ denotes the standard normal distribution function, and $g(s)$ denotes the density function of $\hat{\sigma}/\sigma$. Let ν be the degree of freedom of $\hat{\sigma}^2$. So q_α depends on k, ν and α. We can obtain an interval estimate with the confidence level exactly $1-\alpha$.

Theorem 4.2.2. *For any $\theta_1, ..., \theta_k$ and σ^2, we have*

$$P\{\theta_i - \theta_j \in E_{ij},\ \forall\ i \neq j\} = 1 - \alpha,$$

where E_{ij} is given by Eq. (4.2.10).

Proof.

$$
\begin{aligned}
& P\Big\{\theta_i - \theta_j \in E_{ij},\ \forall i \neq j\Big\} \\
&= \sum_{i=1}^{k} P\Big\{ -\hat{\sigma}\sqrt{2/n_0}q_\alpha < \bar{Y}_i - \bar{Y}_j - (\theta_i - \theta_j) < \hat{\sigma}\sqrt{2/n_0}q_\alpha, \quad \forall j \neq i \\
&\qquad \text{and} \quad \bar{Y}_i - \theta_i = \max(\bar{Y}_l - \theta_l)\Big\} \\
&= \sum_{i=1}^{k} P\Big\{0 < \bar{Y}_i - \bar{Y}_j - (\theta_i - \theta_j) < \hat{\sigma}\sqrt{2/n_0}q_\alpha, \quad \forall j \neq i\Big\} \\
&= \sum_{i=1}^{k} P\Big\{0 < \sqrt{n_0}(\bar{Y}_i - \theta_i)/\sigma - \sqrt{n_0}(\bar{Y}_j - \theta_j)/\sigma < \sqrt{2}q_\alpha\hat{\sigma}/\sigma, \quad \forall j \neq i\Big\} \\
&= kP\Big\{0 < \sqrt{n_0}(\bar{y}_1 - \theta_1)/\sigma - \sqrt{n_0}(\bar{y}_j - \theta_j)/\sigma < \sqrt{2}q_\alpha\hat{\sigma}/\sigma,\ j = 2, \cdots, k\Big\} \\
&= k\int_0^{\infty}\int_{-\infty}^{+\infty} [\phi(z) - \phi(z - \sqrt{2}q_\alpha s)]^{k-1} d\Phi(z)\ g(s)ds \\
&= 1 - \alpha.
\end{aligned}
$$

□

Example 4.2.2. Consider the interval estimate based on the data in Example 4.1.1, where $k = 4$ and $v = 12$. Note that when $\alpha = 0.05$, we have $q_\alpha = 2.97$. Refer to the computational result in Example 4.2.1, $\hat{\sigma}\sqrt{2/n_0}q_\alpha = 44.6$. Then the interval estimate based on Tukey's method is $\bar{y}_i - \bar{y}_j \pm 44.6$. Comparing with (4.2.10), the interval has a shorter length, or equivalently, a higher accuracy.

When the **sample sizes are unequal**, Tukey (1953) and Kramer (1956) constructed an interval estimate similar to that given by (4.2.6) or (4.2.10):

$$D_{ij} = \Big[\bar{y}_i - \bar{y}_j - (\hat{\sigma}/c_{ij})q_\alpha,\ \bar{y}_i - \bar{y}_j + (\hat{\sigma}/c_{ij})q_\alpha\Big],$$

where c_{ij} is given by (4.2.5), and conjectured that it is a confidence interval with confidence level $1 - \alpha$ (see Tukey, 1994). Hayter (1984) proved the following theorem, consequently the Tukey's conjecture.

Theorem 4.2.3. *For any* $\theta_1, \cdots, \theta_k$ *and* σ^2,

$$P\{\theta_i - \theta_j \in D_{ij}, \quad \forall\ i \neq j\} \geq 1 - \alpha,$$

the strict equality holds if and only if $n_1 = \cdots = n_k$.

Now let us go back to the problem of testing (4.2.2), corresponding to interval estimate, the test statistic in the Tukey's method is

$$Q(\boldsymbol{y}) = \max \frac{|\bar{y}_i - \bar{y}_j|}{\sqrt{2/n_0}\ \hat{\sigma}}.$$

If $Q(\boldsymbol{y}) \geq q_\alpha$, reject H_0. We can use c_{ij} instead of $\sqrt{n_0/2}$ if the sample sizes are unequal. From (4.2.5), we can see that the test statistic in the Tukey's method coincides with that in the U-I test (see the discussion following Eq. (2.5.5) in Section 2.5).

4.2.2 *Generalized Multiple Comparisons*

Now consider the problem of testing (4.2.3). Under the assumptions of the one-factor model (4.1.1),

$$\frac{\sum_{i=1}^{k} a_i \bar{y}_i - \sum_{i=1}^{k} a_i \theta_i}{\sqrt{\sum_{i=1}^{k} a_i^2/n_i}\ \sigma} \sim N(0,1),$$

and

$$\frac{(n-k)\hat{\sigma}^2}{\sigma^2} \sim \chi^2(n-k).$$

Furthermore, the two r.v.'s given in the above two equations are mutually independent, thus we can construct a test statistic as follows

$$T_{\boldsymbol{a}}(\boldsymbol{y}) = \frac{\sum_{i=1}^{n} a_i \bar{y}_i}{\sqrt{\sum_{i=1}^{n} a_i^2/n_i}\ \hat{\sigma}}. \tag{4.2.11}$$

If $H_0^{\boldsymbol{a}}$ given by (4.2.3) is true, $T_{\boldsymbol{a}}(\boldsymbol{Y}) \sim t(n-k)$. If $|T_{\boldsymbol{a}}(\boldsymbol{y})| \geq t_{\alpha/2}(n-k)$, then reject $H_0^{\boldsymbol{a}}$.

Let $\mathbf{1} = (1, \cdots, 1)'$ be a vector of all ones in $\boldsymbol{R}^k$, $\boldsymbol{R}_{\mathbf{1}}^{k-1}$ be the orthogonal complement space of $\mathbf{1}$, so $\boldsymbol{a}$ is a vector belonged to $\boldsymbol{R}_{\mathbf{1}}^{k-1}$. Recall Example 3.3.2, it is easy to verify that $T_{\boldsymbol{a}}(\boldsymbol{y})$ is a UMPU test for the problem of testing

$$H_0^{\boldsymbol{a}} : \sum_{i=1}^{k} a_i \theta_i = 0 \quad \text{vs.} \quad H_1^{\boldsymbol{a}} : \boldsymbol{\theta} \in \eta \boldsymbol{a}\ (\eta \neq 0),$$

where $\eta \boldsymbol{a}$ denotes a straight line containing the vector $\boldsymbol{a}$ in $\boldsymbol{R}_{\mathbf{1}}^{k}$.

Example 4.2.3. (Linear combination problem) For the data in Example 4.1.2 in the last section, consider the problem of testing

$$H_0^{\boldsymbol{a}}: \frac{1}{3}(\theta_1+\theta_2+\theta_4)=\theta_3 \quad \text{vs.} \quad H_1^{\boldsymbol{a}}: \frac{1}{3}(\theta_1+\theta_2+\theta_4)\neq\theta_3$$

where $\boldsymbol{a} = (1/3, 1/3, -1, 1/3)'$. As far as the obese Zucker rats are concerned, we can obtain that

$$|T_{\boldsymbol{a}}(\boldsymbol{y})| = 3.46 > 3.055 = t_{0.01/2}(12),$$

which rejects $H_0^{\boldsymbol{a}}$ at a very significant level. This means that the amount of food intake of the obese Zucker rats at the third phase is obviously larger than those at other phases. Hence the third phase is the strongest period in terms of physique for the Zucker rats.

Now let us consider the problem of testing (4.2.4) based on the idea of the U-I test. This kind of problem is proposed originally by Scheffé (1953).

Scheffé's method. Based on the idea of the U-I test, the test statistic for the problem of testing (4.2.4) should be

$$T^2(\boldsymbol{y}) = \max_{\boldsymbol{a}\in A} T_{\boldsymbol{a}}^2(\boldsymbol{y}), \tag{4.2.12}$$

where $T_{\boldsymbol{a}}(\boldsymbol{y})$ is given by Eq. (4.2.12). We must obtain the distribution of $T^2(\boldsymbol{Y})$ when H_0 is true. For this purpose, we give the following lemma at first.

Lemma 4.2.1. *For any given* $(v_1, \cdots, v_k)' \in \boldsymbol{R}^k$, *we have*

$$\max_{\boldsymbol{a}\in A} \frac{(\sum_{i=1}^k a_i v_i)^2}{\sum_{i=1}^k a_i^2/n_i} = \sum_{i=1}^k (v_i - \bar{v})^2, \tag{4.2.13}$$

where $\bar{v}$ *denotes a weighted mean, i.e.* $\bar{v} = \sum_{i=1}^k n_i v_i / \sum_{i=1}^k n_i$.

Proof. Let $b_i = a_i \Big/ \sqrt{\sum_{i=1}^k a_i^2/n_i}$. The left side of (4.2.13) can be written as $(\sum_{i=1}^k b_i v_i)^2$. Meantime, we have $\sum_{i=1}^k b_i = 0$ and $\sum_{i=1}^k b_i^2/n_i = 1$. Furthermore

$$\frac{1}{n^2}\left[\sum_{i=1}^k b_i v_i\right]^2 = \frac{1}{n^2}\left[\sum_{i=1}^k b_i(v_i-\bar{v})\right]^2 = \left[\sum_{i=1}^k \frac{b_i}{n_i}(v_i-\bar{v})\frac{n_i}{n}\right]^2, \tag{4.2.14}$$

here n_i/n, $i = 1, \cdots, k$ can be regarded as a discrete probability distribution, then by the Cauchy-Schwarz Inequality, we have

$$\left[\sum_{i=1}^k \frac{b_i}{n_i}(v_i-\bar{v})\frac{n_i}{n}\right]^2 \le \sum_{i=1}^k \left[\left(\frac{b_i}{n_i}\right)^2 \frac{n_i}{n}\right] \sum_{i=1}^k \left[(v_i-\bar{v})^2 \frac{n_i}{n}\right].$$

From $\sum_{i=1}^k b_i^2/n_i = 1$ and Eq. (4.2.14),

$$\left[\sum_{i=1}^k b_i v_i\right]^2 \le \sum_{i=1}^k n_i(v_i - \bar{v})^2. \tag{4.2.15}$$

Now take $a_i = n_i(v_i - \bar{v})$, then $\sum_{i=1}^k a_i = 0$, namely, $\boldsymbol{a} \in A$. Then we have

$$b_i = \frac{n_i(v_i - \bar{v})}{\sqrt{\sum_{i=1}^k n_i(v_i - \bar{v})^2}},$$

$$\left[\sum_{i=1}^k b_i v_i\right]^2 = \sum_{i=1}^k n_i(v_i - \bar{v})^2,$$

which attains the maximum of right-hand side of (4.2.15). This completes the proof of (4.2.13). □

From the above lemma, the test statistic given by (4.2.12) can be written as

$$T^2(\boldsymbol{y}) = \frac{1}{\hat{\sigma}^2}\sum_{i=1}^k n_i(\bar{y}_i - \bar{y})^2. \tag{4.2.16}$$

This result coincides with the F-test given by Eq. (4.1.10) in the last section.

Theorem 4.2.4. *If the null hypothesis given in* (4.2.4) *is true, then*

$$T^2(\boldsymbol{Y}) \sim (k-1)F(k-1, n-k),$$

where $T^2(\boldsymbol{y})$ *is the U-I test for the problem of testing* (4.2.4).

Similarly, we can also consider the interval estimation problem. For $\boldsymbol{a} \in A$, let

$$f_\alpha = \sqrt{(k-1)F_\alpha(k-1, n-k)},$$

then the interval estimator for $\sum_{i=1}^k a_i\theta_i$ is

$$E_{\boldsymbol{a}} = \left[\sum_{i=1}^k a_i\bar{y}_i - \hat{\sigma} f_\alpha\sqrt{\sum_{i=1}^k \frac{a_i^2}{n_i}},\ \sum_{i=1}^k a_i\bar{y}_i + \hat{\sigma} f_\alpha\sqrt{\sum_{i=1}^k \frac{a_i^2}{n_i}}\right].$$

It can be verified that this interval is also a confidence interval with confidence level $1-\alpha$, *i.e.*

$$P\left\{\sum_{i=1}^k a_i\theta_i \in E_{\boldsymbol{a}},\ \forall \boldsymbol{a} \in A\right\} = 1 - \alpha.$$

Table 4.2.1 Comparing the Tukey's method with the Scheffé's method ($\alpha = 0.05$)

	$k=2$		$k=4$		$k=6$	
v	q_α	f_α	q_α	f_α	q_α	f_α
6	2.45	2.45	3.46	3.79	3.98	4.64
12	2.18	2.18	2.97	3.22	3.36	3.93
18	2.10	2.10	2.83	3.07	3.18	3.68

Comparing with Theorem 4.2.2, as the scope of the Scheffé's method is much more general, the interval given by $E_{\boldsymbol{a}}$ is longer than that given by E_{ij} when we choose $\boldsymbol{a}$ in a paired way. In fact, when $n_1 = \cdots = n_k = n_0$, take $\boldsymbol{a} = (-1, 1, 0, \cdots 0)'$, it suffices to compare q_α with f_α. Table 4.2.1 shows the difference between them.

It can be seen that the length difference between the two kinds of intervals increases gradually with the increasing of k. However, compared with the Tukey's method, the Scheffé's method is applicable for more situations, such as the situation of paired-parameter test with unequal sample sizes.

4.2.3 *The False Discovery Rate Procedure*

As we have known, a Type I error is often considered to be more serious, and therefore more important to avoid, than a Type II error. The hypothesis test procedure is therefore adjusted so that there is a guaranteed 'low' probability of rejecting the null hypothesis wrongly; this probability is never 0. In a multiple hypothesis test, the Type I error can be reduced by selecting an *a-priori* significance value (say, α) to determine statistical significance of individual hypothesis tests. However, the experiment-wise (EW) Type I error is quickly increased at the rate of $1-(1-\alpha)^k$, where k is the number of hypothesis tests performed.

Frequently, Bonferroni corrections (Miller, 1981, pp. 67-70) have been used to control experiment-wise α (denoted by α_{EW}) at a predetermined level using the formula: $\alpha_{EW} = \alpha/k$. Bonferroni corrections are effective at controlling α_{EW}. Furthermore, they are often simple to use in application, require no distributional assumptions and enable individual alternative hypotheses to be identified. Nevertheless, the correction is very conservative and power (the proportion of the false null hypotheses that are correctly rejected) is greatly reduced, especially when several highly-correlated tests are undertaken (Hochberg, 1988; Simes, 1986). While many studies focus on reducing Type I error with Bonferroni corrections, reducing Type II er-

Table 4.2.2 Number of errors committed when testing k null hypotheses

	Decision		Total
	Not significant	Significant	
True null hypotheses	U	V	k_0
Non-true null hypotheses	T	S	$k-k_0$
Total	$k-R$	R	k

ror (or increasing power) is often neglected in multiple comparison tests. Sequential Bonferroni corrections (Holm, 1979; Hochberg, 1988) provide some improvements to power, but assume independence among tests.

Benjamini and Hochberg (1995) presented an alternative multiple testing procedure called False Discovery Rate (FDR). Assume that k_0 of the k null hypotheses are true. Let V denote the number of true null hypotheses rejected (*i.e.*, the number of Type I errors) and R the total number of hypotheses rejected. Table 4.2.2 summarizes the possible outcomes of individual tests. Note that T is the number of Type II errors, and only k, R and $k-R$ are observed while the other variables are unknown. Now, let Q be the unobservable random quotient,

$$Q = \begin{cases} V/R, & \text{if} \quad R > 0, \\ 0, & \text{otherwise.} \end{cases}$$

Then the FDR is simply defined to be the expectation of Q. Their procedure calls for controlling the expected proportion of falsely rejected hypotheses rather than controlling all falsely rejected hypotheses as in α_{EW}.

The FDR procedure presented by Benjamini and Hochberg (1995) is performed as follows:

(1) Order p-values $p_{(1)} \leq \cdots \leq p_{(k)}$, where k is the number of pairwise tests;
(2) Starting with the largest p-value, find the first individual p-value, say $p_{(i)}$, that satisfies: $p_{(i)} \leq i/k \times \alpha$, where i denotes the i-th observation;
(3) The $p_{(i)}$ that satisfies the condition above becomes the critical value for the experiment.

The gain in power with the Benjamini-Hochberg method FDR over the Bonferroni procedure are substantial (Narum, 2006). By now, many modifications of the Benjamini-Hochberg method FDR have been presented (*e.g.*, Storey, 2002; Bickel, 2004; Benjamini and Yekutieli, 2001), some of which have been used successfully in situations with very large numbers of pairwise tests such as microarray gene expression (Reiner *et al.*, 2003).

4.3 One-Factor Model: Trend Tests

From Table 4.1.1, the amount of food intake of the Zucker rats increases gradually with the age increasing in weeks (discussion for the decreasing case with $i = 4$ is delayed to the last part of this section). This implies that the mean of $\bar{y}_i$ may increase gradually. Let us analyze (4.1.1) a bit more closely. Usually, let D_i denote age in the behavior analysis, dose in drug-effect analysis, and time in economic analysis respectively. Hence, under all these situations, they can be expressed in a uniform way, $D_1 < D_2 < \cdots < D_k$, and their corresponding parameters can usually be expressed as

$$H : \theta_1 \leq \cdots \leq \theta_k.$$

In testing analysis, we should consider this information sufficiently. Thus an appropriate testing problem should be formulated as

$$H_0 : \theta_1 = \cdots = \theta_k \quad \text{vs.} \quad H_1 : H - H_0, \tag{4.3.1}$$

where $H - H_0$ denotes that there exist $i < j$ such that $\theta_i < \theta_j$. This problem is usually called **trend test problem** or **order restricted test problem**, especially H is called **simple order restriction**.

First let us show that the testing problem given by (4.3.1) is one-sided, *i.e.* the parameter space corresponding H_1 is a convex set. Let $\Theta = \{\boldsymbol{\theta} \in \boldsymbol{R}^k;\ \theta_1 \leq \cdots \leq \theta_k\}$. Let

$$\begin{aligned} \boldsymbol{a}_1 &= (-1, 1, 0, \cdots, 0, 0)' \\ \boldsymbol{a}_2 &= (0, -1, 1, \cdots, 0, 0)' \\ &\vdots \\ \boldsymbol{a}_{k-1} &= (0, 0, 0, \cdots, -1, 1)', \end{aligned}$$

then Θ can be written as

$$\Theta = \{\boldsymbol{\theta} \in \boldsymbol{R}^k;\ \boldsymbol{a}_i'\boldsymbol{\theta} \geq 0,\ i = 1, \cdots, k-1\}. \tag{4.3.2}$$

Thus Θ is a polyhedral convex cone. We can construct several statistics to study this problem.

4.3.1 *Linearity Tests*

4.3.1.1 *Mantel test*

As the testing problem given by (4.3.1) has too many parameters and is very difficult to analyze, Mantel (1963) proposed a method to reduce parameters: Assume that all the parameters obey a linear increasing law, that is, let

$$\theta_i = \alpha + \beta t_i, \quad i = 1, \cdots, k, \tag{4.3.3}$$

where $\beta \geq 0$, but the t_i are given constants satisfying $t_1 \leq \cdots \leq t_k$. Under this model, the problem can be transformed into

$$H_0 : \beta = 0 \quad \text{vs.} \quad H_1 : \beta > 0. \tag{4.3.4}$$

It is easy to verify that the MLEs of α and β are the solutions of the following equation

$$\min_{\alpha,\beta} \sum_{i=1}^{k} \sum_{j=1}^{n_i} (y_{ij} - \alpha - \beta t_i)^2 = \min_{\alpha,\beta} \sum_{i=1}^{k} (\bar{y}_i - \alpha - \beta t_i)^2 \omega_i$$

where $\omega_i = n_i$ and $\bar{y}_i = \sum_{j=1}^{n_i} y_{ij}/n_i, i = 1, \cdots, k$. Let $\hat{\alpha}$ and $\hat{\beta}$ be the MLEs of α and β, then we have

$$\hat{\alpha} = \bar{y} - \hat{\beta}\bar{t}$$

$$\hat{\beta} = \frac{\sum\limits_{i=1}^{k} (t_i - \bar{t})(\bar{y}_i - \bar{y})\omega_i}{\sum\limits_{i=1}^{k} (t_i - \bar{t})^2 \omega_i},$$

where $\bar{y}$ and $\bar{t}$ are two weighted means, that is, $\bar{y} = \sum_{i=1}^{k} \omega_i \bar{y}_i / \sum_{i=1}^{k} \omega_i$ and $\bar{t} = \sum_{i=1}^{k} \omega_i t_i / \sum_{i=1}^{k} \omega_i$. Then the estimators of (4.1.1) are $\hat{y}_i = \hat{\alpha} + \hat{\beta} t_i, i = 1, \cdots, k$. The sum of residual squares or error squares is

$$\begin{aligned} R^2 &= \sum_{i=1}^{k} (\bar{y}_i - \hat{\alpha} - \hat{\beta} t_i)^2 \omega_i \\ &= \sum_{i=1}^{k} (\bar{y}_i - \bar{y})^2 \omega_i - \hat{\beta}^2 \sum_{i=1}^{k} (t_i - \bar{t})^2 \omega_i. \end{aligned} \tag{4.3.5}$$

For $\theta \in \Theta$, $E_\theta R^2 = (k-1)\sigma^2 - \sigma^2 = (k-2)\sigma^2$, therefore,

$$\hat{\sigma}^2 = \frac{1}{k-2} R^2$$

is an unbiased estimator for σ^2. Note that under H_0 the MLE of α is $\bar{y}$. By (4.3.5),

$$\sum_{i=1}^{k} (\bar{y}_i - \bar{y})^2 \omega_i - \sum_{i=1}^{k} (\bar{y}_i - \hat{\alpha} - \hat{\beta} t_i)^2 \omega_i = \hat{\beta}^2 \sum_{i=1}^{k} (t_i - \bar{t})^2 \omega_i. \tag{4.3.6}$$

Now the LR statistic for the problem of testing (4.3.4) is

$$t(y) = \frac{\hat{\beta}}{\hat{\sigma}} \sqrt{\sum_{i=1}^{k} (t_i - \bar{t})^2 \omega_i}. \tag{4.3.7}$$

Since (4.3.6) denotes a direct sum decomposition, $\hat{\beta}$ and $\hat{\sigma}$ are independent. When H_0 is true, $(\hat{\beta}/\sigma)\sqrt{\sum_{i=1}^{k}(t_i - \bar{t})^2\omega_i} \sim N(0,1)$, but $\hat{\sigma}^2/\sigma^2$ has a χ^2 distribution with $k-2$ degrees of freedom. Therefore, we have the following theorem.

Theorem 4.3.1. *For the problem of testing* (4.3.4), *when* H_0 *is true, the LR statistic given by* (4.3.7), $t(Y)$, *has a Student's t-distribution with* $k-2$ *degrees of freedom. A level-*α *LR test rejects* H_0 *if* $t(y) \geq t_\alpha(k-2)$, *where* $t_\alpha(k-2)$ *is the* α*-quantile of a Student's t-distribution with* $k-2$ *degrees of freedom.*

Example 4.3.1. Let us apply the Mantel test to analyze the problem discussed in Example 4.1.2. Let $t_i = i,\ i = 1, \cdots, 4$. We can obtain that the estimators of β of the obese and lean Zucker rats are $\hat{\beta}^O = 3.2$ and $\hat{\beta}^L = 2.8$ respectively. Then the corresponding t-values are $t^O = 0.32$ and $t^L = 0.28$ respectively. In this case, the degrees of freedom is 2, for size $\alpha = 0.05$, the upper α quantile of a Student's t-distribution is 2.9. So both of them can not reject the null H_0, given by (4.3.4).

Albeit the number of parameters is reduced by applying (4.3.3), it is very difficult to determine the constants $t_i = i, i = 1, \cdots, k$. From (4.3.7) the constants t_i will affect the computation of test statistics. In application, there are usually two methods to treat t_i: in the analysis of drug data we can substitute drug dose for t_i; if there is no more prior information, just like Example 4.3.1, let $t_i = i, i = 1, \cdots, k$. Now we discuss an optimization method.

4.3.1.2 *The best linearity test*

This coincides the method discussed in Section 3.4. For discussion simplicity, consider the case with σ^2 known. Without the loss of generality, let $\sigma^2 = 1$. For $i = 1, \cdots, k$, let

$$c_i = \frac{t_i - \bar{t}}{\sqrt{\sum_{j=1}^{k}(t_j - \bar{t})^2 \omega_j}},$$

so $\sum_{i=1}^k c_i\omega_i = 0$, and $\sum_{i=1}^k c_i^2\omega_i = 1$. Then the LR statistic given by Eq. (4.3.7) can be written as

$$T_{\boldsymbol{c}} = \sum_{i=1}^{k} c_i\bar{y}_i\omega_i, \tag{4.3.8}$$

where $\boldsymbol{c} = (c_1, \cdots, c_k)'$. Note that $c_1 \le \cdots \le c_k$, collect all the $\boldsymbol{c}$'s satisfying the above conditions, we can construct the following set

$$S = \{\boldsymbol{c} \in \boldsymbol{R}^k;\ \sum_{i=1}^{k} c_i\omega_i = 0, \sum_{i=1}^{k} c_i^2\omega_i = 1, c_1 \le \cdots \le c_k\}. \tag{4.3.9}$$

Then the coefficient in Mantel test must be a point in S. Let us look for the optimal coefficient.

Now let us return the problem of testing (4.3.1). It is easy to verify that, for any $\boldsymbol{\theta} \in \boldsymbol{R}^k$ satisfying $\theta_1 = \cdots = \theta_k$, $T_c \sim N(0,1)$. So $\boldsymbol{T}_c$ is the UMP test for

$$H_0 : \theta_1 = \cdots = \theta_k = 0 \quad \text{vs.} \quad H_{\boldsymbol{c}} : \boldsymbol{\theta} \in \eta\boldsymbol{c} \quad (\eta > 0). \tag{4.3.10}$$

For the given level α and $\forall\ \boldsymbol{\theta} \in S$, we can obtain the power function of $T_{\boldsymbol{c}}$

$$\beta(\boldsymbol{c}, \boldsymbol{\theta}) = 1 - \Phi\left(u_\alpha - \sum_{i=1}^{k} c_i\theta_i\omega_i\right).$$

Similar to the discussion in Section 3.4, the linear test statistic $T_{\boldsymbol{c}_0}$ is called as **the best linear test** if its minimal power attains the maximum, that is, $\boldsymbol{c}_0 \in S$ satisfies

$$\min_{\boldsymbol{\theta}\in S} \beta(\boldsymbol{c}_0, \boldsymbol{\theta}) = \max_{\boldsymbol{c}\in S} \min_{\boldsymbol{\theta}\in S} \beta(\boldsymbol{c}, \boldsymbol{\theta}),$$

or equivalently

$$\min_{\boldsymbol{\theta}\in S} \sum_{i=1}^{k} c_{0i}\theta_i\omega_i = \max_{\boldsymbol{c}\in S} \min_{\boldsymbol{\theta}\in S} \sum_{i=1}^{k} c_i\theta_i\omega_i. \tag{4.3.11}$$

Now we employ Eq. (4.3.11) to solve the $\boldsymbol{c}_0$.

Since S can be written as (4.3.2) and form a polyhedral convex cone, the **nonnegative linear combination** of any number of vectors in S still belongs to S, *i.e.* for $\boldsymbol{b}, \boldsymbol{c} \in S$, $\lambda_1 \ge 0$ and $\lambda_2 \ge 0$, $\lambda_1\boldsymbol{b} + \lambda_2\boldsymbol{c} \in S$. In fact, S can be considered to be generated by edge vector. A vector $\boldsymbol{e}$ is called an **edge vector** in S, if $\boldsymbol{e}$ can not be expressed as a nonnegative linear combination of more than two vectors in S that are not collinear with $\boldsymbol{e}$. The half-line in S containing the edge vector is called an **edge**. It can be verified that there are $k-1$ edge vectors in S: let $\boldsymbol{c} \in S$, then there are $k-1$ inequalities among the components of $\boldsymbol{c}$, where the edge vector should satisfy that there are $k-2$ equalities and a strict inequality among the components. By computation, the $k-1$ edge vectors are

$$\boldsymbol{e}^{(m)} = \lambda_m(e_{m1}, \cdots, e_{mk})', \quad m = 1, \cdots, k-1,$$

where $e_{mi} = -1/s_m$ when $i < m$; $e_{mi} = 1/(s_k - s_m)$ when $i > m$; $s_m = \omega_1 + \cdots + \omega_m$ and $s_k = \omega_1 + \cdots + \omega_k$; λ_m is a normalized parameter to ensure that $\boldsymbol{e}^{(m)} \in S$. Let $\boldsymbol{b}, \boldsymbol{c} \in S$, use $A(\boldsymbol{b}, \boldsymbol{c})$ to denote the angle between the vectors $\boldsymbol{b}$ and $\boldsymbol{c}$. Define $\cos A(\boldsymbol{b}, \boldsymbol{c}) = \sum_{i=1}^{k} b_i c_i \omega_i$. Since the cosine function is a monotone decreasing function of angle, for any given vector $\boldsymbol{c}$ in S, the vector $\boldsymbol{b}$ that can ensure the angle to attain its maximum must lie on some edge. Therefore Eq. (4.3.11) can be rewritten as

$$\min_{1 \le m \le k-1} \sum_{i=1}^{k} c_{i0} e_i^{(m)} \omega_i = \max_{c \in S} \min_{1 \le m \le k-1} \sum_{i=1}^{k} c_i e_i^{(m)} \omega_i. \tag{4.3.12}$$

It can be verified that if a vector $\boldsymbol{a} \in S$ and can be expressed as a positive combination of edge vectors, that is,

$$\boldsymbol{a} = r_1 \boldsymbol{e}^{(1)} + \cdots + r_{k-1} \boldsymbol{e}^{(k-1)}, \tag{4.3.13}$$

for $r_i > 0, i = 1, \cdots, k-1$, and satisfies

$$\sum_{i=1}^{k} a_i e_i^{(1)} \omega_i = \cdots = \sum_{i=1}^{k} a_i e_i^{(k-1)} \omega_i, \tag{4.3.14}$$

then the vector $\boldsymbol{a}$ satisfies Eq. (4.3.12). Solving Eqs. (4.3.13) and (4.3.14) we can obtain $\boldsymbol{c}_0 = (c_{01}, \cdots c_{0k})'$,

$$c_{0i} = \lambda \left(\sqrt{s_{i-1}(s_i - s_{i-1})} - \sqrt{s_i(s_k - s_i)} \right) / \omega_i, \tag{4.3.15}$$

$i = 1, \cdots, k$, where $s_0 = 0$ and λ is a normalized coefficient to ensure that $\boldsymbol{c}_0 \in s$. So when σ^2 is unknown we can obtain the following theorem.

Theorem 4.3.2. *For the problem of testing* (4.3.1), *the best linear test is*

$$T_{\boldsymbol{c}_0}(\boldsymbol{y}) = \frac{1}{\hat{\sigma}} \sum_{i=1}^{k} c_{0i} y_i \omega_i, \tag{4.3.16}$$

where $\boldsymbol{c}_0$ is given by Eq. (4.3.15), and $\hat{\sigma}^2$ is an unbiased estimator for σ^2, that is, $\hat{\sigma}^2 = {Q_1}^2/(n-k)$, where Q_1^2 is given by Eq. (4.1.5). When H_0 is true, $T_{\boldsymbol{c}_0}(\boldsymbol{Y})$ has a Student's t-distribution with $n-k$ degrees of freedom.

Example 4.3.2. (Continuation of Example 4.3.1) By Eq. (4.3.15) we can obtain $\boldsymbol{c}_0 = (-0.7, -0.1.0.1, 0.7)'$. By Eq. (4.3.16), $T^O_{\boldsymbol{c}_0} = 0.75$ and $T^L_{\boldsymbol{c}_0} = -0.48$, we cannot reject H_0 given by (4.3.1) in either case. In fact, from Table 4.1.3 we can see that the amount of food intake of both kinds of Zucker rats attains the maximum simultaneously at $i = 3$, namely, at the age of 30 weeks. Therefore the alternative hypothesis in (4.3.1) should be

$$H_1^u : \theta_1 \le \theta_2 \le \theta_3 \ge \theta_4.$$

This kind of restriction is called **umbrella order restriction**. More generally, we introduce

$$H_1^u : \theta_1 \le \cdots \le \theta_p \ge \cdots \ge \theta_k,$$

where the highest point is at p, and θ_p is called a **peak**. For the testing problems under this restriction, the best linear test is given by Shi (1988b), the corresponding optimal coefficient $\boldsymbol{c}_0 = (c_{01}, \cdots, c_{0k})'$ satisfies

$$\begin{aligned} c_{0i} &= \lambda(\sqrt{s_{i-1}(s_k - s_{i-1})} - \sqrt{s_i(s_k - s_i)})/\omega_i \qquad &\text{for } i < p,\\ c_{0p} &= \lambda(\sqrt{s_{p-1}(s_k - s_{p-1})} + \sqrt{s_p(s_k - s_p)})/\omega_p \qquad &\text{for } i = p,\\ c_{0i} &= \lambda(\sqrt{s_i(s_k - s_i)} - \sqrt{s_{i-1}(s_k - s_{i-1})})/\omega_i \qquad &\text{for } i > p, \end{aligned}$$

where $s_0 = 0$ and λ is a normalized coefficient. For this example, we can obtain

$$\boldsymbol{c}_0 = (-0.38, -0.06, 0.82, -0.38)'.$$

By Eq. (4.3.16), we can obtain $T^O_{\boldsymbol{c}_0}(y) = 3.49$ and $T^L_{\boldsymbol{c}_0} = 4.28$. The degrees of freedom are $n - k = 12$. When $\alpha = 0.01$, $t_\alpha(12) = 2.68$, then the two tests reject H_0 given by (4.3.1) significantly. It is meaningful to compare this result with those in Example 4.1.2, 4.3.1 and the first half part of this example. When the alternatives are different, the analysis results will also be different. Consequently, it is very important to construct a "reasonable" testing problem for further statistical inference. The so-called "reasonableness" means that the construction of testing problem must correspond with the background of problems. Furthermore, from this problem we can also see that, the Student's t-distribution in Theorem 4.3.2 has a much bigger degrees of freedom than that in Theorem 4.3.1, this is because Theorem 4.3.2 does not assume the model (4.3.3). **Albeit a simple model will simplify the computation, it can affect the accuracy of analysis.**

4.3.2 *Likelihood Ratio Test*

Consider the LR test of the problem of testing (4.3.1). Assume that the variances are equal. At first, consider the case when the **variances are known**. Then the log-likelihood function is

$$\begin{aligned} l(\boldsymbol{y};\boldsymbol{\theta}) &= -\sum_{i=1}^{k}\sum_{j=1}^{n_i}\frac{1}{2\sigma^2}(y_{ij}-\theta_i)^2+c \\ &= -\frac{1}{2}\sum_{i=1}^{k}(\bar{y}_i-\theta_i)^2\omega_i-\frac{1}{2\sigma^2}\sum_{i=1}^{k}\sum_{j=1}^{n_i}(y_{ij}-\bar{y}_i)^2+c, \end{aligned}$$

where c is a constant independent of $\boldsymbol{\theta}$ and $\omega_i = n_i/\sigma^2$. The MLE of parameter $\boldsymbol{\theta}$ under H_0 is

$$\bar{\boldsymbol{\theta}}_0 = (\bar{y},\cdots,\bar{y}),\ \text{where } \bar{y} = \sum_{i=1}^{k}\bar{y}_i\omega_i/\sum_{i=1}^{k}\omega_i. \tag{4.3.17}$$

Let the MLE of $\boldsymbol{\theta}$ under H_1 be $\hat{\boldsymbol{\theta}} = (\hat{\theta}_1,\cdots,\hat{\theta}_k)'$, then $\hat{\boldsymbol{\theta}} \in C$ and satisfies

$$\sum_{i=1}^{k}(\bar{y}_i-\hat{\theta}_i)^2\omega_i = \min_{\boldsymbol{\theta}\in C}\sum_{i=1}^{k}(\bar{y}_i-\theta_i)^2\omega_i, \tag{4.3.18}$$

where $C = \{\boldsymbol{\theta} \in \boldsymbol{R}^k;\ \theta_1 \leq \cdots \leq \theta_k\}$. Thus, the problem of looking for the MLE of $\boldsymbol{\theta}$ under H_1 is transformed into the limiting problem under the restriction. Kuhn-Tucker method is very efficient for this kind of problem.

4.3.2.1 *Kuhn-Tucker conditions*

Kuhn-Tucker method, is a special type of the Lagrange multiplier method. Let $h(\boldsymbol{y};\boldsymbol{\theta})$ be a function of $\boldsymbol{\theta}$ from $\boldsymbol{R}^n$ to $\boldsymbol{R}$, called **objective function**, where $\boldsymbol{y}$ is a given sample value. Let $h_j(\boldsymbol{\theta})$ be a function of $\boldsymbol{\theta}$, called **restriction function**, $j = 1,\cdots,m$. Now the problem is to solve the following equation

$$\min_{\boldsymbol{\theta}} h(\boldsymbol{y};\boldsymbol{\theta}), \tag{4.3.19}$$

where $\boldsymbol{\theta}$ satisfies restriction condition

$$h_j(\boldsymbol{\theta}) \leq 0, \quad j = 1,\cdots,m. \tag{4.3.20}$$

Kuhn and Tucker (1956) pointed out that the necessary condition for which $\boldsymbol{\theta}_0 \in \boldsymbol{R}^n$ is the solution of (4.3.19) and (4.3.20) is that the following four conditions hold:

(1) $\frac{\partial}{\partial\theta_i}h(\boldsymbol{y};\boldsymbol{\theta}_0)+\sum_{j=1}^{m}\lambda_j\frac{\partial}{\partial\theta_i}h_j(\boldsymbol{\theta}_0)=0, \quad i=1,\cdots,n;$
(2) $\lambda_j h_j(\boldsymbol{\theta}_0)=0, \quad j=1,\cdots,m;$
(3) $h_j(\boldsymbol{\theta}_0)\le 0, \quad j=1,\cdots,m;$
(4) $\lambda_j\ge 0, \quad j=1,\cdots,m,$

where $\boldsymbol{\lambda}=(\lambda_1,\cdots,\lambda_m)'$ is the Lagrange multiplier. Usually the above four conditions are called **Kuhn-Tucker conditions**. Specially, when both the objective and constraint functions are convex, Kuhn-Tucker Conditions are both necessary and sufficient, the solution of $\boldsymbol{\lambda}$ and $\boldsymbol{\theta}$ are unique. We can obtain some good algorithms if we can use the Kuhn-Tucker conditions skilly. Now let us solve (4.3.18).

4.3.2.2 *MLE under H_1*

Corresponding to (4.3.19) and (4.3.20), the objective function is $h(\boldsymbol{y};\boldsymbol{\theta})=\sum_{i=1}^{k}(\bar{y}_i-\theta_i)^2\omega_i$; the restriction function is $h_i(\boldsymbol{\theta})=\theta_i-\theta_{i+1}, i=1,\cdots,k-1$. Then we can obtain the following Lagrangian equation

$$H(\boldsymbol{y};\boldsymbol{\theta},\boldsymbol{\lambda})=\frac{1}{2}\sum_{i=1}^{k}(\bar{y}_i-\theta_i)^2\omega_i+\sum_{i=1}^{k-1}\lambda_i(\theta_i-\theta_{i+1}).$$

And the Kuhn-Tucker conditions are:

(1) $(\bar{y}_i-\theta_i)\omega_i-\lambda_i+\lambda_{i-1}=0, \quad \lambda_0=\lambda_k=0, \quad i=1,\cdots,k;$
(2) $\lambda_i(\theta_i-\theta_{i+1})=0, \quad i=1,\cdots,k-1;$
(3) $\theta_i-\theta_{i+1}\le 0, \quad i=1,\cdots,k-1;$
(4) $\lambda_i\ge 0, \quad i=1,\cdots,k-1.$

Since both the objective and the restriction functions are convex, the above conditions are both necessary and sufficient, and the solution is unique.

For computation simplicity, for $l\le s$, let

$$Av(l,s)=\sum_{i=l}^{s}\bar{y}_i\omega_i\Big/\sum_{i=l}^{s}\omega_i,$$

which is a partly weighted mean. Since the sample here is drawn from a continuous distribution, without the loss of generality, assume that if $s\ne s'$ then $Av(l,s)\ne Av(l,s')$. Let $\hat{\theta}_i$ and $\hat{\lambda}_i$ be the solutions of the Kuhn-Tucker conditions. By Condition (4), there exist $1\le i_1<i_2<\cdots<i_t<k, i_{t+1}=k$ satisfying

$$\begin{cases}\hat{\lambda}_j=0, & \text{for} \quad j=i_1,\cdots,i_t,i_{t+1},\\ \hat{\lambda}_j>0, & \text{otherwise.}\end{cases} \tag{4.3.21}$$

By Conditions (2) and (3) we know that $\hat{\boldsymbol{\theta}} = (\hat{\theta}_1, \cdots, \hat{\theta}_k)'$ satisfies

$$\hat{\theta}_1 = \cdots = \hat{\theta}_{i_1} < \hat{\theta}_{i_1+1} = \cdots = \hat{\theta}_{i_2} < \cdots < \hat{\theta}_{i_t+1} = \cdots = \hat{\theta}_k. \qquad (4.3.22)$$

By solving Condition (1), we can obtain

$$\hat{\theta}_j = Av(i_l + 1, i_{l+1}), \quad i_l + 1 \le j \le i_{l+1}; \; l = 0, 1, \cdots, t; \; i_0 = 0. \quad (4.3.23)$$

Therefore the key is to get a subindex set $\{i_1, \cdots, i_t\}$ satisfying (4.3.21) or (4.3.22).

Lemma 4.3.1. *Assume that i_1 is given by (4.3.21) or (4.3.22), then when $j \neq i_1$, we have*

$$Av(1, i_1) < Av(1, j). \qquad (4.3.24)$$

Proof. We have assumed that (4.3.24) excludes the possibility of equality. By Eq. (4.3.22), let $\hat{\theta}_1 = \cdots = \hat{\theta}_{i_1} = \hat{\theta}^{(1)}$, $\hat{\theta}_{i_1+1} = \cdots = \hat{\theta}_{i_2} = \hat{\theta}^{(2)}$. When $j < i_1$, take the summation of the first j terms in Condition (1), we have

$$\sum_{i=1}^{j} (\bar{y}_i - \hat{\theta}^{(1)})\omega_i - \lambda_j = 0.$$

By $\lambda_j > 0$ we can obtain $Av(1, j) > \hat{\theta}^{(1)} = Av(1, i_1)$; when $j > i_1$, without the loss of generality, let $i_1 < j \le i_2$. Applying the above method and Eq. (4.3.22) we can obtain $Av(i_1 + 1, j) \ge \hat{\theta}^{(2)} > \hat{\theta}^{(1)}$. Note that for positive numbers a, b, c, d, if $c/d > a/b$ then $(a+c)/(b+d) > a/b$. Thus we can obtain $Av(1, j) > Av(1, i_1)$. This completes the proof of (4.3.24). □

Applying the proof method of Lemma 4.3.1, we can determine $i_2, \cdots, i_t$. Thus we can give an algorithm to compute the MLE for θ.

Lower Set Algorithm.

`Step` 1, look for i_1 satisfying $Av(1, i_1) = \min\{Av(1, j); 1 \le j \le k\}$;

`Step` $l+1$, look for i_{l+1} satisfying $Av(i_l+1, i_{l+1}) = \min\{Av(i_l+1; j); i_l + 1 \le j \le k\}$. Repeat until getting the subindex set $\{i_1, \cdots, i_t\}$. The MLE for θ defined by (4.3.23) is $\hat{\theta} = (\hat{\theta}_1, \cdots, \hat{\theta}_k)'$.

For discussing the properties of the MLE, the following algorithm is also useful. Its proof can also be finished by using Lemma 4.3.1.

PAVA (Pool Adjacent Violators Algorithm).

`Step` 1, if $y \in C$, then $\hat{\theta} = y$. Otherwise,

`Step` 2, there exists an integer i such that $\bar{y}_i > \bar{y}_{i+1}$. Let $\bar{y}_{(i)} = Av(i, i+1), \omega_{(i)} = \omega_i + \omega_{i+1}$.

For $\bar{y}_1, \cdots, \bar{y}_{i-1}, \bar{y}_{(i)}, \bar{y}_{i+2}, \cdots, \bar{y}_k$ and the weight functions $\omega_1, \cdots, \omega_{i-1}, \omega_{(i)}, \omega_{i+2}, \cdots, \omega_k$, repeat the above two steps until getting the subindex set $B_1, \cdots, B_l$, satisfying

$$Av(B_1) < \cdots < Av(B_l),$$

where $Av(B_j) = \sum_{i \in B_j} \bar{y}_i \omega_i / \sum_{i \in B_j} \omega_i, j = 1, \cdots, l$. If $i \in B_j$, then

$$\hat{\theta}_i = Av(B_j), \ j = 1, \cdots, l.$$

Theorem 4.3.3. (Projection Theorem) *Let $\hat{\boldsymbol{\theta}} \in \boldsymbol{R}^k$. Then $\hat{\boldsymbol{\theta}}$ is the solution of Eq. (4.3.18) if and only if $\hat{\boldsymbol{\theta}} \in C$, and*

$$\sum_{i=1}^{k} (\bar{y}_i - \hat{\theta}_i)\hat{\theta}_i \omega_i = 0, \tag{4.3.25}$$

$$\sum_{i=1}^{k} (\bar{y}_i - \hat{\theta}_i)\theta_i \omega_i \leq 0, \quad \forall \boldsymbol{\theta} \in C. \tag{4.3.26}$$

Proof. If $\hat{\boldsymbol{\theta}}$ is the solution of Eq. (4.3.18), according to Eq. (4.3.22), $\hat{\theta}_i$ is a constant when $i \in \{i_l + 1, \cdots, i_{l+1}\}$, denoted by $\hat{\theta}^{(l+1)}, l = 0, 1, \cdots, t$, and $i_0 = 0$. Then Eq. (4.3.25) can be written as

$$\begin{aligned} \sum_{i=1}^{k} (\bar{y}_i - \hat{\theta}_i)\hat{\theta}_i \omega_i &= \sum_{l=0}^{t} \sum_{i=i_l+1}^{i_{l+1}} (\bar{y}_i - \hat{\theta}_i)\hat{\theta}_i \omega_i \\ &= \sum_{l=0}^{t} \hat{\theta}^{(l+1)} \sum_{i=i_l+1}^{i_{l+1}} (\bar{y}_i - \hat{\theta}^{(l+1)})\omega_i. \end{aligned}$$

From Eq. (4.3.23), we know that Eq. (4.3.25) holds. Next we prove (4.3.26). For any $\boldsymbol{\theta} \in C$, by the definition of convex cone, for any $\alpha > 0$ we have $\hat{\boldsymbol{\theta}} + \alpha\boldsymbol{\theta} \in C$. By Eq. (4.3.18),

$$\sum_{i=1}^{k} (\bar{y}_i - (\hat{\theta}_i + \alpha\theta_i))^2 \omega_i \geq \sum_{i=1}^{k} (\bar{y}_i - \hat{\theta}_i)^2 \omega_i.$$

Thus we can obtain that

$$\sum_{i=1}^{k} (\bar{y}_i - \hat{\theta}_i)\theta_i \omega_i \leq \frac{\alpha}{2} \sum_{i=1}^{k} \theta_i^2 \omega_i.$$

By the arbitrariness of $\alpha > 0$, (4.3.26) holds.

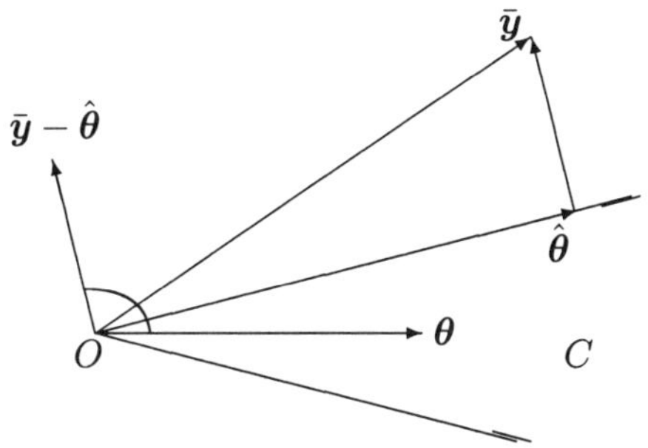

Fig. 4.3.1 Intuitive explanation of the Projection Theorem.

On the contrary, if $\hat{\boldsymbol{\theta}} \in C$ and satisfies (4.3.25) and (4.3.26). For any $\boldsymbol{\theta} \in C$ we have

$$\begin{aligned}\sum_{i=1}^{k}(\bar{y}_i-\theta_i)^2\omega_i &= \sum_{i=1}^{k}(\bar{y}_i-\hat{\theta}_i)^2\omega_i + 2\sum_{i=1}^{k}(\bar{y}_i-\hat{\theta}_i)(\hat{\theta}_i-\theta_i)\omega_i + \sum_{i=1}^{k}(\hat{\theta}_i-\theta_i)^2\omega_i \\ &\geq \sum_{i=1}^{k}(\bar{y}_i-\hat{\theta}_i)^2\omega_i,\end{aligned}$$

i.e. $\hat{\boldsymbol{\theta}}$ is the solution of Eq. (4.3.18). □

The above theorem gives a necessary and sufficient condition for the projection of a vector on to a polyhedral convex cone. Figure 4.3.1 gives its intuitive explanation. Let $\bar{\boldsymbol{y}} = (\bar{y}_1, \cdots, \bar{y}_k)'$. If $\bar{\boldsymbol{y}} \in C$, then $\hat{\boldsymbol{\theta}} = \bar{\boldsymbol{y}}$, and the conclusion holds obviously. If $\bar{\boldsymbol{y}} \notin C$, we can regard $\sum_{i=1}^{k} \bar{y}_i z_i \omega_i$ as an inner product, then $\hat{\boldsymbol{\theta}}$ is the projection of $\bar{\boldsymbol{y}}$ onto C. Hence, $\bar{\boldsymbol{y}}-\hat{\boldsymbol{\theta}}$ and $\hat{\boldsymbol{\theta}}$ are perpendicular which is equivalent to Eq. (4.3.25). Let C^* be a **dual cone** of C, that is,

$$C^* = \left\{ \boldsymbol{a} \in \boldsymbol{R}^k;\ \sum_{i=1}^{k} a_i\theta_i\omega_i \leq 0,\ \forall \boldsymbol{\theta} \in C \right\}.$$

Then $\bar{\boldsymbol{y}}-\hat{\boldsymbol{\theta}}$ is the projection to C^*. Note that the angle between any two vectors belonging to C and C^* respectively is bigger than $\pi/2$, thus (4.3.26) holds.

By (4.3.25) and (4.3.26), it is easy to obtain that for any $\boldsymbol{\theta} \in C$ we have

$$\sum_{i=1}^{k}(\bar{y}_i-\theta_i)^2\omega_i \geq \sum_{i=1}^{k}(\hat{\theta}_i-\theta_i)^2\omega_i.$$

Then for any $\boldsymbol{\theta} \in C$, we have

$$E_{\boldsymbol{\theta}}\left[\sum_{i=1}^{k}(\bar{y}_i - \theta_i)^2\omega_i\right] > E_{\boldsymbol{\theta}}\left[\sum_{i=1}^{k}(\hat{\theta}_i - \theta_i)^2\omega_i\right]. \tag{4.3.27}$$

Albeit $\hat{\boldsymbol{\theta}}$ is not an unbiased estimator for $\boldsymbol{\theta}$, the above equality shows that the MSE of $\hat{\boldsymbol{\theta}}$ is smaller than that of $\bar{y}_i$. Lee (1981) studied this difference and proved that (4.3.27) holds for any i, that is, $E_{\boldsymbol{\theta}}(\bar{y}_i - \theta_i)^2 > E_{\boldsymbol{\theta}}(\hat{\theta}_i - \theta_i)^2, i = 1, \cdots, k$. In many literatures, $\hat{\boldsymbol{\theta}}$ is called as the **isotonic regression** of $\bar{\boldsymbol{y}}$.

4.3.2.3 *Test statistic and its distribution under H_0*

Let us look for the LR test statistic for the problem of testing (4.3.1). As Shown by (4.3.17) and (4.3.18), let $\hat{\boldsymbol{\theta}}_0$ and $\hat{\boldsymbol{\theta}}$ be the MLEs under H_0 and H_1 respectively, then the LR test given by (4.3.16) is equivalent to

$$\begin{aligned}
-2(l(\boldsymbol{y};\hat{\boldsymbol{\theta}}_0) - l(\boldsymbol{y};\hat{\boldsymbol{\theta}})) &= \sum_{i=1}^{k}(\bar{y}_i - \bar{y})^2\omega_i - \sum_{i=1}^{k}(\bar{y}_i - \hat{\theta}_i)^2\omega_i \\
&= \sum_{i=1}^{k}(\hat{\theta}_i - \bar{y})^2\omega_i + 2\sum_{i=1}^{k}(\bar{y}_i - \hat{\theta}_i)(\hat{\theta}_i - \bar{y})\omega_i \\
&= \sum_{i=1}^{k}(\hat{\theta}_i - \bar{y})^2\omega_i,
\end{aligned}$$

where the second term in the second equality above is zero based on Theorem 4.3.3. Usually the LR test statistic is

$$\bar{\chi}^2 = \sum_{i=1}^{k}(\hat{\theta}_i - \bar{y})^2\omega_i, \tag{4.3.28}$$

rejecting H_0 when $\bar{\chi}^2$ is large enough. Now let us compute the distribution of $\bar{\chi}^2$ under H_0.

Let K denote the subindex set, *i.e.* $K = \{1, \cdots, k\}$. By the form of $\hat{\theta}$, there exist a subindex set blocks $B_1, \cdots, B_l$, such that $B_i \bigcap B_j = \phi,\ i \neq j;\ \bigcup B_i = K$. By the PAVA algorithm, we can obtain

$$Av(B_1) < \cdots < Av(B_l).$$

Let L be the r.v. corresponding to the number of sub index set blocks satisfying the above property, then its range is $L = l,\ \ l = 1, \cdots, k$. Let

$$P(l,k) = P\{L = l\} \tag{4.3.29}$$

denote the probability of $L = l$ under H_0. Obviously $\sum P(l,k) = 1$. By the Law of Total Probability, if H_0 is true for $c > 0$ then

$$P\{\bar{\chi}^2 \geq c\} = \sum_{l=2}^{k} P\{\bar{\chi}^2 \geq c | L = l\} P(l,k); \tag{4.3.30}$$

When $c = 0$, we have

$$P\{\bar{\chi}^2 = 0\} = P(1, k).$$

Now let us calculate these probabilities a bit more closely.

Lemma 4.3.2. *Let $\boldsymbol{Z} = (Z_1, \cdots, Z_s)'$ be an r.v. which has s-dimensional standard normal distribution, and $\boldsymbol{A}$ a $t \times s$ real matrix. Then for any $c > 0$, we have*

$$P\{\boldsymbol{Z}'\boldsymbol{Z} \geq c \mid \boldsymbol{A}\boldsymbol{Z} \geq 0\} = P\{\boldsymbol{Z}'\boldsymbol{Z} \geq c\}.$$

Proof. We take the following polar coordinate transformation

$$\begin{cases} Z_1 = R \ \sin\alpha_1 \\ Z_i = R \ \cos\alpha_1 \cdots \cos\alpha_{i-1} \sin\alpha_i, \quad i = 2, \cdots, s-1 \\ Z_s = R \ \cos\alpha_1 \cdots \cos\alpha_{s-1}. \end{cases}$$

It is easy to verify that $\boldsymbol{Z}'\boldsymbol{Z} = R^2$. Since the condition $\boldsymbol{A}\boldsymbol{Z} \geq 0$ depends only on the α_i and R is independent of the α_i's, the result of the lemma holds. □

Lemma 4.3.3. *Let $V_1, \cdots, V_s$ be independent r.v.'s, where V_i has a normal distribution $N(\theta, 1/b_i)$. Then under the condition $V_1 \leq \cdots \leq V_s$, we have*

$$\sum_{i=1}^{s} b_i (V_i - \bar{V})^2 \sim \chi^2_{s-1},$$

where $\bar{V} = \sum_{i=1}^{s} b_i V_i / \sum_{i=1}^{s} b_i$.

Proof. Let $\boldsymbol{V} = (V_1, \cdots, V_s)'$, $\boldsymbol{B} = \mathrm{diag}(\sqrt{b_1}, \cdots, \sqrt{b_s})$, and $\boldsymbol{U} = \boldsymbol{B}\boldsymbol{V}$. Let $\boldsymbol{D} = (d_{ij})$ be an $s \times s$ orthogonal matrix with the last row (s, j)-element $d_{sj} = \sqrt{b_j}/\sqrt{\sum b_i}, j = 1, \cdots, s$. Let $\boldsymbol{Z} = \boldsymbol{D}\boldsymbol{U}$, then we have

$$\sum_{i=1}^{s} b_i (V_i - \bar{V})^2 = \boldsymbol{Z}'\boldsymbol{Z} - Z_s^2 = \sum_{i=1}^{s-1} Z_i^2,$$

where $Z_i \sim N(0, 1), i = 1, \cdots, s$. Let $\boldsymbol{C}$ be an $(s-1) \times s$ matrix,

$$\boldsymbol{C} = \begin{pmatrix} -1 & 1 & 0 \cdots & 0 & 0 \\ 0 & -1 & 1 \cdots & 0 & 0 \\ \vdots & \vdots & \vdots \ddots & \vdots & \vdots \\ 0 & 0 & 0 \cdots & -1 & 1 \end{pmatrix}$$

Let $\boldsymbol{A} = \boldsymbol{C}\boldsymbol{B}^{-1}\boldsymbol{D}'$, then $V_1 \leq \cdots \leq V_s$ is equivalent to that $\boldsymbol{A}\boldsymbol{Z} = \boldsymbol{C}\boldsymbol{B}^{-1}\boldsymbol{D}'\boldsymbol{D}\boldsymbol{U} = \boldsymbol{C}\boldsymbol{V} \geq 0$. By Lemma 4.3.2 the result holds. □

Lemma 4.3.4. *If H_0 is true given $L = l$, then the conditional distribution of $\bar{\chi}^2$ is χ^2_{l-1}.*

Proof. We will use the Mathematical Induction for the number of populations, k. When $k = 1$, the problem is meaningless. When $k = 2$, there are two cases: $l = 1$ and $l = 2$. When $l = 1$, by Eq. (4.3.28) we have $\bar{\chi}^2 = 0$, denoting that it has the distribution χ^2_0, that is, the χ^2 distribution with 0 degree of freedom; When $l = 2$, $\bar{y}_1 \leq \bar{y}_2$, this is the result of Lemma 4.3.3.

Now assume that result holds for $k = m - 1$. Let us discuss the case $k = m$. All the observed data can be divided into the following two cases:

(1) $\bar{y}_1 \leq \cdots \leq \bar{y}_m$;
(2) There exists i such that $\bar{y}_i > \bar{y}_{i+1}$.

For the first case, we can apply Lemma 4.3.3 directly. For the second one, by the PAVA algorithm, let

$$\bar{y}_{(i)} = (\omega_i \bar{y}_i + \omega_{i+1}\bar{y}_{i+1})/(\omega_i + \omega_{i+1}).$$

From Theorem 1.3.5, we know that $\bar{Y}_i - \bar{Y}_{i+1}$ and $\bar{Y}_{(i)}$ are mutually independent. Therefore, under the given condition $\bar{Y}_i - \bar{Y}_{i+1} < 0$, we know that $\bar{Y}_1, \cdots, \bar{Y}_{i-1}, \bar{Y}_{(i)}, \bar{Y}_{i+2}, \cdots, \bar{Y}_m$ are mutually independent. Finally, note that there are only $m - 1$ variables, so the result holds by the induction assumption. □

From Eq. (4.3.30) and Lemma 4.3.4, we have the following theorem.

Theorem 4.3.4. *When H_0 is true for $c > 0$, we have*

$$P\{\bar{\chi}^2 \geq c\} = \sum_{l=2}^{k} P(l,k) P\{\chi^2_{l-1} \geq c\},$$

$$P\{\bar{\chi}^2 = 0\} = P(1,k),$$

where χ^2_{l-1} denotes a r.v. which has the χ^2 distribution with $l - 1$ degrees of freedom, and $P(l,k)$ is given by (4.3.29).

The $P(l,k)$ in the above theorem is called **level probability**, and the corresponding subindex set blocks $B_1, \cdots, B_l$ are called **level sets.** Now we discuss how to compute the level probability. We first study a special case, and then give a general result.

Consider $P(2,4)$. What we need to do is to compute the probability that the four sample means from four normal populations with the same mean are partitioned into two blocks by the low set or PAVA algorithm. Obviously, these two blocks have three cases:

(1) $B_1 = \{1\}, B_2 = \{2, 3, 4\}$;
(2) $B_1 = \{1, 2\}, B_2 = \{3, 4\}$;
(3) $B_1 = \{1, 2, 3\}, B_2 = \{4\}$.

Consider Case (1). By the Lower Set Algorithm, it is made up of two events, that is,

$$D_1 = \{\bar{y}_1 < Av(1,2), \bar{y}_1 < Av(1,2,3), \bar{y}_1 < Av(1,2,3,4)\},$$

and

$$D_2 = \{\bar{y}_2 > \bar{y}_3, Av(2,3) > \bar{y}_4\}.$$

It can be verified that the above two events are mutually independent. Using $P(2,2;\omega_1,\omega_2+\omega_3+\omega_4)$ and $P(1,3;\omega_2,\omega_3,\omega_4)$ to denote the probabilities of the two events respectively, the probability corresponding to Case (1) can be written as

$$P(2,2;\omega_1,\omega_2+\omega_3+\omega_4)P(1,3;\omega_2,\omega_3,\omega_4).$$

Similarly, the probability corresponding to Case (2) is

$$P(2,2;\omega_1+\omega_2,\omega_3+\omega_4)P(1,2;\omega_1,\omega_2)P(1,2;\omega_3,\omega_4).$$

The probability corresponding to Case (3) is

$$P(2,2;\omega_1+\omega_2+\omega_3,\omega_4)P(1,3;\omega_1,\omega_2,\omega_3).$$

Use $c(B_i)$ to denote the number of elements of subindex set B_i. Based on the above analysis, we can get

$$P(2,4) = \sum_{\mathcal{L}_{24}} P(2,2;B_1,B_2)P(1,c(B_1))P(1,c(B_2)),$$

where $\mathcal{L}_{24}$ denotes above three cases. For a generalized case we get

$$P(l,k) = \sum_{\mathcal{L}_{lk}} P(l,l;B_1,\cdots,B_l)\prod_{i=1}^{l} P(1,c(B_i)),$$

where $\mathcal{L}_{lk}$ denotes all possible configurations partitioning dividing k sample means into l blocks, $B_1,\cdots,B_l$, by the Low Set Algorithm. Usually the computation at the probability is very difficult, since it will involve the computation of the probability $P\{Y_1 \le Y_2 \le \cdots \le Y_k\}$. If let $Z_i = Y_{i+1} - Y_i, i = 1,\cdots,k-1$, then the probability can be written as

$$P\{Z_1 \ge 0,\cdots,Z_{k-1} \ge 0\}.$$

This is the probability of a r.v. with a k-dimensional normal distribution, taking values in the first quadrant usually called as **the positive orthant**

probability. However, its numerical computation is also very difficult. Sun (1988) gave an algorithm to calculate the positive orthant probability for the case when $k \leq 10$.

Now consider a special case, that is, under $\omega_1 = \cdots = \omega_k$, when H_0 given by (4.3.1) is true, to calculate $P(l, k)$. Here we can regard $y_1, \cdots, y_k$ as an i.i.d. sample. Without the loss of generality, suppose that $y_1 < \cdots < y_k$. Now we use the Low Set or PAVA algorithm to obtain k subindex sets, *i.e.* $l = k$. Let $\pi(1), \cdots, \pi(k)$ be a permutation of $1, \cdots, k$. By the given condition we know that the probabilities of obtaining the samples $y_{\pi(1)}, \cdots, y_{\pi(k)}$ respectively are equal, however, we must have $l < k$ by the algorithm. Let $r(l, k)$ denote the number of l subindex sets obtained by running through all the permutations, then we have

$$\sum_{l=1}^{k} r(l,k)s^l = s(s+1)\cdots(s+k-1). \tag{4.3.31}$$

We will prove (4.3.31) by a mathematical induction.

When $k = 2$, for $y_1 < y_2, \pi(1) = 2, \pi(2) = 1 \Leftrightarrow l = 1; \pi(1) = 1, \pi(2) = 2 \Leftrightarrow l = 2$. So (4.3.31) holds.

Now suppose that (4.3.31) holds for the case $k = m - 1$, we need to prove that it holds for the case $k = m$. It can be divided into two cases: $\pi(m) = m$ and $\pi(m) = i$, then we have $i \in \{1, 2, \cdots, m-1\}$. Let $r_1(l, m)$ and $r_2(l, m)$ denote the number of permutations corresponding to the two cases respectively,

$$r(l,m) = r_1(l,m) + r_2(l,m). \tag{4.3.32}$$

For the first case, the permutations only occur among the first $m-1$ indices, hence by induction assumption

$$\begin{aligned}\sum_{l=1}^{m} r_1(l,m)s^l &= s\sum_{l=1}^{m} r(l-1,m-1)s^{l-1} \\ &= s\sum_{l=1}^{m-1} r(l,m-1)s^l \\ &= s^2(s+1)\cdots(s+m-2).\end{aligned}$$

For the second case, for the fixed $i \in \{1, \cdots, m-1\}$, the permutations also occur among these $m-1$ indices. Use $r_i^*(l, m-1)$ to denote the number of all the permutations. By induction assumption, then

$$\begin{aligned}\sum_{l=1}^{m} r_2(l,m)s^l &= \sum_{i=1}^{m-1}\sum_{l=1}^{m-1} r_i^*(l,m-1)s^l \\ &= (m-1)s(s+1)\cdots(s+m-2).\end{aligned}$$

By (4.3.32) we can obtain (4.3.31).

Theorem 4.3.5. *Let $y_1, \cdots, y_k$ be an i.i.d. sample from a continuous distribution. Let $P(l,k)$ denote the probability of obtaining l level sets by the Low Set algorithm. If $\omega_1 = \cdots = \omega_k$, then the probability generating function of $P(l,k)$ is*

$$P_k(s) = \sum_{l=1}^{k} P(l,k)s^l = s(s+1)\cdots(s+k-1)/k! \tag{4.3.33}$$

Proof. Let $\boldsymbol{Y} = (Y_1, \cdots, \cdots, Y_k)$. Use $P_{\boldsymbol{Y}}$ to denote its joint distribution. From Example 1.3.5, this distribution is symmetric about any coordinate. Use $\hat{\boldsymbol{Y}}^*$ to denote the order statistics of $\boldsymbol{Y}$, and $P_{\boldsymbol{Y}^*}$ the probability distribution of $\boldsymbol{Y}^*$. Let $C = \{\boldsymbol{y} \in \boldsymbol{R}^k;\ y_1 < y_2 < \cdots < y_k\}$. Obviously, $P_{\boldsymbol{Y}^*}(C) = 1$. Let $m(y_1, y_2, \cdots, y_k)$ denote the number of level sets, then we have

$$\begin{aligned}
P_k(s) &= \int s^{m(y_1,y_2,\cdots,y_k)} dP_{\boldsymbol{Y}} \\
&= \frac{1}{k!}\sum_{\pi}\int_C s^{m(y_{\pi(1)},y_{\pi(2)},\cdots,y_{\pi(k)})} dP_{\boldsymbol{Y}^*} \\
&= \frac{1}{k!}\int_C \sum_{\pi} s^{m(y_{\pi(1)},y_{\pi(2)},\cdots,y_{\pi(k)})} dP_{\boldsymbol{Y}^*} \\
&= \frac{1}{k!}\int_C s(s+1)\cdots(s+k-1) dP_{\boldsymbol{Y}^*} \\
&= \frac{1}{k!} s(s+1)\cdots(s+k-1),
\end{aligned}$$

where the two summations run over all the permutations, and the fourth equality follows from Eq. (4.3.31). □

By Eq. (4.3.33), it is easy to compute $P_k(s) = sP_{k-1}(s)/k + (k-1)P_{k-1}(s)/k$, then we can obtain the following result.

Corollary 4.3.1. *If $\omega_1 = \omega_2 = \cdots = \omega_k$, then the level probabilities can be denoted as*

$$\begin{aligned}
&P(1,k) = \frac{1}{k}, \quad P(k,k) = \frac{1}{k!}, \\
&P(l,k) = \frac{1}{k}P(l-1,k-1) + \frac{k-1}{k}P(l,k-1).
\end{aligned} \tag{4.3.34}$$

When the **variances are unknown**, we still assume that $\omega_1 = \omega_2 = \cdots = \omega_k$, see Eq. (4.3.16). Let $\hat{\sigma}_0^2$ and $\hat{\sigma}_1^2$ be the MLEs of σ^2 under H_0 and $H_0 \cup H_1$, respectively. Then

$$\hat{\sigma}_0^2 = \frac{1}{n}\sum_{i=1}^{k}\sum_{j=1}^{n_i}(y_{ij} - \bar{y})^2,$$

$$\hat{\sigma}_1^2 = \frac{1}{n}\sum_{i=1}^{k}\sum_{j=1}^{n_i}(y_{ij} - \hat{\theta}_i)^2,$$

where $n = n_1 + n_2 + \cdots + n_k$, and $\hat{\boldsymbol{\theta}} = (\hat{\theta}_1, \hat{\theta}_2, \cdots, \hat{\theta}_k)'$ is the MLE of θ under $H_0 \cup H_1$. Then the LR test statistic is $\Lambda = (\hat{\sigma}_1^2/\hat{\sigma}_0^2)^{\frac{n}{2}}$, which is equivalent to

$$1 - \Lambda^{\frac{2}{n}} = \frac{\sum_{i=1}^{k}\sum_{j=1}^{n_i}(y_{ij} - \bar{y})^2 - \sum_{i=1}^{k}\sum_{j=1}^{n_i}(y_{ij} - \hat{\theta}_i)^2}{\sum_{i=1}^{k}\sum_{j=1}^{n_i}(y_{ij} - \bar{y})^2}.$$

By Theorem 4.3.3, we can obtain

$$\sum_{i=1}^{k}\sum_{j=1}^{n_i}(y_{ij} - \bar{y})^2 = \sum_{i=1}^{k}\sum_{j=1}^{n_i}(y_{ij} - \hat{\theta}_i)^2 + \sum_{i=1}^{k} n_i(\hat{\theta}_i - \bar{y})^2. \tag{4.3.35}$$

Thus the LR test statistic is equivalent to

$$\bar{E}^2 = \frac{\sum_{i=1}^{k} n_i(\hat{\theta}_i - \bar{y})^2}{\sum_{i=1}^{k}\sum_{j=1}^{n_i}(y_{ij} - \bar{y})^2}. \tag{4.3.36}$$

Comparing with Eq. (4.3.28), we can see that the numerator of the above equation is $\bar{\chi}^2$. Note that (4.3.35) denotes a direct sum decomposition, so the two right-hand terms are independent. By using Example 1.1.12, we can obtain the following theorem whose proof is similar to that of Theorem 4.3.4.

Theorem 4.3.6. *When H_0 is true for $c > 0$, we have*

$$P\{\bar{E}^2 \geq c\} = \sum_{l=2}^{k} P(l,k)P\{B_{\frac{l-1}{2},\frac{n-l}{2}} \geq c\},$$
$$P\{\bar{E}^2 = 0\} = P(1,k),$$

where $B_{s,t}$ denotes the r.v. corresponding to the beta distribution with s and t degrees of freedom, and $P(l,k)$ is given by Corollary 4.3.1.

Example 4.3.3. Continue to analyze the data in Example 4.1.1, which have already been analyzed in Example 4.1.2, 4.2.3, 4.3.1 and 4.3.2. Recall the computation result provided by Table 4.1.3 in Example 4.1.2. Suppose $\bar{y}_i^O$ and $\bar{y}_i^L$ have normal distributions $N(\theta_i^0, \sigma_0^2)$ and $N(\theta_i^L, \sigma_L^2), i = 1, 2, 3, 4$, respectively. For the obese rats, the testing problem is

$$H_0^O : \theta_1^O = \theta_2^O = \theta_3^O = \theta_4^O \quad \text{vs.} \quad H_1^O : \theta_1^O \le \theta_2^O \le \theta_3^O \le \theta_4^O, \qquad (4.3.37)$$

For the lean rats, the testing problem is

$$H_0^L : \theta_1^L = \theta_2^L = \theta_3^L = \theta_4^L \quad \text{vs.} \quad H_1^L : \theta_1^L \le \theta_2^L \le \theta_3^L \le \theta_4^L. \qquad (4.3.38)$$

Calculate the LR statistics respectively. By the low set algorithm, the MLEs of θ_i^O and θ_i^L under H_1 are respectively

$\hat{\theta}_i^O$:	634.75	660.42	660.42	660.42	$\bar{y}^O = 654.00$
$\hat{\theta}_i^L$:	467.75	479.25	508.00	508.00	$\bar{y}^L = 490.75$

The MLEs of σ_O^2 and σ_L^2 can be found in Tables 4.1.4 and 4.1.5 of Example 4.1.2. By Eq. (4.3.36), we can obtain that the LR test statistics are $\bar{E}_O^2 = 0.12$ and $\bar{E}_L^2 = 0.35$ respectively. Since $\omega_1 = \omega_2 = \omega_3 = \omega_4$, we can use Theorems 4.3.5 and 4.3.6 to calculate the p-value. Now $k = 4$, by Eq. (4.3.34) we can obtain

$$\begin{aligned}
P(1,4) &= \frac{1}{4}, \quad P(4,4) = \frac{1}{24}, \\
P(2,4) &= \frac{1}{4}P(1,3) + \frac{3}{4}P(2,3) = \frac{1}{4}\cdot\frac{1}{3} + \frac{3}{4}\cdot\frac{1}{2} = \frac{11}{24}, \\
P(3,4) &= 1 - \frac{1}{4} - \frac{1}{24} - \frac{11}{24} = \frac{1}{6}.
\end{aligned}$$

So the p-values are respectively

$$\begin{aligned}
p^O &= P\{\bar{E}_0^2 \ge 0.12\} \\
&= \frac{11}{24}P\{B_{\frac{1}{2},\frac{14}{2}} \ge 0.12\} + \frac{1}{6}P\{B_{\frac{2}{2},\frac{13}{2}} \ge 0.12\} + \frac{1}{24}P\{B_{\frac{3}{2},\frac{12}{2}} \ge 0.12\} \\
&= 0.186, \\
p^L &= P\{\bar{E}_L^2 \ge 0.35\} \\
&= \frac{11}{24}P\{B_{\frac{1}{2},\frac{14}{2}} \ge 0.35\} + \frac{1}{6}P\{B_{\frac{2}{2},\frac{13}{2}} \ge 0.35\} + \frac{1}{24}P\{B_{\frac{3}{2},\frac{12}{2}} \ge 0.35\} \\
&= 0.023.
\end{aligned}$$

Since $p^O > 0.05$, as far as the obese rats are concerned, the age difference has no significant effect on the amount of food intake; however, it has significant effect as far as the lean rats are concerned. Because the lean rats are used as the control group, the results suggest that some recessive gene plays a role in the treatment group.

4.3.3 *Linear Rank Tests*

In this subsection, we consider a more general model

$$y_{ij} = \theta_i + \varepsilon_{ij}, \quad i = 1, \cdots, k; \ j = 1, \cdots, n_i. \tag{4.3.39}$$

Now, we assume that the ε_{ij}'s are i.i.d. from a continuous distribution function, $F(x)$, and that $F(x)$ is symmetric about 0. From (4.3.39), we can regard $y_{i1}, \cdots, y_{in_i}$ as an i.i.d. sample from $F(x - \theta_i)$. Obviously the normal distribution belongs to the above distribution family. Consider the trend test given by (4.3.1), that is,

$$H_0 : \theta_1 = \cdots = \theta_k \quad \text{vs.} \quad H_1 - H_0, \tag{4.3.40}$$

where $H_1 : \theta_1 \leq \cdots \leq \theta_k$. Refer to Section 3.5, we need to construct a test statistic based on rank.

Let r_{ij} denote the rank of y_{ij} in the mixed sample

$$y_{11}, \cdots, y_{1n_1}, \cdots, y_{k1}, \cdots, y_{kn_k},$$

and $\bar{r}_i = \sum_j r_{ij}/n_i$ the rank mean. Let R_{ij} and $\bar{R}_i$ be the r.v.'s corresponding to r_{ij} and $\bar{r}_i$. Similar to (2.4.18) in Chapter 2, when H_0 is true, we can obtain

$$\begin{aligned} &E_0\bar{R}_i = (N+1)/2, \\ &V_0\bar{R}_i = (N-n_i)(N+1)/(12n_i), \\ &CV_0(\bar{R}_i, \bar{R}_j) = -(N+1)/12, \end{aligned} \tag{4.3.41}$$

where $N = n_1 + \cdots + n_k$. Now consider the limiting case and suppose that the k sample sizes tend to infinity at the same speed, that is, for any $i \in \{1, \cdots, k\}$, there exists $\lambda_i \in (0,1)$, when $N \to \infty$, we have $n_i/N \to \lambda_i$. For the given $\boldsymbol{a} \in \boldsymbol{R}^k$ satisfying $\sum_{i=1}^k \lambda_i a_i = 0$ and $\sum_{i=1}^k \lambda_i a_i^2 = 1$,

$$T_{\boldsymbol{a}} = \sum_{i=1}^{k} \lambda_i a_i \bar{R}_i \tag{4.3.42}$$

is called a **linear rank test**.

Theorem 4.3.7. *If H_0 is true, then*

$$\sqrt{\frac{12}{N}} T_{\boldsymbol{a}} \xrightarrow{L} N(0\ ,1) \tag{4.3.43}$$

as $N \to \infty$.

Proof. For $i = 1, \cdots, k$, let

$$T_i = \frac{a_i}{\sqrt{N}}(\bar{R}_i - \frac{N+1}{2}).$$

From Eq. (4.3.41), $E_0 T_i = 0$. When N is large enough,

$$V_0 T_i = \frac{a_i^2}{12\lambda_i}(1 - \lambda_i),$$

$$CV_0(T_i, T_j) = -\frac{1}{12} a_i a_j, \quad \text{if } i \neq j.$$

Since $\sqrt{12/N} T_{\boldsymbol{a}} = \sqrt{12} \sum \lambda_i T_i$, we have $E_0 \sqrt{12/N} T_{\boldsymbol{a}} = 0$.

$$V_0(\sqrt{12/N} T_{\boldsymbol{a}}) \approx \sum_{i=1}^{k} \lambda_i a_i^2 - (\sum_{i=1}^{k} \lambda_i a_i)^2 = 1.$$

By the Central Limiting Theorem, (4.3.43) holds. □

Recall the discussion in Section 3.5. Consider the Pitman efficiency of the statistic $T_{\boldsymbol{a}}$. For given $\beta > 0$ and $c \in R^k$ satisfying $\sum_{i=1}^k \lambda_i c_i = 0$ and $\sum_{i=1}^k \lambda_i c_i^2 = 1$, construct the following hypothesis:

$$H(\boldsymbol{c}): \; \theta_i = \beta c_i / \sqrt{N}, \quad i = 1, \cdots, k.$$

Similar to the proof of Theorem 4.3.7 and the discussion in Example 3.5.3, we can obtain that if $H(\boldsymbol{c})$ is true, then

$$\sqrt{\frac{12}{N}} T_{\boldsymbol{a}} \xrightarrow{L} N(\beta e(\boldsymbol{a}, \boldsymbol{c}), 1)$$

as $N \to \infty$, where

$$e(\boldsymbol{a}, \boldsymbol{c}) = \sum_{i=1}^{k} \lambda_i a_i c_i \sqrt{12} \int f^2(x) d\mu(x), \tag{4.3.44}$$

satisfying $e(\boldsymbol{a}, \boldsymbol{c}) < \infty$, and $f(x)$ is the density of $F(x)$. From Theorem 3.5.1, $e(\boldsymbol{a}, \boldsymbol{c})$ is the Pitman efficiency of $T_{\boldsymbol{a}}$ at the point $\boldsymbol{c}$. Next we let

$$C = \{\boldsymbol{c} \in \boldsymbol{R}^k; \; c_1 \leq \cdots \leq c_k, \sum_{i=1}^{k} \lambda_i c_i = 0, \sum_{i=1}^{k} \lambda_i c_i^2 = 1\}.$$

Corresponding to the problem of testing (4.3.40), both the $\boldsymbol{a}$ in the statistic $T_{\boldsymbol{a}}$ and the $\boldsymbol{c}$ in $H(\boldsymbol{c})$ satisfy $\boldsymbol{a} \in C$ and $\boldsymbol{c} \in C$ obviously. Let B denote the set of all linear rank tests whose coefficients belong to C, *i.e.*

$$B = \{T_{\boldsymbol{b}}; \; T_{\boldsymbol{b}} = \sum_{i=1}^{k} \lambda_i b_i \bar{R}_i, \boldsymbol{b} \in C\}.$$

$T_{\boldsymbol{a}}$ is called **the best linear rank test**, if $T_{\boldsymbol{a}} \in B$, and satisfies

$$\min_{\boldsymbol{c}\in C} e(\boldsymbol{a}, \boldsymbol{c}) = \max_{T_{\boldsymbol{b}}\in B} \min_{\boldsymbol{c}\in C} e(\boldsymbol{b}, \boldsymbol{c}).$$

It coincides with the idea of the best linear test discussed in Section 4.3.1. So the best coefficient $\boldsymbol{a}$ satisfies $\boldsymbol{a} \in C$, and

$$\min_{\boldsymbol{c}\in C} \sum_{i=1}^{k} \lambda_i a_i c_i = \max_{\boldsymbol{b}\in C} \min_{\boldsymbol{c}\in C} \sum_{i=1}^{k} \lambda_i b_i c_i.$$

Hence $\boldsymbol{a} = (a_1, \cdots, a_k)'$ satisfies

$$a_i = \beta(\sqrt{s_{i-1}(s_k - s_{i-1})} - \sqrt{s_i(s_k - s_i)})/\lambda_i,$$

where $s_m = \lambda_1 + \cdots + \lambda_m$, $m = 1, \cdots, k$ and $s_0 = 0$; β is a normalized coefficient to ensure that $\boldsymbol{a} \in C$.

As discussed in Section 3.5, the best linear rank test is very robust. For more detailed discussion see Shi (1988).

In summary, all the problems discussed Section 4.3 are usually called **statistical inferences under order restrictions**, whose primary goal is to study the estimation and hypotheses testing problem of k independent population parameters under order restrictions. Take the normal distribution as an example. Suppose there are k populations $N(\theta_i, a_i\sigma^2)$, with known a_i, σ^2 is known or unknown, $i = 1, \cdots, k$. According to different practical problems, where we can consider the following restriction about the means, respectively:

$$\begin{aligned} \text{Simple order}: &\ \theta_1 \le \theta_2 \le \cdots \le \theta_k; \\ \text{Umbrella order}: &\ \theta_1 \le \cdots \le \theta_p \ge \cdots \ge \theta_k; \\ \text{Simple tree order}: &\ \theta_1 \le \theta_i, \quad i = 2, \cdots, k; \\ \text{Simple loop order}: &\ \theta_1 \le \theta_i \le \theta_k, \quad i = 2, \cdots, k-1. \end{aligned}$$

Bartholomew (1959a, 1959b) considered originally the LR test under the simple order restriction. The result in Theorem 4.3.5 was given by Barton and Mallows (1961). Shi (1988a) discussed the LR test under the umbrella order restriction and provided the level prabobilities when $k \le 10$. Abelson and Tukey (1963), and Schaafsma (1968) discussed the best linear test under the simple order restriction. Shi (1988b) discussed the best linear rank test under the umbrella order restriction.

For an overview about the LR test, see Barlow *et al.* (1972); Robertson *et al.* (1988); Silvapulle and Sen (2004); for the study about the linear test statistic, see Schaafsma and Smid (1966); Snijders (1979); Shi (1987); Akkerboom (1990).

4.4 Multi-Factor Model: Analysis of Variance

Now let us go back to the model discussed in Section 4.1 and suppose that the data are affected not only by Factor A_i. but also by Factor B_j. Through generalizing the saturated model given by (4.1.3), we can obtain the following two classes of models. This kind of problem is usually called **Two-factor (or Twoway) ANOVA**.

4.4.1 *Model Without Interaction Effects*

If the interaction effects between A_i and B_j are not considered, then the model is usually denoted by

$$y_{ijl} = \mu + \theta_i + \nu_j + \epsilon_{ijl}, \quad i = 1, \cdots, k; \; j = 1, \cdots, r; \; l = 1, \cdots, c, \quad (4.4.1)$$

where μ, θ_i, ν_j are parameters: θ_i corresponding to factor A_i is called as **row effects**, ν_j corresponding to B_j is called as **column effects**; y_{ijl} denotes the l-th observation in the i-th row and the j-th column; the ϵ_{ijl}'s are i.i.d. $N(0, \sigma^2)$ observational errors. Similar to the discussion about the model (4.1.3), we can assume that

$$\sum_{i=1}^{k} \theta_i = 0, \quad \sum_{j=1}^{r} \nu_j = 0.$$

Here, we are interested in the problem of testing

$$\begin{aligned} H_{0\theta} &: \theta_1 = \cdots = \theta_k = 0, \\ H_{0\nu} &: \nu_1 = \cdots = \nu_r = 0. \end{aligned} \quad (4.4.2)$$

Example 4.4.1. (Establishment of model) Recall the behavior of food intake of Zucker rats discussed in Example 4.1.1. We used A_i to denote the age in weeks, and studied the amount of food intake for the Obese and Lean rats, respectively. The corresponding data were given Tables by 4.1.1 and 4.1.2. Now we add the type of rats as a new factor, using B_1 to denote the Obese rats and B_2 to denote the lean rats. For any $A_i \bigcup B_j$ there are four samples. So $k = 4, r = 2, c = 4$. Extract the data of A+B in Tables 4.1.1 and 4.1.2, *i.e.* the amount of food intake, we can obtain Table 4.1.1.

Let the y_{ijl}'s denote the data in Table 4.1.1. we can obtain the model (4.4.1). Note that the order of the data in Table 4.1.1 cannot be changed arbitrarily, since the row data denote the amount of food intake of each Zucker rat at different ages in weeks. Now we want to make a comparison analysis. The testing problems are

$$H_{0\theta} : \theta_1 = \cdots = \theta_4 = 0; \text{ and } H_{0\nu} : \nu_1 = \nu_2 = 0.$$

Table 4.4.1 Amount of food intake of Zucker rats.

	A_1	A_2	A_3	A_4
B_1	666	659	706	642
	598	634	678	598
	654	710	670	694
	621	655	645	634
B_2	473	467	496	463
	479	457	497	444
	463	525	528	439
	456	468	469	403

Obviously, $H_{0\theta}$ concerns the effect of the age in weeks on the amount of food intake, and $H_{0\nu}$ concerns the effect of the type of rats on the amount of food.

Let $n = krc$, then the MLEs of parameters μ, θ_i, and ν_j are

$$\begin{aligned}\hat{\mu} &= \bar{y} = 1/n \sum_{i=1}^{k} \sum_{j=1}^{r} \sum_{l=1}^{c} y_{ijl},\\ \hat{\theta}_i &= \bar{y}_{i++} - \bar{y} = 1/(rc) \sum_{j=1}^{r} \sum_{l=1}^{c} y_{ijl} - \bar{y},\\ \hat{\nu}_j &= \bar{y}_{+j+} - \bar{y} = 1/(kc) \sum_{i=1}^{k} \sum_{l=1}^{c} y_{ijl} - \bar{y},\end{aligned} \tag{4.4.3}$$

respectively. Under $H_{0\theta}$ the MLE of θ_i is 0, while others do not change; Under $H_{0\nu}$ the MLE of ν_i is 0, while others do not change. Because

$$\begin{aligned}\sum_{i=1}^{k}\sum_{j=1}^{r}\sum_{l=1}^{c}(y_{ijl} - \bar{y})^2 &= \sum_{i=1}^{k}\sum_{j=1}^{r}\sum_{l=1}^{c}[(y_{ijl} - \bar{y}_{i++} - \bar{y}_{+j+} + \bar{y}) + (\bar{y}_{i++} - \bar{y})\\ &\quad +(\bar{y}_{+j+} - \bar{y})]^2\\ &= rc\sum_{i=1}^{k}(\bar{y}_{i++} - \bar{y})^2 + kc\sum_{j=1}^{r}(\bar{y}_{+j+} - \bar{y})^2\\ &\quad +\sum_{i=1}^{k}\sum_{j=1}^{r}\sum_{l=1}^{c}(y_{ijl} - \bar{y}_{i++} - \bar{y}_{+j+} + \bar{y})^2\\ &= Q_1 + Q_2 + Q_3.\end{aligned} \tag{4.4.4}$$

Let Q denote the left-hand side of (4.4.4), then $Q = Q_1 + Q_2 + Q_3$, which is also a direct sum decomposition. It is easy to verify that if $H_{0\theta}$ and $H_{0\nu}$

are true, $Q/\sigma^2 \sim \chi^2(n-1)$, $Q_1/\sigma^2 \sim \chi^2(k-1)$, $Q_2/\sigma^2 \sim \chi^2(r-1)$, and $Q_3/\sigma^2 \sim \chi^2(n-k-r+1)$.

In this chapter, we have used the concept of the direct sum decomposition several times. Its core involves the relationship between the χ^2 distribution and quadratic form of independent r.v.'s from a normal distribution. Now we analyze this concept in detail.

Let $X_1, \cdots, X_n$ be i.i.d. r.v.s from the standard normal $N(0,1)$. Let $\boldsymbol{X} = (X_1, \cdots, X_n)'$, and $\boldsymbol{A}$ be an $n \times n$ nonnegative-definite matrix, then a quadratic form is defined as $Q = \boldsymbol{X}'\boldsymbol{A}\boldsymbol{X}$, and the rank of $\boldsymbol{A}$ is called as the rank of Q. By Lemma 2.3.1 $Q \sim \chi_r^2$ if and only if $\boldsymbol{A}$ is an idempotent matrix and $\text{rank}(\boldsymbol{A}) = r$. The following theorem provides a basic result for the direct sum decomposition.

Theorem 4.4.1. (Basic Theorem of Quadratic Form) *Let $Q_i, i = 1, \cdots, s$, be a quadratic form in $\boldsymbol{X}$ with rank n_i, satisfying*

$$Q_1 + \cdots + Q_s = \sum_{i=1}^{n} X_i^2.$$

Then the Q_i's are mutually independent and Q_i has a χ^2 distribution with n_i degrees of freedom if and only if $n_1 + \cdots + n_s = n$.

Proof. **Necessarity**. If $Q_i \sim \chi^2_{n_i}, i = 1, \cdots, s$, then the moment generating function is

$$M_{Q_i}(t) = Ee^{tQ_i} = (1-2t)^{-n_i/2}.$$

By the independence, if $Z = Q_1 + \cdots + Q_s$, then the moment generating function of Z is

$$M_Z(t) = M_{Q_1}(t) \cdots M_{Q_s}(t) = (1-2t)^{-(n_1+\cdots+n_s)/2},$$

which is the moment generating function of the χ^2 distribution with $n = n_1 + \cdots + n_s$ degrees of freedom.

Sufficiency. Let $Q_i = \boldsymbol{X}'\boldsymbol{A}_i\boldsymbol{X}$, where $\boldsymbol{A}_i$ is a nonnegative-definite matrix with rank n_i. There exists an $n_i \times n$ matrix $\boldsymbol{B}_i$, such that $\boldsymbol{B}_i'\boldsymbol{B}_i = \boldsymbol{A}_i$. Let $\boldsymbol{Y}_i = \boldsymbol{B}_i\boldsymbol{X}$ be an n_i-dimensional vector, then $Q_i = \boldsymbol{Y}_i'\boldsymbol{Y}_i$. Let

$$\boldsymbol{Y} = \begin{pmatrix} \boldsymbol{Y}_1 \\ \vdots \\ \boldsymbol{Y}_s \end{pmatrix}, \quad \boldsymbol{B} = \begin{pmatrix} \boldsymbol{B}_1 \\ \vdots \\ \boldsymbol{B}_s \end{pmatrix},$$

then $\boldsymbol{Y} = \boldsymbol{B}\boldsymbol{X}$, where $\boldsymbol{B}$ is an $n \times n$ matrix, $\boldsymbol{X}'\boldsymbol{X} = Q_1 + \cdots + Q_s = \boldsymbol{Y}'\boldsymbol{Y} = \boldsymbol{X}'\boldsymbol{B}'\boldsymbol{B}\boldsymbol{X}$, *i.e.* $\boldsymbol{B}$ is an orthogonal matrix, hence $\boldsymbol{Y} \sim N(\boldsymbol{0}, \boldsymbol{I}_n)$. By the definition of Q_i, we know that the Q_i's are mutually independent, and that Q_i has a χ^2 distribution with n_i degrees of freedom. □

Table 4.4.2 Variance calculation table for Zucker rats.

$i =$	1	2	3	4
$\bar{y}_{i++} =$	551.25	571.88	626.75	539.63

$\bar{y}_{+1+} = 654, \quad \bar{y}_{+2+} = 490.75, \quad \bar{y} = 572.38$

Table 4.4.3 ANOVA table for the amount of food intake of Zucker rats.

Source of variation	Sum of squares	Degrees of freedom	F-value	F_α
Age in weeks (A)	$Q_1 = 35803.12$	3	5.84	$F^A_{0.01}$=4.60
Type of rats (B)	$Q_2 = 213178.24$	1	104.28	$F^B_{0.01}$=7.68
Random error	$Q_3 = 55196.14$	27		
Total	$Q = 304177.5$	31		

By calculation, we can obtain that the LR tests for $H_{0\theta}$ and $H_{0\nu}$ are

$$F_A = \frac{Q_1/(k-1)}{Q_3/(n-k-r+1)} \quad \text{and} \quad F_B = \frac{Q_2/(r-1)}{Q_3/(n-k-r+1)}, \tag{4.4.5}$$

respectively. If we consider Q_1 as error caused by Factor A_i, *i.e.* row effects, Q_2 as error caused by Factor B_j, *i.e.* column effects, and Q_3 as random error, then reject $H_{0\theta}$ when F_A is large and reject $H_{0\nu}$ when F_B is large. It agrees with the intuitive explanation of one-way ANOVA discussed in Section 4.1. From Example 1.1.10, if $H_{0\theta}$ is true, $F_A \sim F(k-1, n-k-r+1)$. If $H_{0\nu}$ is true, $F_B \sim F(r-1, n-k-r-1)$. Consequently we can use F_A and F_B to test $H_{0\theta}$ and $H_{0\nu}$, respectively. Now we discuss the testing method by the following example.

Example 4.4.2. (Two-factor ANOVA table) Continue to analyze the problem given in Example 4.4.1. We can obtain Table 4.4.2 by using the data given by Table 4.4.1 of Example 4.4.1. Based on the calculating results given in Table 4.4.2, we can get an ANOVA table for the data (see Table 4.4.3).

Analyzing Table 4.4.3 we can observe that, no matter whether the age A or the type of rats B affects the amount of food intake of Zucker rats significantly, especially the effect of the type of rats B is extremely significant. Comparing with the analysis of Example 4.1.2, When the type of rats was studied as a unique factor, we find that age in weeks of the obese rats has no effect on the amount of the food intake. Obviously, the reason for the change of the analysis results is that the two types of rats are put into a model together. Hence it is important to choose a suitable model in practice. For this problem, if we do not want to study the difference between the two types of rats, we should not use the two-factor model. We shall analyze this problem more detailed in the next subsection.

By virtue of the method discussed in Section 4.2, we can consider trend test for two-factor model. In our example, the corresponding testing problems are

$$H_{1\theta} : \theta_1 \leq \cdots \leq \theta_k,$$
$$H_{1\nu} : \nu_1 \leq \cdots \leq \nu_r.$$

Some methods to obtain the MLEs of parameters can be found in Robertson *et al.* (1988). Unfortunately, it is very difficult to calculate the null distribution of the LR test.

4.4.2 *Model with Interaction Effects*

In the ANOVA of two-factor model, we will pay more attention to the interaction effects. Then the model is

$$y_{ijl} = \mu + \theta_i + \nu_j + \omega_{ij} + \varepsilon_{ijl}, \quad i = 1, \cdots, k; \ j = 1, \cdots, r; \ l = 1, \cdots, c, \tag{4.4.6}$$

where ω_{ij} denote the common effects of A_i and B_j, called **interaction effects**. Without the loss of generality, we can suppose that

$$\sum_{i=1}^{k} \omega_{ij} = \sum_{j=1}^{r} \omega_{ij} = 0.$$

For the model (4.4.6), what we are most interested in is the problem of testing

$$H_{0\omega} : \omega_{ij} = 0, \quad i = 1, \cdots k; \ j = 1, \cdots, r.$$

we can obtain the MLE of ω_{ij}

$$\hat{\omega}_{ij} = \bar{y}_{ij+} - \bar{y}_{i++} - \bar{y}_{+j+} + \bar{y}, \tag{4.4.7}$$

in which $\bar{y}_{i++}$, $\bar{y}_{+j+}$ and $\bar{y}$ are given by (4.4.3), and

$$\bar{y}_{ij+} = \frac{1}{c} \sum_{l=1}^{c} y_{ijl}.$$

Table 4.4.4 Calculation results of the interaction effects.

	A_1	A_2	A_3	A_4
B_1	643.75	664.5	674.75	642.00
B_2	467.75	479.25	578.75	437.25

Now we can decompose the last term of (4.4.4) further

$$\begin{aligned}\sum_{i=1}^{k}\sum_{j=1}^{r}\sum_{l=1}^{c}(y_{ijl}-\bar{y}_{i++}-\bar{y}_{+j+}+\bar{y})^2 =& c\sum_{i=1}^{k}\sum_{j=1}^{r}(\bar{y}_{ij+}-\bar{y}_{i++}-\bar{y}_{+j+}+\bar{y})^2 \\ &+\sum_{i=1}^{k}\sum_{j=1}^{r}\sum_{l=1}^{c}(y_{ijl}-\bar{y}_{ij+})^2 \\ =& Q_I+Q_E.\end{aligned} \tag{4.4.8}$$

In conjunction with (4.4.4), we can get $Q = Q_1 + Q_2 + Q_I + Q_E$. It is a direct sum decomposition. From Theorem 4.4.1, when $H_{0\omega}$ is true, $Q_I/\sigma^2 \sim \chi^2(k-1)(r-1)$ and $Q_E/\sigma^2 \sim \chi^2(kr(c-1))$. Then the LR test for $H_{0\omega}$ is

$$F_{AB} = \frac{Q_I/[(k-1)(r-1)]}{Q_E/[kr(c-1)]}. \tag{4.4.9}$$

If F_{AB} is large then reject $H_{0\omega}$. Still from Example 1.1.10, when $H_{0\omega}$ is true, $F_{AB} \sim F((k-1)(r-1), kr(c-1))$. As an intuitive explanation, we can consider Q_I as the error caused by interaction effects, Q_E as random error. By the discussion from model (4.1.1) to model (4.4.1), then to model (4.4.6), our basic researching idea is based on the direct sum decomposition, *i.e.* separating the effects of Factor B and the interaction effect of $A\bigcap B$ from the error terms, then considering the residual term as a new error term, and finally making a comparison.

Example 4.4.3. (ANOVA table for interaction effects) Continue to analyze the data in Example 4.4.1, the y_{ij+}'s are given by Table 4.4.4.

Note that the data in Table 4.4.4 conform to those in Table 4.1.1. Using Eq. (4.4.8) we can calculate Q_I and Q_E, and F_{AB} accordingly.

Based on Table 4.4.5, we can observe that the interaction effects don't affect the amount of food intake of Zucker rats, that is, we should study the change of the amount of food intake separately for two types of rats. In a sense, we should adopt model (4.1.1) and then support the conclusion given in Example 4.1.2.

Table 4.4.5 ANOVA table for the amount of food intake of Zucker rats.

Source of variation	Sum of squares	Degrees of freedom	F-value	F_α
Interaction effects	$Q_I = 9050.14$	3	1.57	$F_{0.05}$=3.01
Random error	$Q_E = 46146.00$	24		$F_{0.05}$=4.72
Total	$Q_3 = 55196.14$	27		

Based on the analysis of the above example, we can see that it is usually very complicated to study a practical problem. Therefore it is very important to select a suitable model.

4.5 Multi-Factor Model: Log-linear Models

In this section we will discuss the testing problem of contingency table based on log-linear models, we mainly focus on the two and three-dimensional contingency tables, and discuss how to deepen our research by using ancillary parameters.

4.5.1 *Two-Dimensional Contingency Table*

Use X and Y to define an $I \times J$ two-dimensional contingency table. For $i \in \{1, \cdots, I\}$ and $j \in \{1, \cdots, J\}$, n_{ij} denote the frequency of cell (i, j) , and N_{ij} the corresponding random variable, and p_{ij} the probability of event occurrence, that is, $p_{ij} = P\{N_{ij} = n_{ij}\} = P\{X = i, Y = j\}$. Furthermore, let $\mu_{ij} = EN_{ij}$. Suppose that

$$\ln \mu_{ij} = \lambda + \lambda_i^X + \lambda_j^Y + \lambda_{ij}^{XY}, \quad i = 1, \cdots, I; \quad j = 1, \cdots, J. \qquad (4.5.1)$$

The model (4.5.1) is called **log-linear model**, where $\sum_i \lambda_i^X = 0, \sum_j \lambda_j^Y = 0, \sum_i \lambda_{ij}^{XY} = \sum_j \lambda_j^{XY} = 0$. Comparing with (4.4.6), the λ_i^X's, λ_j^Y's, and λ_{ij}^{XY}'s respectively represent the row effects the column effects and the interaction effects. Sometimes (4.5.1) is also called a **saturated model**. If all $\lambda_{ij}^{XY} = 0$, then (4.5.1) can be reduced to the following form:

$$\ln \mu_{ij} = \lambda + \lambda_i^X + \lambda_j^Y, \quad i = 1, \cdots, I; \ j = 1, \cdots, J, \qquad (4.5.2)$$

which is called an **incomplete model**. Recall the discussion on contingency table in Section 2.3, for the two-dimensional case, what we are concerned is the independence between X and Y, and we have

$$X \perp\!\!\!\perp Y \Longleftrightarrow p_{ij} = p_{i+}p_{+j}, \quad \forall i, j,$$

where $p_{i+} = \sum_j p_{ij}$ and $p_{+j} = \sum_i p_{ij}$. Recall the odds ratio we have ever discussed in Section 2.1, for the model (4.5.1), we define

$$\theta_{ij} = \frac{p_{ij}p_{i+1,j+1}}{p_{i,j+1}p_{i+1,j}}, \quad i = 1, \cdots, I-1; \; j = 1, \cdots, J-1,$$

which is called a **local odds ratio**. It is easy to verify that if $\theta_{ij} = 1$ for $\forall i, j$, then

$$\frac{p_{11}}{p_{i1}} = \frac{p_{12}}{p_{i2}} = \cdots = \frac{p_{1J}}{p_{iJ}}, \quad i = 1, \cdots, I,$$

so $p_{ij} = p_{i+}p_{+j}$. Similarly we can obtain the following result

$$p_{ij} = p_{i+}p_{+j}, \forall i, j \Longleftrightarrow \theta_{ij} = 1, \quad \forall i, j.$$

Now let us go back to the discussion about model (4.5.1). Assume that N_{ij} has a **multinomial distribution** when n is fixed, then $\mu_{ij} = nP_{ij}$, so $\theta_{ij} = (\mu_{ij}\mu_{i+1j+1})/(\mu_{ij+1}\mu_{i+1j})$. Hence $\theta_{ij} = 1$ is equivalent to $\lambda_{ij}^{XY} + \lambda_{i+1j+1}^{XY} = \lambda_{ij+1}^{XY} + \lambda_{i+1j}^{XY}$ for $i = 1, \cdots, I-1, j = 1, \cdots, J-1$. Thus we can obtain

$$\theta_{ij} = 1, \quad \forall i, j \Longleftrightarrow \lambda_{ij}^{XY} = 0, \quad \forall i, j.$$

Consequently, as far as the model (4.5.1) is concerned, testing whether X and Y are independent is equivalent to the problem of testing

$$H_0 : \lambda_{ij}^{XY} = 0, \quad i = 1, \cdots, I; \; j = 1, \cdots, J. \tag{4.5.3}$$

Obviously, if H_0 is true, then the model (4.5.1) can be reduced to the model (4.5.2). So the testing problem of log-linear model for contingency table **depends on the model selection** in a great degree.

4.5.1.1 *AIC and model selection*

Let M denote a model, and θ^M denote the MLE of parameter θ under the model M. For the multinomial distribution (see Example 1.4.4), the log-likelihood function is

$$L(M) = C + \sum_{i=1}^{I} \sum_{j=1}^{J} n_{ij} \ln p_{ij}^M, \tag{4.5.4}$$

where $C = \ln(n!/\prod_{i=1}^I \prod_{j=1}^J n_{ij})$. If we use M_1 and M_0 to denote the saturated model (4.5.1) and the model (4.5.2) respectively, then we have

$$\begin{aligned} L(M_1) &= C + \sum_{i=1}^{I} \sum_{j=1}^{J} n_{ij} \ln n_{ij} - n \ln n, \\ L(M_0) &= C + \sum_{i=1}^{I} n_{i+} \ln n_{i+} + \sum_{j=1}^{J} n_{+j} \ln n_{+j} - 2n \ln n, \end{aligned} \tag{4.5.5}$$

where $n_{i+} = \sum_{j=1}^{J} n_{ij}$ and $n_{+j} = \sum_{i=1}^{I} n_{ij}$. According to the maximum likelihood criterion, the larger the maximum log-likelihood value is, the better the corresponding model. But the maximum log-likelihood value depends on the size of parameter space, such as if $M_0 \subset M_1$, then $L(M_0) < L(M_1)$. Akaike information criterion (AIC) is a model selection criterion proposed for eliminating this kind of influence (see Burnham and Anderson (1998); Sakamoto *et al.* (1999)). It is defined as follows:

$$\mathrm{AIC}(M) = -2(L(M) - d(M)), \tag{4.5.6}$$

where $d(M)$ is the dimension of parameter space under the model M. For example, $d(M_1) = I \times J - 1 = n - 1$ and $d(M_0) = I - 1 + J - 1 = I + J - 2$. According to the maximum likelihood criterion again, the smaller $\mathrm{AIC}(M)$, the better the model obviously.

Example 4.5.1. Let us analyze the data given in Table 2.3.1 further. The data were used to analyze whether there exists racial discrimination in the death penalty verdict in USA. Based on the data set, we can obtain the maximum likelihood values in (4.5.5) as follows:

$$\begin{aligned} L(M_1) &= C + \sum_{i=1}^{2}\sum_{j=1}^{2} n_{ij}\ln n_{ij} - n\ln n \\ &= C + 53\ln 53 + 430\ln 430 + 15\ln 15 + 176\ln 176 - 674\ln 674 \\ &= C + 3768.5 - 4389.9 \\ &= C - 621.4, \\ L(M_0) &= C + \sum_{i=1}^{2} n_{i+}\ln n_{i+} + \sum_{j=1}^{2} n_{+j}\ln n_{+j} - 2n\ln n \\ &= C + 68\ln 68 + 606\ln 606 + 483\ln 483 + 191\ln 191 - 2\times 674\ln 674 \\ &= C + 3767.7 - 4389.9 \\ &= C - 622.2. \end{aligned}$$

Furthermore, $d(M_1) = 4 - 1 = 3$ and $d(M_0) = 4 - 2 = 2$. Thus by (4.5.5) we can obtain

$$\begin{aligned} \mathrm{AIC}(M_1) &= -2(C - 621.4 - 3) = 1248.8 - 2C, \\ \mathrm{AIC}(M_0) &= -2(C - 622.2 - 2) = 1248.4 - 2C. \end{aligned}$$

Since $\mathrm{AIC}(M_0) < \mathrm{AIC}(M_1)$, model M_0 would be better, that is, X and Y can be considered to be independent, which coincides with the analysis result in Section 2.3. From this example we can see that the dimension of parameter space will affect the analysis result.

4.5.1.2 *Testing model*

We cannot essentially test a saturated model because we have no way to determine random error. What we are discussing is to test whether model (4.5.2) is true or not under the assumption that (4.5.1) holds, that is, if (4.5.2) is not true, then (4.5.1) holds. This is equivalent to testing H_0 given by (4.5.3), and the corresponding alternative hypothesis is

$$H_1: \ \lambda_{ij}^{XY} \neq 0, \ \ \exists\, i, j.$$

For the given model M, let

$$G^2(M) = -2L(M),$$

where $L(M)$ is given by (4.5.4). Let M_0 and M_1 denote the models given by (4.5.1) and (4.5.2) respectively, the log-likelihood difference between these two models is

$$\begin{aligned} G^2(M_0|M_1) &= G^2(M_0) - G^2(M_1) \\ &= 2\sum_{i=1}^{I}\sum_{j=1}^{J} n_{ij} \ln \frac{n\, n_{ij}}{n_{i+}n_{+j}}. \end{aligned} \tag{4.5.7}$$

This is just the LR test. Reject M_0 when $G^2(M_0|M_1)$ is large. By Theorem 2.2.2, when M_0 is true, $G^2(M_0|M_1) \xrightarrow{L} \chi^2_{(I-1)(J-1)}$. This result coincides with Theorem 2.3.3 since the Pearson χ^2 test is an approximation to the LR test (recall the discussion in Section 2.3). By virtue of Eq. (4.5.7), we can construct the following AIC criterion to compare two models:

$$\mathrm{AIC}(M_0|M_1) = G^2(M_0|M_1) - 2(d(M_1) - d(M_0)). \tag{4.5.8}$$

If it is negative then M_0 is better than M_1, otherwise M_1 is better than M_0. We have ever discussed that H_0 is usually protected in testing problems, that is, it will only be rejected if there is strong evidence against it in favor of H_1. Therefore, the model selection criterion is inclined to treat the two models with equal emphasis. Let us see the following example.

Example 4.5.2. (Continuation of Example 4.5.1) Based on the calculation result in Example 4.5.1, we can obtain the following result by (4.5.7)

$$G^2(M_0|M_1) = 2 \times (3768.5 - 3767.7) = 1.6.$$

Compute the p-value, $P\{\chi_1^2 \geq 1.6\} \approx 0.19$. As this probability is large, we cannot reject H_0, that is, consider that the death penalty verdict and the defendant's race is independent. This result is almost the same as that of the goodness-of-fit test in Section 2.3. Furthermore, we can see that the approximation is very satisfactory.

Next let us calculate the AIC criterion. From Eq. (4.5.8)

$$\mathrm{AIC}(M_0|M_1) = 1.6 - 2 \times (3 - 2) = -0.4.$$

Since it is negative, we can consider that the model M_0 is better than M_1. In this example, we can see that we should select the model M_1 based on the AIC criterion. On the other hand, since $P\{\chi^2 \geq 3.8\} = 0.05$, we reject H_0 and select the model M_1 only when the log-likelihood $G^2(M_0|M_1)$ is larger than 3.8. Hence we can see that H_0 is protected in the testing problem.

4.5.2 *Three-Dimensional Contingency Table*

Use X, Y and Z to define an $I \times J \times K$ three-dimensional contingency table. Let n_{ijk} denote the frequency of cell (i, j, k). Let N_{ijk} denote the corresponding r.v., and p_{ijk} be the probability of event occurrence. Let $\mu_{ijk} = EN_{ijk}$. Similar to model (4.5.1), here the saturated log-linear model is

$$\ln \mu_{ijk} = \lambda + \lambda_i^X + \lambda_j^Y + \lambda_k^Z + \lambda_{ij}^{XY} + \lambda_{ik}^{XZ} + \lambda_{jk}^{YZ} + \lambda_{ijk}^{XYZ}. \qquad (4.5.9)$$

We still assume that

$$\begin{aligned}
\sum_i \lambda_i^X &= \sum_i \lambda_{ij}^{XY} = \sum_i \lambda_{ik}^{XZ} = \sum_i \lambda_{ijk}^{XYZ} = 0, \\
\sum_j \lambda_j^Y &= \sum_j \lambda_{ij}^{XY} = \sum_j \lambda_{jk}^{YZ} = \sum_j \lambda_{ijk}^{XYZ} = 0, \\
\sum_k \lambda_k^Z &= \sum_k \lambda_{ik}^{XZ} = \sum_k \lambda_{ik}^{YZ} = \sum_k \lambda_{ijk}^{XYZ} = 0.
\end{aligned}$$

As in the two-way case, what we are interested in is the independence of X, Y and Z. Obviously, the mutual independence of the three variables can be expressed as:

$$X \perp\!\!\!\perp Y \perp\!\!\!\perp Z \Leftrightarrow \lambda_{ij}^{XY} = \lambda_{ik}^{XZ} = \lambda_{jk}^{YZ} = \lambda_{ik}^{XZ} = \lambda_{ijk}^{XYZ} = 0, \qquad (4.5.10)$$

usually denoted as (X, Y, Z).

Independence of one variable from the other two can be formulated as

$$\begin{aligned}
X \perp\!\!\!\perp Y, X \perp\!\!\!\perp Z &\Leftrightarrow \lambda_{ij}^{XY} = \lambda_{ik}^{XZ} = \lambda_{ijk}^{XYZ} = 0; \\
Y \perp\!\!\!\perp X, Y \perp\!\!\!\perp Z &\Leftrightarrow \lambda_{ij}^{XY} = \lambda_{jk}^{YZ} = \lambda_{ijk}^{XYZ} = 0; \\
Z \perp\!\!\!\perp X, Z \perp\!\!\!\perp Y &\Leftrightarrow \lambda_{ik}^{XZ} = \lambda_{jk}^{YZ} = \lambda_{ijk}^{XYZ} = 0,
\end{aligned} \qquad (4.5.11)$$

usually denoted as (X, YZ), (Y, XZ), and (Z, XY), respectively. In addition to that, we will pay more attention to the conditional independence for a three-way contingency table. X and Y are called to be conditionally independent given Z, if for $\forall Z = k$ and $\forall i, j$ we have

$$P(X = i, Y = j|Z = k) = P(X = i|Z = k)P(Y = j|Z = k),$$

denoted as $X \perp\!\!\!\perp Y|Z$. If we treat the above case as a two-way contingency table given that $Z = k$, then $p_{ijk} = p_{i+k}p_{+jk}$. In this case, the **conditional local odds ratio** can be defined as

$$\theta_{ij(k)} = \frac{p_{ijk}p_{i+1,j+1,k}}{p_{i,j+1,k}p_{i+1,j,k}}, \quad i = 1, \cdots, I-1; j = 1, \cdots, J-1; k = 1, \cdots, K.$$

Similarly we can define $\theta_{(i)jk}$ and $\theta_{i(j)k}$. Recall the discussion on odds ratio in two-dimensional contingency tables. We can obtain

$$\begin{aligned} X \perp\!\!\!\perp Y|Z &\Leftrightarrow \theta_{ij(k)} = 1, \quad \forall i, j, k; \\ X \perp\!\!\!\perp Z|Y &\Leftrightarrow \theta_{i(j)k} = 1, \quad \forall i, j, k; \\ Y \perp\!\!\!\perp Z|X &\Leftrightarrow \theta_{(i)jk} = 1, \quad \forall i, j, k. \end{aligned} \tag{4.5.12}$$

Next let us go back to the log-linear model (4.5.9). Consider a multinomial distribution model, *i.e.* $N_{ijk} \sim N(n, p_{ijk})$, where n denotes the total sum of frequency. Then we have $\mu_{ijk} = EN_{ijk} = np_{ijk}$. So we can use the means to express the conditional local odds ratio. Similar to the two-way contingency table, by (4.5.12), the conditional independence of two variables given a third one can be formulated as:

$$\begin{aligned} X \perp\!\!\!\perp Y|Z &\Leftrightarrow \lambda_{ij}^{XY} = \lambda_{ijk}^{XYZ} = 0, \quad \forall i, j, k; \\ X \perp\!\!\!\perp Z|Y &\Leftrightarrow \lambda_{ik}^{XZ} = \lambda_{ijk}^{XYZ} = 0, \quad \forall i, j, k; \\ Y \perp\!\!\!\perp Z|X &\Leftrightarrow \lambda_{jk}^{YZ} = \lambda_{ijk}^{XYZ} = 0, \quad \forall i, j, k, \end{aligned} \tag{4.5.13}$$

usually denoted as (XZ, YZ), (XY, ZY), and $YX, ZX)$, respectively. Thus we have given three kinds of independence, and related them to the parameters in the log-linear model. Since every kind of independence demands that $\lambda_{ijk}^{XYZ} = 0$, so we will assume that this condition holds when we study the independence problems.

We can utilize the criterion or G^2 to select an appropriate model. Both methods need the calculation of maximum likelihood. Here, we give some calculation details.

For the model (X, Y, Z), since $p_{ijk} = p_{i++} \cdot p_{+j+} \cdot p_{++k}$, by (4.5.10), we have

$$\begin{aligned} L((X, Y, Z)) = c &+ \sum_i n_{i++} \ln n_{i++} + \sum_j n_{+j+} \ln n_{+j+} \\ &+ \sum_k n_{++k} \ln n_{++k} - 3n \ln n, \end{aligned}$$

where $c = \ln(n!/\Pi_i\Pi_j\Pi_k n_{ijk})$. For the model (X, YZ), since

$$\begin{aligned} p_{ijk} &= P\{X = i, Y = j, Z = k\} \\ &= P\{Y = j, Z = k|X = i\}P\{X = i\} \\ &= p_{i++}p_{+jk}, \end{aligned}$$

by (4.5.11), we have

$$L((X, YZ)) = c + \sum_i n_{i++} \ln n_{i++} + \sum_j \sum_k n_{+jk} \ln n_{+jk} - 2n \ln n.$$

For the model (XY, YZ), since

$$\begin{aligned} p_{ijk} &= P\{X = i, Y = j, Z = k\} \\ &= P\{X = i|Z = k\}P\{Y = j|Z = k\}P\{Z = k\} \\ &= p_{i+k}p_{+jk}/p_{++k}, \end{aligned}$$

by (4.5.14) we have

$$\begin{aligned} L((XZ, YZ)) = c &+ \sum_i \sum_k n_{i+k} \ln n_{i+k} + \sum_j \sum_k n_{+jk} \ln n_{+jk} \\ &- \sum_k n_{++k} \ln n_{++k} - n \ln n. \end{aligned}$$

Finally, for the three-way saturated model (4.5.9), denoted as (XYZ), the maximum likelihood is

$$L((XYZ)) = c + \sum_i \sum_j \sum_k n_{ijk} \ln n_{ijk} - n \ln n.$$

Next we discuss the problem of **degrees of freedom** of the models. Based on the assumption of the multinomial distribution, it can be verified that for the parameters in the saturated model (4.5.9), its corresponding degrees of freedom can be written as

$$(I - 1) + (J - 1) + (K - 1) + (I - 1)(J - 1) + (I - 1)(K - 1)$$

$$+(J - 1)(K - 1) + (I - 1)(J - 1)(K - 1) = IJK - 1.$$

Thus, we can determine the degrees of freedom of the models based on (4.5.10), (4.5.11) and (4.5.13). Now let's discuss the method of computing maximum likelihood and the technique of making statistical inference by the following example.

Example 4.5.3. (Continuation of Examples 4.5.1 and 4.5.2) Discuss the data given in Tables 2.3.1 and 2.3.2. Let X and Z respectively represent the races of defendant and victim, and Y the death penalty verdict. We can obtain the following table.

Table 4.5.1 Model selection.

Model	$L(M)$	$d(M)$	$G^2(M)$	AIC(M)
(X, Y, Z)	−990.40	3	1980.80	1986.80
(X, YZ)	−981.99	4	1963.98	1971.98
(Y, XZ)	−1532.08	4	3064.16	3072.16
(Z, XY)	−989.63	4	1979.26	1987.26
(XZ, YZ)	−791.70	5	1583.40	1593.40
(XY, ZY)	−981.22	5	1962.44	1972.44
(XYZ)	−789.00	7	1578.00	1592.00

From Table 4.5.1, we can see that the two models (XYZ) and (XZ, YZ) yield smaller values of AIC. This result suggests that the two models may be appropriate. The former is a saturated models, showing that there does not exist any kind of independent relationship among X, Y, and Z, in other words, the death penalty verdict is related to the race. The latter shows that X and Y are conditionally independent given Z, thus Z (*i.e.* the victim's race) is a key variable. Next we judge whether there is an essential difference between the two models by testing methods. By Eq. (4.5.8) we can obtain

$$\begin{aligned} G^2(XZ, YZ|XYZ) &= G^2(XZ, YZ) - G^2(XYZ) \\ &= 1583.40 - 1578.00 = 5.4. \end{aligned}$$

From Table 4.5.1, we can get that the degrees of freedom are $d(\text{XYZ}) - d(XZ, YZ) = 7 - 5 = 2$ and the p-value is $P\{\chi_2^2 \geq 5.4\} \approx 0.07$. This probability neither rejects nor supports the model (XZ, YZ) strongly. Therefore, the final conclusion is that the death penalty verdict is related to the race, especially the victim's race.

4.5.3 *Ancillary Parameter Model*

When in the discussion of linearity test in Section 4.3.1, we have ever used an ancillary parameter to reduce the number of parameters. Now for model (4.5.1), we discuss this method further. Since what we are concerned is the independence problem between X and Y, so it can be rewritten as

$$\ln \mu_{ij} = \lambda + \lambda_i^X + \lambda_i^Y + \beta c_i d_j, \; i = 1, \cdots, I; j = 1, \cdots, J, \tag{4.5.14}$$

where c_i and d_j are given constants. Without the loss of generality, let $\sum_i c_i = \sum_j d_j = 0$. Sometimes to study the trend of change we can assume that $c_1 \leq \cdots \leq c_I$ and $d_1 \leq \cdots \leq d_J$. Here we consider the problem of testing

$$H_0 : \beta = 0 \quad \text{vs.} \quad H_1 : \beta > 0. \tag{4.5.15}$$

Consider a most general model, *i.e.* Poisson model, the corresponding log-likelihood function is

$$\begin{aligned} L(\mu) &= \sum_i \sum_j n_{ij} \ln \mu_{ij} - \sum_i \sum_j n_{ij} \mu_{ij} + c \\ &= n\lambda + \sum_i n_{i+}\lambda_i^X + \sum_j n_{+j}\lambda_j^Y + \beta \sum_i \sum_j c_i d_j n_{ij} \\ &\quad - \sum_i \sum_j \exp(\lambda + \lambda_i^X + \lambda_j^Y + \beta c_i d_j) + c, \end{aligned} \tag{4.5.16}$$

where c is a constant independent of the parameters. Recall the definition of local odds ratio,

$$\ln \theta_{ij} = \beta(c_i - c_{i+1})(d_j - d_{j+1}), \quad i = 1, \cdots, I-1; j = 1, \cdots, J-1.$$

If assume that both the constants c_i and d_j are equally spaced, such as $c_i = i$ and $d_j = j$, we can see that model (4.5.14) is equivalent to that the local odds ratio is a constant. Hence the problem of testing (4.5.16) becomes the discussion of the independence between X and Y.

Let $\mu_{i+} = \sum_j \mu_{ij}$ and $\mu_{+j} = \sum_i \mu_{ij}$. By differentiating the log-likelihood function, we can obtain that $\hat{\mu}_{ij}$ is the MLE of μ_{ij} if and only if $\hat{\mu}_{ij}$ is the solution of the following equation

$$\begin{aligned} \mu_{i+} &= n_{i+}, \quad i = 1, \cdots, I-1, \\ \mu_{+j} &= n_{+j}, \quad j = 1, \cdots, J-1, \\ \sum_i \sum_j c_i d_j \mu_{ij} &= \sum_i \sum_j c_i d_j n_{ij}. \end{aligned} \tag{4.5.17}$$

We can use some methods, such as the Newton-Raphson method, to get $\hat{\mu}_{ij}$, then the LR test statistic can be written as

$$-2 \ln \Lambda = 2 \sum_i \sum_j n_{ij} \ln \frac{n_{ij}}{\hat{\mu}_{ij}}.$$

By Theorem 2.2.2, when H_0 is true, the limiting distribution of the above equation is a χ^2 distribution with $d = IJ - I - J$ degrees of freedom. For more detailed discussion see Agresti (2002).

4.6 Problems

4.1. Let $\boldsymbol{x} = (x_1, \cdots, x_k)'$ be a sample from a normal population $N(\boldsymbol{\mu}, \boldsymbol{I})$, where $\boldsymbol{\mu} = (\mu_1, \cdots, \mu_k)'$ is an unknown k-dimensional parameter vector. Let $T(\boldsymbol{x}) = \sum_{i=1}^{k} c_i x_i$ denote a linear test statistic for testing

$$H_0 : \mu_1 = \cdots = \mu_k \quad \text{vs.} \quad H_1 : \mu_1 \geq \cdots \geq \mu_k,$$

where $\sum_{i=1}^{j} c_i \geq 0$ $(j = 1, \cdots, k-1)$, and $\sum_{i=1}^{k} c_i = 0$.

(a) Prove that for any $\boldsymbol{\mu} \in H_1$, we have $T(\boldsymbol{\mu}) \geq 0$;
(b) Find the rejection region for the above test.

4.2. Suppose that $\boldsymbol{X}$ has a normal distribution $N(\mathbf{0}, \boldsymbol{\Sigma})$ with density function $f(\boldsymbol{x})$, where

$$\boldsymbol{\Sigma} = \begin{bmatrix} 1 & \rho_{12} & \cdots & \rho_{1k} \\ \rho_{21} & 1 & \cdots & \rho_{2k} \\ \vdots & \vdots & \ddots & \vdots \\ \rho_{k1} & \rho_{k2} & \cdots & 1 \end{bmatrix}.$$

Prove that

$$\frac{\partial^2 f(\boldsymbol{x})}{\partial x_i \partial x_j} = \frac{\partial f(\boldsymbol{x})}{\partial \rho_{ij}}.$$

4.3. Suppose that $\boldsymbol{X}_k$ has a k-dimensional normal distribution $N(\mathbf{0}, \boldsymbol{\Sigma})$. Let $p_k = P\{X_1 \geq 0, \cdots, X_k \geq 0\}$, *i.e.* p_k is the positive orthant probability. Prove that

$$p_1 = \frac{1}{2}; \quad p_2 = \frac{1}{4} + \frac{1}{2\pi} \arcsin \rho_{12};$$

$$p_3 = \frac{1}{8} + \frac{1}{4\pi}(\arcsin \rho_{12} + \arcsin \rho_{13} + \arcsin \rho_{23}),$$

where ρ_{ij} is the correlation coefficient between X_i and X_j.

4.4. Suppose that X is a random variable with mean θ and variance $g(\theta)$. Verify that

(a) Let $f(x)$ be a differentiable function of x, we approximately have $V_\theta[f(x)] = [\frac{d}{d\theta} f(\theta)]^2 g(\theta)$.
(b) Let $f(x) = \int 1/\sqrt{g(x)}dx$, then the approximate variance of $f(x)$ is independent of θ.
(c) If x has a Poisson distribution $P(\theta)$, and $f(x) = \sqrt{g(x)}$, then the variance of $f(x)$ is independent of θ.
(d) If x has a binomial distribution $Bi(n;p)$, and $f(x) = \arcsin(\sqrt{x/n})$, then the variance of $f(x)$ is independent of p.

The above results show that, for some distributions whose mean and variance are correlative, we can make the mean and the variance become approximately independent by choosing a suitable transformation (see Bar-Lev and Enis, 1988, 1990).

4.5. For k binomial populations, $Bi(n_i, p_i)$, $i = 1, \cdots, k$, let x_i be a sample drawn from the i-th population. Using the transformation given in Problem 4.4, construct a test statistic for testing the null hypothesis $H_0 : p_1 = \cdots = p_k$ and compute the distribution of the statistic when H_0 is true.

4.6. For a 2×2 contingency table, given the marginal totals,

(a) Prove that n_{11} has a hypergeometric distribution, *i.e.*

$$P(n_{11}|n_{1+}, n_{2+}, n_{+1}, n_{+2}) = \frac{\binom{n_{1+}}{n_{11}}\binom{n_{2+}}{n_{+1}-n_{11}}}{\binom{n}{n_{+1}}},$$

where $n_{11} = \max(0, n_{1+} + n_{+1} - n), \cdots, \min(n_{1+}, n_{+1})$;

(b) Find a UMP test for the problem of testing $H_0 : \theta = 1$ vs. $H_1 : \theta > 1$, where θ is the odds ratio.

4.7. For model (4.5.9), if $\lambda_{ijk}^{XYZ} \equiv 0$ holds for all i, j, k, then the conditional local odds ratio satisfies:

$$\theta_{ij(1)} = \theta_{ij(2)} = \cdots = \theta_{ij(K)};$$

$$\theta_{i(1)k} = \theta_{i(2)k} = \cdots = \theta_{i(J)k};$$

$$\theta_{(1)jk} = \theta_{(2)jk} = \cdots = \theta_{(I)jk}.$$

4.8. Let $\boldsymbol{p} = (p_1, \cdots, p_k)'$ and $\boldsymbol{q} = (q_1, \cdots, q_k)'$ be two probability vectors, verify the following information inequality

$$\sum_{i=1}^{k} |p_i - q_i| \le \left[2\sum_{i=1}^{k} p_i \ln \frac{p_i}{q_i}\right]^{1/2}.$$

4.9. For model (4.5.9), if $\lambda_{ijk}^{XYZ} \equiv 0$ holds for all i, j, k, then the MLE of μ_{ijk} exists and is unique satisfying

$$\mu_{ij+} = n_{ij+}; \quad \mu_{i+k} = n_{i+k}; \quad \mu_{+jk} = n_{+jk}$$

for $i = 1, \cdots, I$, $j = 1, \cdots, J$, and $k = 1, \cdots, K$.

4.10. For the above problem the MLEs of the μ_{ijk}'s can be obtained by an iterative algorithm, IPFP (Iterative Proportional Fitting Procedure):

$$\mu_{ijk}^{(m,1)} = \mu_{ijk}^{(m-1,3)}\left[n_{ij+}/\mu_{ij+}^{(m-1,3)}\right];$$
$$\mu_{ijk}^{(m,2)} = \mu_{ijk}^{(m,1)}\left[n_{i+k}/\mu_{i+k}^{(m,1)}\right];$$
$$\mu_{ijk}^{(m,3)} = \mu_{ijk}^{(m,2)}\left[n_{+jk}/\mu_{+jk}^{(m,2)}\right],$$

where the initial values for iteration are chosen satisfying that $\lambda_{ijk}^{XYZ} \equiv 0$ (the simplest method is that all of the initial values are 1). Verify that when $m \to \infty$, the given sequence converges to the MLE.

4.11. For model (4.5.9), using the data set given by Table 2.3.2 to test H_0: $\lambda_{ijk}^{XYZ} \equiv 0$.

4.12. For an $r \times c$ contingency table, suppose that the local odds ratio is

$$\frac{p_{ij}p_{i+1,\,j+1}}{p_{i,\,j+1}p_{i+1,\,j}} = \psi_{ij}.$$

Find the maximum likelihood estimation equation of p_{ij} and construct an iterative algorithm similar to that given in Problem 4.10 when ψ_{ij} is given.

4.13. **(Kruskal-Wallis Test)** It is an analogue of the F-test used in ANOVA. However, ANOVA tests depend on the assumption that all populations under comparison are normally distributed, while the Kruskal-Wallis Test places no such restriction. When we use it, rank all of the observations y_{ij} from smallest to largest firstly, then replace each observations g_{ij} by its rank R_{ij}. In event of tie, each of the tied observations gets the average rank. Let R_{i+} be the sum of the ranks for the i-th level, then the test statistic is

$$T = \frac{1}{S^2}\left[\sum_{i=1}^{k}\frac{R_{i+}^2}{n_i} - \frac{n(n+1)^2}{4}\right],$$

where k is the number of levels, n_i is the number of observations at the i-th level, n is the total number of observations and

$$S^2 = \frac{1}{n-1}\left[\sum_{i=1}^{k}\sum_{i=1}^{n_i}R_{ij}^2 - \frac{n(n+1)^2}{4}\right].$$

If n_i is large enough, then under the null hypothesis H_0 as (4.1.2), T has approximately a χ^2_{k-1} distribution. If $T > \chi^2_{\alpha,k-1}$ then reject H_0. Using the data of $A+B$ given in Tables 4.1.1 and 4.1.2 to test H_0 by the Kruskal-Wallis Test, and compare the testing result with that given in Example 4.1.2.

4.14. **(Random effects model)** If the ν_i's in model (4.1.3) are random variables with normal distribution $N(0,\sigma_\nu^2)$ and independent of the ε_{ij}'s, then (4.1.3) is called a random effects model. The testing problem becomes

$$H_0 : \sigma_\nu^2 = 0 \quad \text{vs.} \quad H_1 : \sigma_\nu^2 > 0.$$

Find an LR test.

4.15. In a class survey, students were asked "how many alcoholic beverages do you consume each week?" Data were classified by gender and seat location (Front, Middle, Back)

	Gender	
Seat location	Male	Female
Front	0	0
	2	1
	1	0
Middle	3	2
	2	1
	1	0
Back	3	2
	5	1
	7	6

(a) Describe the two-way ANOVA design associated to this experiment. (Response, factors, levels, replications)

(b) Give the table of group means, marginal means and the overall mean. Plot the group means and comment on the plot.

(c) Show that the sums of squares for main effects are 36.1 and 6.72, and indicate which factor goes with which sum of squares.

(d) Using previous results find the sums of squares for interaction given that the residuals sum of squares is 28.66 and the total sum of squares is 72.93.

(e) Use results of (c) and (d) to complete the two-way ANOVA table. Are the main effects and interaction significant?

(f) In view of the results, explain why it could be useful to increase the sample size.

4.16. In a study on Occupational choice and Testosterone level, a researcher has recorded Testosterone level in men who were ministers, actors and football players.

Occupation	obs. 1.	obs. 2.
Ministers	0.82	0.75
Actors	1.75	2.50
Football players	2.30	2.72

(a) State the ANOVA (H) for this experiment and find the group sample means and the group sample variances.

(b) Find the Pooled estimate of the variance.

(c) Show that the group sum of squares is 3.27 given that the total sums of squares is 3.65.

(d) Write up the complete ANOVA table.

(e) Does the data support (H) or not? (Justify your answer)

(f) Does the Pooled estimate of the variance reflect the variation between groups or within groups?

4.17. An economist for government is constructing a study of vegetable prices in various metropolitan areas for the purpose of estimating the cost to an individual of maintaining a diet containing all the recommended daily amounts of nutrients. The following data give the retail prices per kilogram of four basic vegetables in seven major metropolitan areas:

	Lettuce	Tomatoes	Carrots	Onions	Average
Adelaide	81	87	87	85	85
Brisbane	84	95	92	79	87.5
Canberra	89	104	97	94	96
Melbourne	87	110	89	84	92.5
Darwin	75	99	83	95	88
Perth	79	92	81	88	85
Sydney	86	99	87	91	90.75
Average	83	98	88	88	89.25

(a) Complete the following Two-Way ANOVA table

Source of variation	Sum of squares	Degrees of freedom	Mean square	F-value
Vegetables				
Cities	396.5			
Error				
Total	1699.25	27		

(b) Do the results of the survey show that vegetable prices differ among the metropolitan areas studied in terms of the p-value?

(c) Does the study indicate that there are substantial price differences among the various vegetables in terms of the p-value?

(d) Describe the statistical model underlying the analysis and indicate the null hypothesis used in (b) and (c).

4.18. A department store is considering building a new store at one of four different sites. One of the main factors in the decision is the annual income of the households in the four areas. In a preliminary study, a random sample of residents in each area is asked to state their annual income. The results and the totals for each column are given below:

	Annual income		
Area 1	Area 2	Area 3	Area 4
42	18	27	30
26	12	51	29
29	20	19	17
23	25	22	31
18	48	32	27
	32	46	25
	27	30	
		35	
		32	
138	182	294	159

(a) Given that $\sum\sum x_{ij} = 773$ and $\sum\sum x_{ij}^2 = 24493$, perform an analysis of variance to test the null hypothesis of no difference in mean incomes of four areas. Present your results as an ANOVA table. Use level $\alpha = 0.05$.

(b) The analysis of part (a) assumes the observations come from populations with the same variance σ^2. What other assumptions are necessary?

(c) Show that a point estimate of σ^2 is 92.8. Give a 90% confidence interval for the common variance, σ^2.

4.19. A two factor experiment (factors are A and B) has been conducted to investigate the effect of interaction between the levels of each factor. Three different response levels for each factor have been used and three observations taken from each cell (that is, there is a total of 27 observations). The following part ANOVA table has been prepared and you are asked to conduct the test.

Source of variation	Sum of squares	Degrees of freedom	Mean square	F-value
Factor A	20.22	2	10.11	4.20
Factor B	8.22	*	*	1.71
Interaction A×B	46.22	*	*	*
Error	*	*	*	
Total	118	26		

Fill the missing entries * of the above table and test the null hypothesis of no interaction at 1% level of significance.

Chapter 5

Parametric Tests for Regression Models

The previous chapter discussed the problems of testing hypotheses about model parameters. However, in essence, we just use parameters to denote the means of random variables, and explore the laws of random variables by testing parameters. In this chapter we will establish some correlations among variables and explore the change laws by testing parameters. In the first section, we still discuss the parameter expression of the means of random variables which has a bit more complicated form as well as a preparation for future discussion.

5.1 Linear Models

5.1.1 *Establishment of Model*

First, we give an example to present the idea of constructing models. In economics, there is a well-known law of consumption known as Keynes' Psychological Law which states that: "households increase their consumption as their income increases, but not as much as their income increases." (Keynes, 1936, p. 96). We use the symbols x and y to denote income and expenditure, respectively. Then we should have $y = f(x)$, where f is an increasing function. Obviously, both x and y are random, thus the statistical model should be

$$y = f(x) + \varepsilon, \tag{5.1.1}$$

where ε denotes the random error. Here the random error mainly contains two parts: first, the observation error which is the same as the situation discussed in Chapter 4; and the second part is the disturbance to the model caused by some factors which can not be identified clearly. Suppose all the errors can be counteracted as a whole, thus we have

$$E\varepsilon = 0, \text{ and } V\varepsilon = \sigma^2. \tag{5.1.2}$$

If we consider the simplest function, *i.e.* linear function, then (5.1.1) reduces to

$$y = \beta_0 + \beta_1 x + \varepsilon. \tag{5.1.3}$$

For the convenience of investigation problem, let us suppose, at first, that x **is a given number**, then model (5.1.3) is called a **linear model**, where β_0 and β_1 are the model parameters. By the Keynes' Psychological Law, $\beta_1 \in (0, 1)$.

Next it is necessary to infer the rationality of model based on data. Suppose that we have obtained n pairs of data, $(x_1, y_1), \cdots, (x_n, y_n)$. From (5.1.3),

$$y_i = \beta_0 + \beta_1 x_i + \varepsilon_i, \quad i = 1, \cdots, n, \tag{5.1.4}$$

where the ε_i's are i.i.d. Now model (5.1.4) forms the basis for making statistical inference, called as a **data Model** corresponding to the linear model (5.1.3).

Example 5.1.1. (Income and expenditure) The data in Table 5.1.1 is cited from China Statistical Yearbook (Chinese Statistical Publishing Houses, 2000), recording the per capita income and expenditure of China rural and urban households during the 1978-2002 period. Since the value of urban Per Capita consumption quantity in 1979 is missing, we use the EM algorithm to yield an estimate about it, *i.e.* 350. Furthermore, the per capita gross domestic products (GDPs) during the 1978-2002 period are listed in the last column of Table 5.1.1. We shall analyze these data in Sections 2 and 3. Let y^X and x respectively the rural per capita living expenditure and net income. Let y^Z and z respectively the urban per capita living expenditure and disposable income. Similar to (5.1.4), we establish the following two data models corresponding to the linear model based on Keynes' Psychological Law,

$$y_i^X = \alpha_0 + \alpha_1 x_i + \varepsilon_i^X, \quad i = 1, \cdots, 25,$$

$$y_i^Z = \beta_0 + \beta_1 z_i + \varepsilon_i^Z, \quad i = 1, \cdots, 25,$$

where α_1 and β_1 denote the income/expenditure unit ratios. We can consider the problem of testing

$$H_0 : \alpha_1 = \beta_1. \tag{5.1.5}$$

The ratio in the United States of America is 0.72 (Gujarati, 1995, p. 9). If we want to compare with that of the United States of America, we should test

$$H_{01} : \alpha_1 = 0.72,$$

$$H_{02} : \beta_1 = 0.72. \tag{5.1.6}$$

In Figs. 5.1.1 and 5.1.2, we give, respectively, two linear models which show the relation between the per capita net income and living expenditure in rural households and the corresponding relation in urban households. We will further consider some testing problems about the models in Example 5.1.3.

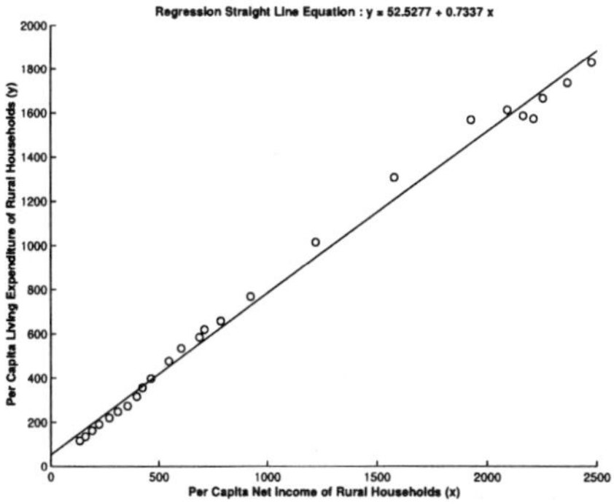

Fig. 5.1.1 Per capita net income vs. living expenditure of rural households during the 1978-2002 period

Now we generalize model (5.1.3) into the following form

$$\boldsymbol{y} = \beta_0 + \beta_1 \boldsymbol{x}_1 + \cdots + \beta_{p-1} \boldsymbol{x}_{p-1} + \varepsilon. \tag{5.1.7}$$

The data model corresponding to (5.1.4) is

$$\boldsymbol{Y} = \boldsymbol{X}\boldsymbol{\beta} + \boldsymbol{\varepsilon}, \tag{5.1.8}$$

where $\boldsymbol{Y}$ is an n-dimensional vector, called observation vector; $\boldsymbol{X}$ is an $n \times p$ known matrix, called design matrix; $\boldsymbol{\beta}$ is a p-dimensional parameter vector; $\boldsymbol{\varepsilon}$ is an n-dimensional random vector satisfying

$$E\boldsymbol{\varepsilon} = \boldsymbol{0}, \quad \text{and} \quad CV(\boldsymbol{\varepsilon}, \boldsymbol{\varepsilon}) = \sigma^2 \boldsymbol{I}_n. \tag{5.1.9}$$

Usually (5.1.9) is called the **Gauss-Markov assumptions**. Note that many nonlinear models, especially the widely-used exponential-linear

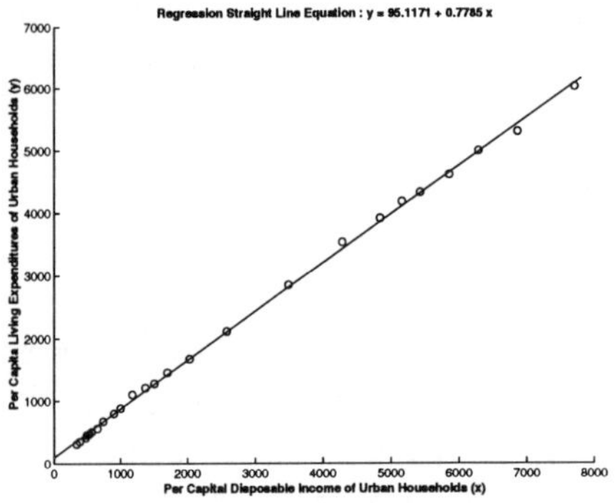

Fig. 5.1.2 Per capita disposable income vs. living expenditure of urban households during the 1978-2002 period

model, can be linearized by appropriate transformations. Here is an example.

Example 5.1.2. (Production function) In production theory, a random form of the famous Cobb-Douglas production function is

$$y = ax_1^b x_2^c \eta, \tag{5.1.10}$$

where y, x_1, x_2, and η respectively denote the production, labor force, capital, and random disturbance. Since the random function is monotone, the above equation is equivalent to

$$\ln y = \beta_0 + \beta_1 \ln x_1 + \beta_2 \ln x_2 + \varepsilon. \tag{5.1.11}$$

Usually $\beta_0 + \beta_1$ is called **returns to scale**. If it is small than 1, then the returns to scale decreases, and the output is less than the input; if it is larger than 1, then the returns to scale increases, and the output is more than the input. Table 5.1.2 offers the data of the output and input of Taiwan's agriculture during the 1958-1972 period (Gujarati, 1995, p. 216). We shall consider the problem of testing

$$H_0 : \beta_0 + \beta_1 = 1. \tag{5.1.12}$$

From the above discussion we can see that a test of the log-linear model usually boils down to a test of the parameter β. In general, there are three cases as follows:

(1) **Model simplification.** Consider model (5.1.7) with some β_i's equal to 0. Let $q \leq p - 1$, then the testing problem can be written as

$$H_0 : \beta_i = 0, \ i = 1, \cdots, q. \tag{5.1.13}$$

Table 5.1.1 Per capita income and expenditure in China during the 1978-2002 period (RMB: YUAN).

		Rural households per capita		Urban households per capita		Per capita
Number	Year	Income	Expenditure	Income	Expenditure	GDP
1	1978	134	116	343	311	273
2	1979	160	135	387	350	300
3	1980	191	162	478	412	348
4	1981	223	191	492	457	416
5	1982	270	220	527	471	457
6	1983	310	248	564	506	487
7	1984	355	274	651	559	591
8	1985	398	317	739	673	737
9	1986	424	357	900	799	809
10	1987	463	398	1002	884	999
11	1988	545	477	1181	1104	1349
12	1989	602	535	1376	1211	1589
13	1990	686	585	1510	1279	1763
14	1991	709	620	1701	1454	2041
15	1992	784	659	2027	1672	2557
16	1993	922	770	2577	2111	3633
17	1994	1221	1017	3496	2851	5355
18	1995	1578	1310	4283	3538	6787
19	1996	1926	1572	4839	3919	7990
20	1997	2090	1617	5160	4186	9179
21	1998	2162	1590	5425	4332	10066
22	1999	2210	1577	5852	4616	10797
23	2000	2253	1670	6280	4998	11601
24	2001	2366	1741	6860	5309	12362
25	2002	2476	1834	7703	6030	13497

If we let M_1 denote (5.1.7) and M_0 denote the model when H_0 is true, from the discussion of Eq. (4.5.8) in Section 4.5, we can use the model selection method again. Our principle is that we should use the model selection method if we treat the two models with equal emphasis; we should use some testing methods if we emphasize particularly on model M_0.

(2) **When the parameters are constant.** Similar to (5.1.6), for a given constant β_0, test

$$H_0 : \beta = \beta_0, \tag{5.1.14}$$

which covers (5.1.13) as far as testing methods are concerned, however, has a different significance.

(3) **When the parameters are restricted.** Similar to (5.1.12), for a $(p-1) \times q$ matrix $\boldsymbol{A}$ $(q \leq p-1)$ and a q-dimensional vector $\boldsymbol{b}$, test

$$H_0 : \boldsymbol{A\beta} = \boldsymbol{b}, \tag{5.1.15}$$

Table 5.1.2 The agricultural total output values (in million Taiwan dollars), labor working days (in million days), and capital inputs (in million Taiwan dollars) in Taiwan during the 1958-1972 period.

Year	Total output value Y	Labor working day X_1	Capital input X_2
1958	16607.7	275.5	17803.7
1959	17511.3	274.4	18096.8
1960	20171.2	269.7	18271.8
1961	20932.9	267.0	19167.3
1962	20406.0	267.8	19647.6
1963	20831.6	275.0	20803.5
1964	24806.3	283.0	22076.6
1965	26465.8	300.7	23445.2
1966	27403.0	307.5	24939.0
1967	28628.7	303.7	26713.7
1968	29904.5	304.7	29957.8
1969	27508.2	298.6	31585.9
1970	29035.5	295.5	33474.5
1971	29281.5	299.0	34821.8
1972	31535.8	288.1	41794.3

which is the most general case.

5.1.2 *Estimating and Testing the Parameters*

Parameter estimation. By (5.1.8), define the residual error as $\boldsymbol{\varepsilon} = \boldsymbol{Y} - \boldsymbol{X\beta}$. Usually $\boldsymbol{\varepsilon}'\boldsymbol{\varepsilon}$ is called as the residual sum of squares (RSS). A good estimator, $\hat{\beta}$, should attain the minimum of the RSS, that is, a solution to

$$\min_{\boldsymbol{\beta}} \|\boldsymbol{Y} - \boldsymbol{X\beta}\|^2.$$

Take the partial derivative of the above expression with respect to $\boldsymbol{\beta}$, set it equal to zero, and we can obtain

$$\boldsymbol{X}'\boldsymbol{X\beta} = \boldsymbol{X}'\boldsymbol{Y}. \tag{5.1.16}$$

This equation is called as a **regularity equation**. Thus we can obtain the unique solution

$$\hat{\boldsymbol{\beta}} = (\boldsymbol{X}'\boldsymbol{X})^{-1}\boldsymbol{X}'\boldsymbol{Y}. \tag{5.1.17}$$

$\hat{\boldsymbol{\beta}}$ is called the **least squares estimate (LSE)** of $\boldsymbol{\beta}$ which is the projection coefficient of $\boldsymbol{Y}$ onto a p-dimensional linear subspace generated by $\boldsymbol{X}$, the projection is $\hat{\boldsymbol{Y}} = \boldsymbol{X}\hat{\boldsymbol{\beta}} = \boldsymbol{X}(\boldsymbol{X}'\boldsymbol{X})^{-1}\boldsymbol{X}'\boldsymbol{Y}$. It is easy to obtain that the

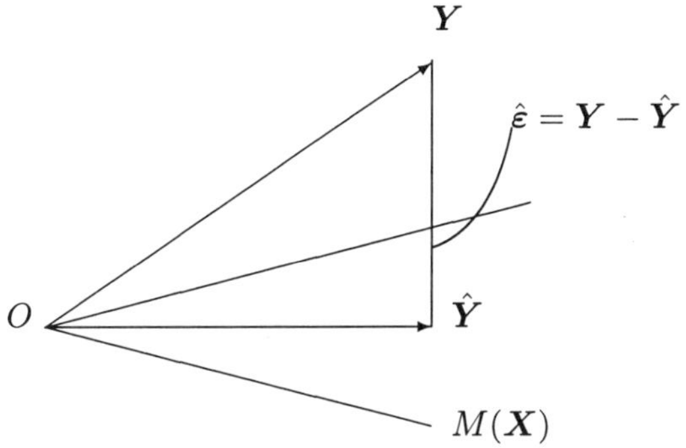

Fig. 5.1.3 Intuitive explanation of the LSE

residual error is $\hat{\boldsymbol{\varepsilon}} = \boldsymbol{Y} - \hat{\boldsymbol{Y}} = (\boldsymbol{I}_n - \boldsymbol{X}(\boldsymbol{X}'\boldsymbol{X})^{-1}\boldsymbol{X}')\boldsymbol{Y}$. Therefore, we can get the following direct sum decomposition (see Fig. 5.1.3)

$$\boldsymbol{Y} = \hat{\boldsymbol{Y}} + \hat{\boldsymbol{\varepsilon}}. \tag{5.1.18}$$

If $\boldsymbol{\varepsilon} \sim N(\boldsymbol{0}, \boldsymbol{I}_n)$, from Example 2.2.5, we know that $\hat{\boldsymbol{\beta}}$ is the **MLE** of $\boldsymbol{\beta}$.

Based on model (5.1.8), it is easy to obtain the following results

$$\begin{aligned} E\boldsymbol{Y} = \boldsymbol{X}\boldsymbol{\beta}, \quad E\hat{\boldsymbol{Y}} &= \boldsymbol{X}\boldsymbol{\beta}, \quad E\hat{\boldsymbol{\varepsilon}} = 0; \\ CV(\boldsymbol{Y}, \boldsymbol{Y}) &= \sigma^2 \boldsymbol{I}_n, \\ CV(\hat{\boldsymbol{Y}}, \hat{\boldsymbol{Y}}) &= \sigma^2 \boldsymbol{X}(\boldsymbol{X}'\boldsymbol{X})^{-1}\boldsymbol{X}', \\ CV(\hat{\boldsymbol{\varepsilon}}, \hat{\boldsymbol{\varepsilon}}) &= \sigma^2(\boldsymbol{I}_n - \boldsymbol{X}(\boldsymbol{X}'\boldsymbol{X})^{-1}\boldsymbol{X}'). \end{aligned} \tag{5.1.19}$$

Thus we have RSS=$\hat{\boldsymbol{\varepsilon}}'\hat{\boldsymbol{\varepsilon}}$, based on Fig. 5.1.3, we can give an estimate of σ^2 as follows

$$\hat{\sigma}^2 = \text{RSS}/(n-p). \tag{5.1.20}$$

Theorem 5.1.1. *The estimates $\hat{\beta}$ and $\hat{\sigma}^2$ have the following properties*

$$E\hat{\boldsymbol{\beta}} = \boldsymbol{\beta}, \quad CV(\hat{\boldsymbol{\beta}}, \hat{\boldsymbol{\beta}}) = \sigma^2(\boldsymbol{X}'\boldsymbol{X})^{-1}, \quad E\hat{\sigma}^2 = \sigma^2.$$

Proof. The first two equalities hold obviously. Let us prove the last one. Since

$$\begin{aligned} E(\text{RSS}) &= E\hat{\boldsymbol{\varepsilon}}'\hat{\boldsymbol{\varepsilon}} = \text{tr}\{CV(\hat{\boldsymbol{\varepsilon}}, \hat{\boldsymbol{\varepsilon}})\} \\ &= \sigma^2 \text{tr}\{\boldsymbol{I}_n - \boldsymbol{X}(\boldsymbol{X}'\boldsymbol{X})^{-1}\boldsymbol{X}'\} \\ &= \sigma^2(n - \text{tr}\{(\boldsymbol{X}'\boldsymbol{X})^{-1}\boldsymbol{X}'\boldsymbol{X}\}) \\ &= (n-p)\sigma^2, \end{aligned}$$

thus we have $E\hat{\sigma}^2 = \sigma^2$, where $\text{tr}\{\boldsymbol{A}\}$ is the trace of matrix $\boldsymbol{A}$: $\text{tr}\{\boldsymbol{A}\} = \sum_i a_{ii}$, satisfying $\text{tr}\{\boldsymbol{A}+\boldsymbol{B}\} = \text{tr}\{\boldsymbol{A}\} + \text{tr}\{\boldsymbol{B}\}$ and $\text{tr}\{\boldsymbol{AB}\} = \text{tr}\{\boldsymbol{BA}\}$.

If $\boldsymbol{\varepsilon} \sim N(\boldsymbol{0}, \sigma^2 \boldsymbol{I}_n)$, then by Theorem 5.1.1, it is easy to obtain that

$$\hat{\boldsymbol{\beta}} \sim N(\boldsymbol{\beta}, \sigma^2(\boldsymbol{X}'\boldsymbol{X})^{-1}). \tag{5.1.21}$$

By Eq. (5.1.18) and Theorem 4.3.1, we can obtain

$$\hat{\boldsymbol{\beta}} \perp\!\!\!\perp \text{RSS}, \quad \text{and} \quad \text{RSS}/\sigma^2 \sim \chi^2_{n-p}, \tag{5.1.22}$$

and this completes the proof of the theorem. □

Testing parameters. Consider the testing problem (5.1.15) which is the most general case. Let $\hat{\boldsymbol{\beta}}_0$ denote the estimate of $\boldsymbol{\beta}$ under H_0. Based on the idea of LSE, $\hat{\boldsymbol{\beta}}_0$ should be the solution to

$$\min_{\boldsymbol{\beta}} \|\boldsymbol{Y} - \boldsymbol{X\beta}\|^2,$$

where the parameters satisfy the following condition

$$\boldsymbol{A\beta} = \boldsymbol{b}.$$

It is a problem of solving the minimum value under restriction. Here the Lagrangian equation is

$$L(\boldsymbol{\beta}, \boldsymbol{\lambda}) = (\boldsymbol{Y} - \boldsymbol{X\beta})'(\boldsymbol{Y} - \boldsymbol{X\beta}) + 2\boldsymbol{\lambda}(\boldsymbol{A\beta} - \boldsymbol{b}),$$

where $\boldsymbol{\lambda} = (\lambda_1, \cdots, \lambda_m)'$ is the Lagrange multiplier. Take the partial derivatives of $L(\boldsymbol{\beta}, \boldsymbol{\lambda})$ with respect to $\boldsymbol{\beta}$ and $\boldsymbol{\lambda}$, set these derivatives equal to zero, we can obtain the following equations

$$\boldsymbol{X}'\boldsymbol{X\beta} + \boldsymbol{A}'\boldsymbol{\lambda} = \boldsymbol{X}'\boldsymbol{Y}, \tag{5.1.23}$$

$$\boldsymbol{A\beta} = \boldsymbol{b}. \tag{5.1.24}$$

By Eq. (5.1.23) we can get

$$\begin{aligned} \hat{\boldsymbol{\beta}}_0 &= (\boldsymbol{X}'\boldsymbol{X})^{-1}\boldsymbol{X}'\boldsymbol{Y} - (\boldsymbol{X}'\boldsymbol{X})^{-1}\boldsymbol{A}'\hat{\boldsymbol{\lambda}} \\ &= \hat{\boldsymbol{\beta}} - (\boldsymbol{X}'\boldsymbol{X})^{-1}\boldsymbol{A}'\hat{\boldsymbol{\lambda}}. \end{aligned} \tag{5.1.25}$$

Substituting it into Eq. (5.1.24), we get

$$\hat{\boldsymbol{\lambda}} = [\boldsymbol{A}(\boldsymbol{X}'\boldsymbol{X})^{-1}\boldsymbol{A}']^{-1}(\boldsymbol{A}\hat{\boldsymbol{\beta}} - \boldsymbol{b}).$$

Substituting $\hat{\boldsymbol{\lambda}}$ into (5.1.25), we get

$$\hat{\boldsymbol{\beta}}_0 = \hat{\boldsymbol{\beta}} - (\boldsymbol{X}'\boldsymbol{X})^{-1}\boldsymbol{A}'[\boldsymbol{A}(\boldsymbol{X}'\boldsymbol{X})^{-1}\boldsymbol{A}']^{-1}(\boldsymbol{A}\hat{\boldsymbol{\beta}} - \boldsymbol{b}). \tag{5.1.26}$$

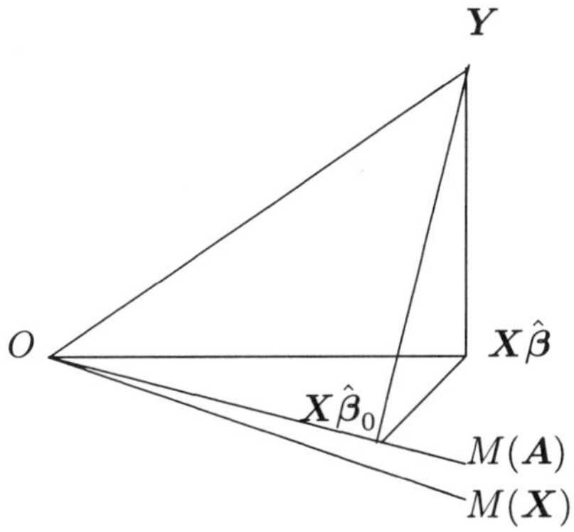

Fig. 5.1.4 Intuitive explanation of the projection under restriction when $\boldsymbol{b} = \mathbf{0}$

It can be verified that $\hat{\boldsymbol{\beta}}_0$ is the MLE of $\boldsymbol{\beta}$ under H_0 when $\boldsymbol{\varepsilon} \sim N(\mathbf{0}, \boldsymbol{I}_n)$. For the residual error, we can get the following decomposition

$$\|\boldsymbol{Y} - \boldsymbol{X}\hat{\boldsymbol{\beta}}_0\|^2 = \|\boldsymbol{Y} - \boldsymbol{X}\hat{\boldsymbol{\beta}}\|^2 + \|\boldsymbol{X}(\hat{\boldsymbol{\beta}} - \hat{\boldsymbol{\beta}}_0)\|^2. \tag{5.1.27}$$

Figure 5.1.4 offers an intuitive explanation of (5.1.27) when $\boldsymbol{b} = \mathbf{0}$.

Under the restriction, the residual error is $\hat{\boldsymbol{\varepsilon}}_0 = \boldsymbol{Y} - \boldsymbol{X}\hat{\boldsymbol{\beta}}_0$. Let $\mathrm{RSS}_0 = \hat{\boldsymbol{\varepsilon}}_0'\hat{\boldsymbol{\varepsilon}}_0$, by (5.1.27),

$$\mathrm{RSS}_0 = \mathrm{RSS} + \|\boldsymbol{X}(\hat{\boldsymbol{\beta}} - \hat{\boldsymbol{\beta}}_0)\|^2. \tag{5.1.28}$$

So the estimate of σ^2 under the restriction may have the following form

$$\hat{\sigma}_0^2 = \mathrm{RSS}_0/(n - p + q). \tag{5.1.29}$$

Theorem 5.1.2. *When H_0 is true, we have*

$$E\hat{\sigma}_0^2 = \sigma^2.$$

Proof. From Theorem 5.1.1, $E\hat{\boldsymbol{\beta}} = \boldsymbol{\beta}$, then we have

$$E[\boldsymbol{A}(\hat{\boldsymbol{\beta}} - \boldsymbol{\beta})]'[\boldsymbol{A}(\boldsymbol{X}'\boldsymbol{X})^{-1}\boldsymbol{A}']^{-1}(\boldsymbol{A}\boldsymbol{\beta} - \boldsymbol{b}) = 0.$$

Thus we can consider

$$\begin{aligned}\|X(\hat{\boldsymbol{\beta}} - \hat{\boldsymbol{\beta}}_0)\|^2 = {} & [\boldsymbol{A}(\hat{\boldsymbol{\beta}} - \boldsymbol{\beta})]'[\boldsymbol{A}(\boldsymbol{X}'\boldsymbol{X})^{-1}\boldsymbol{A}']^{-1}[\boldsymbol{A}(\hat{\boldsymbol{\beta}} - \boldsymbol{\beta})] \\ & + (\boldsymbol{A}\boldsymbol{\beta} - \boldsymbol{b})'[\boldsymbol{A}(\boldsymbol{X}'\boldsymbol{X})^{-1}\boldsymbol{A}']^{-1}(\boldsymbol{A}\boldsymbol{\beta} - \boldsymbol{b}).\end{aligned}$$

Utilizing the properties of the trace, we can calculate the mean of the first term in the right side of the above equation as follows:

$$\begin{aligned}
&E[\boldsymbol{A}(\hat{\boldsymbol{\beta}}-\boldsymbol{\beta})]'[\boldsymbol{A}(\boldsymbol{X}'\boldsymbol{X})^{-1}\boldsymbol{A}']^{-1}[\boldsymbol{A}(\hat{\boldsymbol{\beta}}-\boldsymbol{\beta})]\\
&=\mathrm{tr}\{\boldsymbol{A}'[\boldsymbol{A}(\boldsymbol{X}'\boldsymbol{X})^{-1}\boldsymbol{A}']^{-1}\boldsymbol{A}\cdot CV(\hat{\boldsymbol{\beta}},\hat{\boldsymbol{\beta}})\}\\
&=\sigma^2\mathrm{tr}\{\boldsymbol{A}'[\boldsymbol{A}(\boldsymbol{X}'\boldsymbol{X})^{-1}\boldsymbol{A}']^{-1}\boldsymbol{A}(\boldsymbol{X}'\boldsymbol{X})^{-1})\}\\
&=\sigma^2\mathrm{tr}\{[\boldsymbol{A}(\boldsymbol{X}'\boldsymbol{X})^{-1}\boldsymbol{A}']^{-1}\boldsymbol{A}(\boldsymbol{X}'\boldsymbol{X})^{-1})\boldsymbol{A}'\}\\
&=\sigma^2\mathrm{tr}\{\boldsymbol{I}_q\}=q\sigma^2.
\end{aligned}$$

By (5.1.28) and the proof of Theorem 5.1.1,

$$\begin{aligned}
E(\mathrm{RSS}_0)&=E(\mathrm{RSS})+E(\|\boldsymbol{X}(\hat{\boldsymbol{\beta}}-\hat{\boldsymbol{\beta}}_0)\|^2)\\
&=(n-p)\sigma^2+q\sigma^2+E(\boldsymbol{A\beta}-\boldsymbol{b})'[\boldsymbol{A}(\boldsymbol{X}'\boldsymbol{X})^{-1}\boldsymbol{A}']^{-1}(\boldsymbol{A\beta}-\boldsymbol{b}).
\end{aligned}$$

If H_0 is true, that is, $\boldsymbol{A\beta}=\boldsymbol{b}$, then the above equation reduces to

$$E(\mathrm{RSS}_0)=(n-p+q)\sigma^2.$$

By Eq. (5.1.29), $\hat{\sigma}_0^2$ is an unbiased estimate of σ^2. □

Therefore, for the testing problem (5.1.15), we can construct a test statistic

$$F(y)=\frac{(\mathrm{RSS}_0-\mathrm{RSS})/q}{\mathrm{RSS}/(n-p)}. \tag{5.1.30}$$

Recall that RSS $=\hat{\boldsymbol{\varepsilon}}'\hat{\boldsymbol{\varepsilon}}$ and $\mathrm{RSS}_0=\hat{\boldsymbol{\varepsilon}}_0'\hat{\boldsymbol{\varepsilon}}_0$, thus we reject H_0 when $F(y)$ is large. When $\boldsymbol{\varepsilon}\sim N(\boldsymbol{0},\boldsymbol{I}_n)$, $F(y)$ is an LR test for the testing problem (5.1.15).

Theorem 5.1.3. *If $\boldsymbol{\varepsilon}\sim N(\boldsymbol{0},\boldsymbol{I}_n)$, when H_0 is true, then we have*

$$F(y)\sim F(q,n-p).$$

Proof. By Eq. (5.1.28),

$$\begin{aligned}
\mathrm{RSS}_0-\mathrm{RSS}&=\|\boldsymbol{X}(\hat{\boldsymbol{\beta}}-\hat{\boldsymbol{\beta}}_0)\|^2\\
&=(\boldsymbol{A}\hat{\boldsymbol{\beta}}-\boldsymbol{b})'[\boldsymbol{A}(\boldsymbol{X}'\boldsymbol{X})^{-1}\boldsymbol{A}']^{-1}(\boldsymbol{A}\hat{\boldsymbol{\beta}}-\boldsymbol{b}).
\end{aligned} \tag{5.1.31}$$

It is a function of $\hat{\boldsymbol{\beta}}$. By (5.1.22), we have $(\mathrm{RSS}_0-\mathrm{RSS})\perp\!\!\!\perp\mathrm{RSS}$. If H_0 is true, then

$$A\hat{\boldsymbol{\beta}}\sim N(\boldsymbol{b},\sigma^2\boldsymbol{A}(\boldsymbol{X}'\boldsymbol{X})^{-1}\boldsymbol{A}').$$

By (5.1.31), $(\mathrm{RSS}_0-\mathrm{RSS})/\sigma^2\sim\chi^2_q$. By (5.1.22), $\mathrm{RSS}/\sigma^2\sim\chi^2_{n-p}$. The conclusion follows from Example 1.1.10. □

5.1.3 *Centralization*

In many practical testing problems (cf. Examples 5.1.1 and 5.1.2), we usually do not consider the problem of testing the constant term β_0 in model (5.1.7). Therefore, β_0 becomes a nuisance parameter. From the discussion in Chapters 2 and 3, we know that it is not easy to solve this kind of problems. But for linear models, we can use the Centralization Method. Recall that the data model corresponding to model (5.1.7) is

$$y_i = \beta_0 + \beta_1 x_{i1} + \cdots + \beta_{p-1} x_{ip-1} + \varepsilon_i, \quad i = 1, \cdots, n. \tag{5.1.32}$$

Now let $\alpha = \beta_0 + \beta_1 \bar{x}_{i1} + \cdots + \beta_{p-1}\bar{x}_{p-1}$, where $\bar{x}_j = \Sigma_i x_{ij}/n, \quad j = 1, \cdots, p-1$. Then model (5.1.32) can be rewritten as

$$y_i = \alpha + \beta_1(x_{i1} - \bar{x}_1) + \cdots + \beta_{p-1}(x_{ip-1} - \bar{x}_{p-1}) + \varepsilon_i, \quad i = 1, \cdots, n,$$

which also can be written as the following matrix form

$$\boldsymbol{Y} = \alpha \mathbf{1}_n + \boldsymbol{X}_c \boldsymbol{\beta} + \boldsymbol{\varepsilon}, \tag{5.1.33}$$

where $\mathbf{1}_n$ denotes an n-dimensional vector with all 1's, $\boldsymbol{X}_c$ is an $n \times (p-1)$ matrix with (i,j) element $(x_{ij} - \bar{x}_j)$, and $\boldsymbol{\beta} = (\beta_1, \cdots, \beta_{p-1})'$ are called **model parameters**. Since $\mathbf{1}_n' \boldsymbol{X}_c = 0$, the regularity equation corresponding to Eq. (5.1.16) is

$$\begin{pmatrix} n & 0 \\ 0 & \boldsymbol{X}_c'\boldsymbol{X}_c \end{pmatrix} \begin{pmatrix} \alpha \\ \boldsymbol{\beta} \end{pmatrix} = \begin{pmatrix} \mathbf{1}_n'\boldsymbol{Y} \\ \boldsymbol{X}_c'\boldsymbol{Y} \end{pmatrix}. \tag{5.1.34}$$

It is easy to obtain the following LSEs of $(\alpha, \boldsymbol{\beta})$

$$\begin{aligned} \hat{\alpha} &= \bar{y}, \\ \hat{\boldsymbol{\beta}} &= (\boldsymbol{X}_c'\boldsymbol{X}_c)^{-1}\boldsymbol{X}_c'\boldsymbol{Y}. \end{aligned} \tag{5.1.35}$$

It can be seen that, after the centralization, we have separated the constant term parameters from the model parameters in the linear model, so that their LSEs are mutually independent, that is, $\hat{\alpha} \perp\!\!\!\perp \hat{\boldsymbol{\beta}}$. If return to the original model, then we can easily get that

$$\beta_0 = \bar{y} - \beta_1 \bar{x}_1 - \cdots - \beta_{p-1}\bar{x}_{p-1}. \tag{5.1.36}$$

Next let us consider the testing problem. For the problem of testing (5.1.15), *i.e.*

$$H_0 : \boldsymbol{A}\boldsymbol{\beta} = \boldsymbol{b}. \tag{5.1.37}$$

Since we do not consider the constant term β_0, we need to assume that all the elements in the first column of $\boldsymbol{A}$ are zero. For examples, $\boldsymbol{A} = (0,1)$ and $b = 0.72$ for the problem of testing (5.1.6); $\boldsymbol{A} = (0,1,1)$ and $b = 1$

for (5.1.12). After the centralization, there has no any constant term in $\boldsymbol{\beta}$, consequently the first column of $\boldsymbol{A}$ should be erased, and denote the resulting matrix by $\boldsymbol{A}_c$. Therefore, (5.1.37) can be written as

$$H_0 : \boldsymbol{A}_c\boldsymbol{\beta} = \boldsymbol{b}. \tag{5.1.38}$$

For examples, $\boldsymbol{A}_c = (1)$ for the problem of testing (5.1.16); $\boldsymbol{A}_c = (1, 1)$ for (5.1.12). Then we should construct test statistics on the basis of $\boldsymbol{X}_c$ and $\boldsymbol{A}_c$, and the degrees of freedom under H_0 should also be adjusted accordingly.

Example 5.1.3. (Income and expenditure) Let us calculate the data given by Example 5.1.1. To simplify notations, we rewrite the urban and rural linear models in the following uniform data model

$$y_i = \alpha + \beta x_i + \epsilon_i, \quad i = 1, \cdots, 25, \tag{5.1.39}$$

where $\epsilon_i \sim N(0, \sigma^2)$. Let $\bar{x}$ and $\bar{y}$ be the means of the x_i's and the y_i's respectively. Suppose

$$S_{xx} = \sum_{i=1}^{n} (x_i - \bar{x})^2 \quad \text{and} \quad S_{xy} = \sum_{i=1}^{n} (x_i - \bar{x})(y_i - \bar{y}). \tag{5.1.40}$$

Solving (5.1.34) and (5.1.36) yields the following LSEs of α and β

$$\hat{\alpha} = \bar{y} - \hat{\beta}\bar{x},$$

$$\hat{\beta} = S_{xy}/S_{xx}. \tag{5.1.41}$$

By Table 5.1.1, we can obtain that for the rural households,

$$\begin{array}{ll} \bar{x} = 1018.32, & \bar{y} = 799.68; \\ S_{xx} = 16571781.44, & S_{xy} = 12158893.56; \\ \hat{\beta} = 0.7337, & \hat{\alpha} = 52.5277. \end{array}$$

Thus the linear model on the relationship between income and expenditure is

$$y = 52.5277 + 0.7337x.$$

For the urban households,

$$\begin{array}{ll} \bar{x} = 2654.12, & \bar{y} = 2161.28; \\ S_{xx} = 137338552.68, & S_{xy} = 106914464.16; \\ \hat{\beta} = 0.7785, & \hat{\alpha} = 95.1171. \end{array}$$

The corresponding linear model is

$$y = 95.1171 + 0.7785x.$$

Next consider the problem of testing (5.1.6), that is, testing

$$H_0: \ \beta = 0.72.$$

An approximate equation for (5.1.39) can be written as

$$\hat{y}_i = \hat{\alpha} + \hat{\beta}x_i, i = 1, \cdots, 25.$$

By Eq. (5.1.20), the residual sum of squares is

$$\text{RSS} = \sum_{i=1}^{n}(y_i - \hat{y}_i)^2.$$

The residual sums of squares corresponding to the rural and urban households are

$$\text{RSS} = 59802.99, \quad \text{and} \quad \text{RSS} = 66634.47$$

respectively. Notice that (5.1.28), we can obtain that

$$\begin{aligned}\text{RSS}_0 - \text{RSS} &= \|X_c(\hat{b} - 0.72)\|^2 \\ &= (\hat{b} - 0.72)^2 S_{xx}\end{aligned}$$

By Eq. (5.1.30) we can obtain the following LR test statistic

$$F(y) = \frac{(\hat{b} - 0.72)^2 S_{xx}}{\text{RSS}/23}.$$

The values of $F(y)$ for the rural and urban households are 1.196 and 162.228 respectively. Furthermore, we know that the critical values for F-distribution with 1 and 23 degrees of freedom and at significance levels of 5% and 1% are 4.28 and 7.88, respectively. We can see that we can not reject H_0 for the rural households. However, we strongly reject H_0 for the urban households. These results show that if the average income is used a benchmark of comparison, the expenditure proportion for the urban households is higher than that for the rural households.

By the way, we consider the corresponding interval estimation. Since the square of a t random variable has an F-distribution with 1 numerator degree of freedom, so a 95% confidence interval for β is

$$\left(\hat{\beta} - t_{0.025}(n-2)\sqrt{\frac{\text{RSS}}{(n-2)S_{xx}}}, \ \hat{\beta} + t_{0.025}(n-2)\sqrt{\frac{\text{RSS}}{(n-2)S_{xx}}}\right).$$

We can obtain that the confidence intervals of β for the rural and urban households are

$$(0.7122, \ 0.7552) \ \text{and} \ (0.7706, \ 0.7864),$$

respectively. We observe that they are disjoint, which coincides with the conclusion drawn by the testing method.

5.2 Regression Models

In the previous section, we have discussed the relationship between the two variables Y and X. However, essentially, we treat X as a constant. In this section, we will consider both Y and X as random variables. Among all these kinds of models, Regression Model is the simplest one.

The name "regression" comes from a paper by Francis Galton (Galton, 1886). Galton found that, although there was a tendency for tall parents to have tall children and for short parents to have short children, the average height of children of both parents of a given height tended to move or "regress" toward the average height in the population as a whole. This was Galton's law of universal regression. This law was then confirmed by Galton's friend Karl Pearson (Pearson and Lee, 1903). If we let X and Y respectively denote the parents' height and children's height, then the law stated that $E(Y|X = x)$ tends to a constant value, this conditional mean is called a **regression model**. In modern statistics, what we pay more attention to is how to use the X to predict or explain the value of Y, and X is called the **predictor variable** and Y is called as the **response variable**.

Suppose that (X, Y) is a random vector from a statistical space $(\mathcal{X} \times \mathcal{Y}, \mathcal{A}\times\mathcal{B}, \mathcal{P}_\theta)$. Without the loss of generality, assume that its joint density is $f_\theta(x, y)$. Let h be a measurable mapping from $(\mathcal{X}, \mathcal{A})$ to $(\mathcal{Y}, \mathcal{B})$. Given that $X = x$, $y = h(x)$ is called as a **predictor model** of Y. We will indicate that the regression model is a predictor model with some good properties, and has a profound relation with the linear model. Furthermore, we shall discuss a type of regression model whose parameters have some functional relations.

5.2.1 *Regression Model and Linear Model*

Obviously, we can construct a lot of predictor models. We wish to select the "best" in some sense. An intuitive idea is to make the MSE as small as possible. If for $\forall\theta \in \Theta$ we have

$$E_\theta(Y - h_0(x))^2 = \min_h E_\theta(Y - h(x))^2, \tag{5.2.1}$$

then h_0 is called as the **least MSE predictor model**, where the minimum is taken over all measurable mappings.

Theorem 5.2.1. *A measurable mapping $h_0(x)$ is the least MSE predictor* *model if and only if*

$$h_O(x) = E_\theta(Y|X = x), \tag{5.2.2}$$

where $h_O(x)$ *has a maximum* **correlation coefficient** *with* y, *that is,*

$$\rho_\theta(Y, h_O(X)) = \max_h \rho_\theta(Y, h(X))), \tag{5.2.3}$$

where $\rho_\theta(X,Y) = CV_\theta(X,Y)/[V_\theta(X)V_\theta(Y)]^{\frac{1}{2}}$.

Proof. If we use $q_\theta(x)$ to denote the marginal distribution density of x, and $f_\theta(y|x)$ the conditional distribution density of Y given that $X = x$, then by Eq. (1.2.19) in Section 1.2, we have

$$f_\theta(x,y) = f_\theta(y|x)q_\theta(x).$$

Thus, for any measurable mapping $h(x)$,

$$\begin{aligned}
&\int_{\mathcal{X}}\int_{\mathcal{Y}}(y - h_O(x))(h_O(x) - h(x))f_\theta(x,y)d\nu(y)d\mu(x)\\
&= \int_{\mathcal{X}}\int_{\mathcal{Y}}(y - h_O(x))(h_O(x) - h(x))f_\theta(y|x)q_\theta(x)d\nu(y)d\mu(x)\\
&= \int_{\mathcal{X}}(h_O(x) - h(x))q_\theta(x)\left[\int_{\mathcal{Y}} y f_\theta(y|x)d\nu(y) - h_O(x)\right]d\mu(x)\\
&= 0.
\end{aligned}$$

Therefore, we have the following direct sum decomposition,

$$E_\theta(Y - h(X))^2 = E_\theta(Y - h_O(X))^2 + E_\theta(h_O(X) - h(X))^2, \tag{5.2.4}$$

which attains its minimum value if and only if $h(x) = h_O(x)$ almost surely, and this completes the first part of the theorem.

By Theorem 1.2.1 we have $E_\theta h_O(X) = E_\theta Y$. Similar to the proof of Eq. (5.2.4), we can get the following direct sum decomposition, for any measurable mapping,

$$CV_\theta(Y, h(X)) = CV_\theta(h_O(X), h(X)). \tag{5.2.5}$$

Especially, take $h(x) = h_O(x)$, then

$$CV_\theta(Y, h_O(X)) = V_\theta(h_O(X)) \geq 0. \tag{5.2.6}$$

Thus the correlation coefficient is

$$\begin{aligned}
\rho_\theta(Y, h_O(X)) &= CV_\theta(Y, h_\theta(X))/[V_\theta(Y)V_\theta(h_O(X))]^{\frac{1}{2}}\\
&= [V_\theta(h_O(X))/V_\theta(Y)]^{\frac{1}{2}}.
\end{aligned}$$

So by Eqs. (5.2.5) and (5.2.6), for any measurable mapping $h(x)$, we have

$$\begin{aligned}\rho_\theta^2(Y,h(X)) &= \frac{CV_\theta^2(Y,h(X))}{V_\theta(Y)V_\theta(h(X))}\\ &= \frac{CV_\theta^2(h_0(X),h(X))}{V_\theta(h_0(X))V_\theta(h(X))}\cdot\frac{V_\theta(h_0(X))}{V_\theta(Y)}\\ &= \rho_\theta^2(h_0(X),h(X))\cdot\rho_\theta^2(Y,h_0(X))\\ &\le \rho_\theta^2(Y,h_0(X)).\end{aligned}$$

By Eq. (5.2.6), we have $|\rho_\theta(Y,h(X))| \le \rho_\theta(Y,h_0(X))$, and this completes the second part of the theorem. □

From the Cauchy-Schwarz Inequality, the equality holds if and only if $|\rho_\theta(h_0(X),h(X))| = 1$, that is, $h(x)$ is a linear function of $h_0(x)$.

From the above theorem, we know that the regression model has some good statistical properties. Especially, when (X,Y) has a multivariate normal distribution, it has a profound relation with the linear model.

Let $(\boldsymbol{X}',\boldsymbol{Y}')$ have a multivariate distribution $N(\boldsymbol{\mu},\boldsymbol{\Sigma})$, then $\boldsymbol{\theta}=(\boldsymbol{\mu},\boldsymbol{\Sigma})$. Let $\boldsymbol{\mu}^{(1)}$ and $\boldsymbol{\mu}^{(2)}$ denote the means of $\boldsymbol{X}$ and $\boldsymbol{Y}$ respectively, and make a corresponding partition of $\boldsymbol{\Sigma}$ as follows

$$\boldsymbol{\Sigma} = \begin{pmatrix}\boldsymbol{\Sigma}_{11} & \boldsymbol{\Sigma}_{12}\\ \boldsymbol{\Sigma}_{21} & \boldsymbol{\Sigma}_{22}\end{pmatrix}.$$

Obviously, both $\boldsymbol{\Sigma}_{11}$ and $\boldsymbol{\Sigma}_{22}$ are positive definite matrices, and $\boldsymbol{\Sigma}_{21}=\boldsymbol{\Sigma}_{12}'$. Here the conditional distribution of $\boldsymbol{Y}$ given that $\boldsymbol{X}=\boldsymbol{x}$ is still a normal distribution, and we have (see Problem 5.5)

$$E_{\boldsymbol{\theta}}(\boldsymbol{Y}|\boldsymbol{X}=\boldsymbol{x}) = \boldsymbol{\mu}^{(2)} + \boldsymbol{\Sigma}_{21}\boldsymbol{\Sigma}_{11}^{-1}(\boldsymbol{x}-\boldsymbol{\mu}^{(1)}) \tag{5.2.7}$$

$$V_{\boldsymbol{\theta}}(\boldsymbol{Y}|\boldsymbol{X}=\boldsymbol{x}) = \boldsymbol{\Sigma}_{22} - \boldsymbol{\Sigma}_{21}\boldsymbol{\Sigma}_{11}^{-1}\boldsymbol{\Sigma}_{12}. \tag{5.2.8}$$

Then the regression model is given by Eq. (5.2.7), which is a linear function of $\boldsymbol{x}$. Especially, when Y is a random variable and $\boldsymbol{X}$ is a $(p-1)$-dimensional random vector, the regression model (5.2.7) can be rewritten as

$$E_{\boldsymbol{\theta}}(Y|\boldsymbol{X}=\boldsymbol{x}) = \alpha + \beta_1(x_1-\mu_1^{(1)}) + \cdots + \beta_{p-1}(x_{p-1}-\mu_{p-1}^{(1)}), \tag{5.2.9}$$

where $\alpha=\mu^{(2)}$. If we let $\boldsymbol{\beta}=(\beta_1,\cdots,\beta_{p-1})'$, then

$$\boldsymbol{\beta}' = \boldsymbol{\Sigma}_{21}\boldsymbol{\Sigma}_{11}^{-1}. \tag{5.2.10}$$

Usually β is called a **regression coefficient**. We can see that model (5.2.9) coincides with the centralized linear model given by (5.1.33) formally. Especially, when $\boldsymbol{\mu}$ and $\boldsymbol{\Sigma}$ are unknown, and if there are n samples, then $\bar{\boldsymbol{x}}$ and $\bar{\boldsymbol{y}}$ are the MLEs of $\boldsymbol{\mu}^{(1)}$ and $\boldsymbol{\mu}^{(2)}$ respectively. Refer to Eq. (5.1.34),

$$\frac{1}{n}\boldsymbol{X}_c'\boldsymbol{X}_c = \frac{1}{n}\left(\sum_{i=1}^{n}(x_{ij}-\bar{x}_j)(x_{il}-\bar{x}_l)\right)_{(p-1)\times(p-1)}$$

and

$$\frac{1}{n}\boldsymbol{X}_c'\boldsymbol{y} = \frac{1}{n}\left(\sum_{i=1}^{n}(x_{ij} - \bar{x}_j)(y_i - \bar{y})\right)_{(p-1)\times 1}$$

are the MLEs of $\boldsymbol{\Sigma}_{11}$ and $\boldsymbol{\Sigma}_{21}$ respectively. By Eq. (5.2.10) we can obtain the MLEs of α and β given by Eq. (5.1.35). Therefore, under the assumption of normality, the data model corresponding to the linear model is also that corresponding to the regression model again. By (5.2.9), the following model

$$Y = \alpha + \beta_1 X_1 + \cdots + \beta_{p-1}X_{p-1} + \varepsilon \qquad (5.2.11)$$

is called as a **linear regression model**, where $Y, X_1, \cdots, X_p$ are random variables satisfying $Y \perp\!\!\!\perp (X_1, \cdots, X_{p-1})$, and ε is the random error satisfying $E\varepsilon = 0$.

5.2.2 *Errors-in-Variables Model*

For model (5.2.11), consider the case $p-1 = 1$. Assume that we have drawn n pairs of sample values (x_1, y_1), $\cdots$,(x_n, y_n), then we have

$$y_i = \alpha + \beta x_i + \varepsilon_i, \varepsilon_i \sim N(0, \sigma_\varepsilon^2). \qquad (5.2.12)$$

We wish to separate out the observation error from X. Consider the following error model for X (cf. model (5.1.1) in Section 5.1)

$$x_i = \theta_i + \delta_i, \quad \delta_i \sim N(0, \sigma_x^2). \qquad (5.2.13)$$

If $EY_i = \eta_i, i = 1, \cdots, n$, by (5.2.12) and (5.2.13), we have

$$\eta_i = \alpha + \beta\theta_i, \quad i = 1, \cdots, n.$$

Thus we have established a function-relation between the mean parameters. So far, our basic idea about the method of studying the model parameters is to **investigate the change of the mean based on the variance.** Now (5.2.12) and (5.2.13) can be written as

$$\begin{aligned} y_i &= \alpha + \beta\theta_i + \varepsilon_i, \quad \varepsilon_i \sim N(0, \sigma_y^2), \\ x_i &= \theta_i + \delta_i, \quad \delta_i \sim N(0, \sigma_x^2). \end{aligned} \qquad (5.2.14)$$

Therefore, in the above model, ε_i and δ_i respectively denote the observation errors of y_i and x_i. So (5.2.14) is called as **errors-in-variables model**, or EIV model (Fuller, 1987; Carroll *et al.*, 1995; Casella and Berger, 2002).

Example 5.2.1. (Analyzing the relation between income and GDP) We will use the data in Example 5.1.1 to investigate the relation between the per capita income of Chinese households (rural and urban) and the per capita gross domestic product (GDP). Let Y_i and X_i respectively denote the per capita income and the per capita GDP, $i = 1, \cdots, 25$. Obviously, we should treat the data errors with equal emphasis, consequently we get model (5.2.14). Based on the idea of Keynes' Psychological Law, the model parameter β would represent the ratio of the per capita income to the per unit per capita GDP, so we can imagine that $\beta \in (0,1)$. For the given $\beta_0 \in (0,1)$, consider the problem of testing

$$H_0: \quad \beta = \beta_0. \tag{5.2.15}$$

Alternatively, we can also give a confidence interval for β based on testing methods, which will be beneficial to analyzing the difference between the rural and the urban income.

There are two variances in model (5.2.14), when they are both unknown and unequal, the testing problem under consideration boils down to the famous Behrens-Fisher problem. Now suppose that the ratio between the variances is known, *i.e.* $\sigma_x^2 = \lambda\sigma_y^2$, where $\lambda > 0$ is known. Then the parameters in the model are α, β, θ_i, and σ_x^2. From the normal distribution density, the likelihood function can be written as

$$L(\alpha, \beta, \theta_i, \sigma_x^2) = c \cdot \exp\left\{-\frac{1}{2\sigma_x^2}\sum_{i=1}^{n}\left[(x_i - \theta_i)^2 + \lambda(y_i - \alpha - \beta\theta_i)^2\right]\right\},$$

where $c = \lambda^{\frac{n}{2}}/(2\pi\sigma_x^2)^n$. It is easy to verify that θ_i attains the maximum value of the above equation if and only if θ_i attains the minimum value of

$$f(\alpha, \beta, \theta_i) = \sum_{i=1}^{n}\left[(x_i - \theta_i)^2 + \lambda(y_i - \alpha - \beta\theta_i)^2\right],$$

and solving it yields $\theta_i^* = (x_i + \lambda\beta(y_i - \alpha))/(1 + \lambda\beta^2)$. Substituting θ_i^* into the above function, we have

$$f(\alpha, \beta, \theta_i^*) = \frac{\lambda}{1 + \lambda\beta^2}\sum_{i=1}^{n}(y_i - \alpha - \beta x_i)^2. \tag{5.2.16}$$

To minimize the above function, take its partial derivatives with respect to each of the parameters and set to zero. Finally, we can obtain the following MLEs of α and β

$$\begin{aligned} \hat{\alpha} &= \bar{y} - \hat{\beta}\bar{x}, \\ \hat{\beta} &= \left[-(S_{xx} - \lambda S_{yy}) + \sqrt{(S_{xx} - \lambda S_{yy})^2 + 4\lambda S_{xy}^2}\right] \Big/ (2\lambda S_{xy}), \end{aligned} \tag{5.2.17}$$

respectively, where S_{xx}, S_{yy}, and S_{xy} are given by (5.1.40). Substituting them into θ_i^* yields the MLE of θ_i as follows

$$\hat{\theta}_i = \frac{x_i + \lambda\hat{\beta}(y_i - \hat{\alpha})}{1 + \lambda\hat{\beta}^2}.$$

Next, by changing the variances, we reveal some characteristics of the parameter estimates, and analyze the model (5.2.14). Note that $\lambda \to 0$ implies $\sigma_X^2 \to 0$, then x_i tends to a one-point distribution, that is, model (5.2.14) becomes model (5.2.12). Thus we have $\hat{\theta}_i \to x_i$ by (5.2.17). Furthermore, by (5.2.16), $\hat{\beta} \to S_{xy}/S_{xx}$, which coincides with (5.1.40), and called as a regression of y on x. Note that $\lambda \to \infty$ implies $\sigma_Y^2 \to 0$, then y_i tends to a one-point distribution. We can get a regression of x on y. When $\lambda = 1$, we need to make a further analysis. In this case,

$$\theta_i^* = \frac{x_i + \beta(y_i - \alpha)}{1 + \beta^2}.$$

Let $\tilde{x}_i = \theta_i^*$ and $\tilde{y}_i = \alpha + \beta\theta_i^*$, then $(\tilde{x}_i, \tilde{y}_i)$ is the projection of (x_i, y_i) on the straight line $y = \alpha + \beta x$. Thus, it is justified for us to find the solution of

$$\min \sum_{i=1}^{n} \left[(x_i - \tilde{x}_i)^2 + (y_i - \tilde{y}_i)^2\right]. \tag{5.2.18}$$

The solution of (5.2.18) is called as a **minimum total sum of squares estimate**. Since

$$\sum_{i=1}^{n} \left[(x_i - \tilde{x}_i)^2 + (y_i - \tilde{y}_i)^2\right] = \frac{(y_i - (\alpha + \beta)x_i)^2}{1 + \beta^2},$$

it is easy to verify that the $\hat{\alpha}$ and $\hat{\beta}$ given by Eq. (5.2.17) are exactly the minimum total sum of squares estimates when $\lambda = 1$. This demonstrates from another point of view that (5.2.14) treats the errors of x_i and y_i with equal emphasis.

By Example 1.4.7, $\hat{\alpha}$, $\hat{\beta}$, and $\hat{\theta}_i$ are consistent estimators for α, β, and θ_i, respectively. Based on the likelihood function, we can obtain

$$\hat{\sigma}_x^2 = \frac{\lambda}{2n(1 + \lambda\hat{\beta}^2)} \sum_{i=1}^{n} (y_i - \hat{\alpha} - \hat{\beta}x_i)^2.$$

Since σ_x^2 appears in the likelihood function repeatedly, $\hat{\sigma}_x^2$ is a consistent estimator for $\sigma_x^2/2$. Therefore

$$\hat{\sigma}_\beta^2 = \frac{(1 + \lambda\hat{\beta}^2)^2(S_{xx}S_{yy} - S_{xy}^2)}{(S_{xx} - \lambda S_{yy})^2 + 4\lambda S_{xy}^2} \tag{5.2.19}$$

is a consistent estimator for the variance of $\hat{\beta}$ (Kendall and Stuart, 1979, Chapter 29). For the problem of testing (5.2.15), the LR test statistic is

$$Z_n = \frac{\sqrt{n}(\hat{\beta} - \beta_0)}{\hat{\sigma}_\beta}.$$

When H_0 is true, $Z_n \xrightarrow{L} Z \sim N(0, 1)$. Furthermore, we consider the following modified LR test based on the Student's t-distribution (see, e.g., Gleser and Hwang, 1987; Casella and Berger, 2002, Chapter 12)

$$Z_n = \frac{\sqrt{n-2}(\hat{\beta} - \beta_0)}{\hat{\sigma}_\beta}.$$

When H_0 is true and $n \to \infty$, the limiting distribution of Z_n follows a Student's t-distribution with $n - 2$ degrees of freedom. As a result, an interval estimate for β is

$$(\hat{\beta} - \hat{\sigma}_\beta t_{\alpha/2}/\sqrt{n-2}, \hat{\beta} + \hat{\sigma}_\beta t_{\alpha/2}/\sqrt{n-2}). \tag{5.2.20}$$

Example 5.2.2. (Calculating the relation between income and GDP) Continue to consider the problem proposed in Example 5.2.1. Recall some notations, we have used x_i to denote the per capita GDP and y_i the per capita income of households (rural and urban). Now let $\lambda = 1$, by the data given in Table 5.1.1 in Example 5.1.1,

$$S_{xx} = 4958136651.44.$$

For the rural households and (5.2.16),

$$S_{yy} = 16571781.44, \quad S_{xy} = 89854207.44,$$

$$\hat{\alpha} = 249.6139, \quad \hat{\beta} = 0.1813.$$

For the urban households and (5.2.16),

$$S_{yy} = 137338552.64, \quad S_{xy} = 260156064.04,$$

$$\hat{\alpha} = 426.7918, \quad \hat{\beta} = 0.5254.$$

Return to model (5.2.12), we obtain that the regression lines of the per capita income of the rural and urban households on the per capita GDP are

$$y = 249.6139 + 0.1813x \quad \text{and} \quad y = 426.7918 + 0.5254x,$$

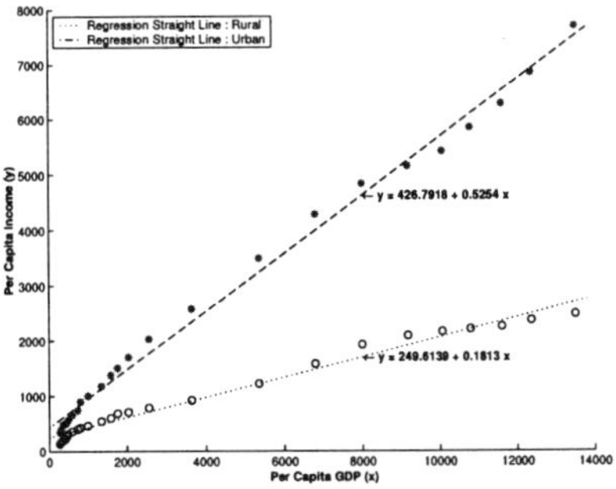

(a) $\lambda = 1$

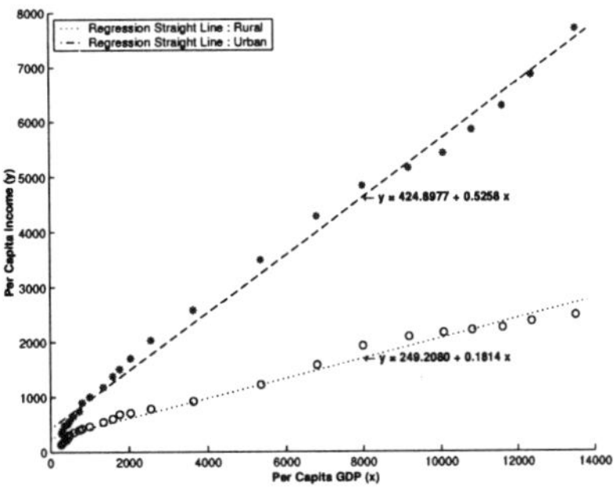

(b) $\lambda = 2$

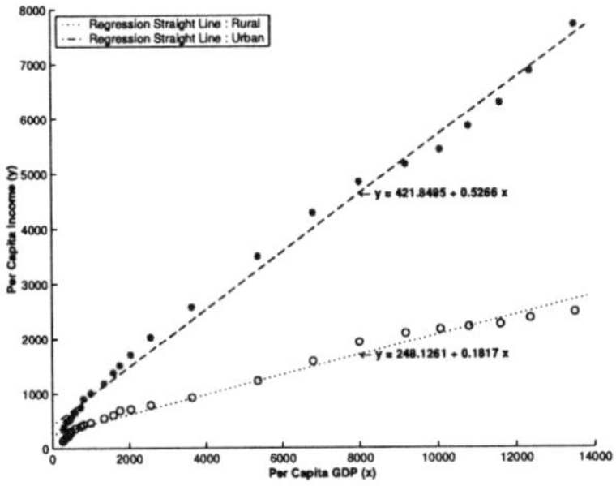

(c) $\lambda = 5$

Fig. 5.2.1 Relationship between the per capita GDP and the per capita net income of the rural and urban households during the 1978-2002 period

respectively (cf. Fig. 5.2.1 (a)). Additionally, Figs. 5.2.1 (b) and (c) show the regression lines of the per capita income of the rural and urban households on the per capita GDP when $\lambda = 2$ and 5 respectively. As shown in Figs. 5.2.1 (a)-(c), we can observe that the change of λ has no significant effect on the analysis of the problem.

By Eq. (5.2.19), the variance estimators of $\hat{\beta}$ of the rural and urban households are 0.00058128, 0.001685, respectively. By (5.2.20) in conjunction with the Student's t-distribution, we can obtain the interval estimator for β of the rural and urban households. The following table reports some estimation results at different levels.

α	Rural	Urban
0.01	(0.1672, 0.1954)	(0.5014, 0.5494)
0.05	(0.1709, 0.1917)	(0.5077, 0.5431)
0.10	(0.1729, 0.1899)	(0.5107, 0.5401)

5.3 Logistic Regression Models

We have known that the linear model describes the relationship between the mean of a random variable y and $\boldsymbol{x} = (x_1, \cdots, x_p)'$. Let $\theta = EY$, then the linear model (5.1.7) can be written as follows:

$$\theta = \beta_0 + \beta_1 x_1 + \cdots + \beta_{p-1} x_{p-1}. \tag{5.3.1}$$

Let g be a function of θ, model (5.3.1) can be generalized as follows

$$g(\theta) = \beta_0 + \beta_1 x_1 + \cdots + \beta_{p-1} x_{p-1}, \tag{5.3.2}$$

which is called a **generalized linear model (GLM)**, and g is called a **link function**. Obviously, it reduces to the ordinary linear model when $g(\theta) = \theta$.

In this section, we will discuss a class of simple but widely used type GLM. Let y be a binomial $Bi(n, \theta)$ random variable with success probability θ. The expected number of successes in n trials (*i.e.* the mean value of the binomial distribution) $EY = n\theta$, and the expected number of failures is $n - n\theta = n(1-\theta)$. Recall the discussion in Example 1.4.2, the ratio of the expected number of success to the expected number of failure $\theta/(1-\theta)$, is called an odds ratio. Now we assume that the logarithm of odds ratio is a linear function of x:

$$\ln\left(\frac{\theta}{1-\theta}\right) = \beta_0 + \beta_1 x_1 + \cdots + \beta_{p-1} x_{p-1}, \tag{5.3.3}$$

which is usually called a **logistic regression model**, sometimes is written as

$$\text{logit}(\theta) = \beta_0 + \beta_1 x_1 + \cdots + \beta_{p-1} x_{p-1}.$$

Obviously it is a generalized linear model. We can see that it is convenient to use the Logistic regression model to analyze discrete data, especially contingency tables. In recent years, this model has been used to analyze genetic data and some meaningful results have been obtained (see, e.g., Henshall and Goddard, 1999; Levinson *et al.*, 2000).

First we discuss some properties and practical significance of the one-variable logistic regression model, *i.e.* the case with $p = 2$ in Eq. (5.3.3).

5.3.1 *One-Variable Case*

In this case, Eq. (5.3.3) reduces to

$$\ln\left(\frac{\theta(x)}{1-\theta(x)}\right) = \alpha + \beta x. \tag{5.3.4}$$

It is easy to obtain that

$$\theta(x) = \frac{\exp(\alpha+\beta x)}{1+\exp(\alpha+\beta x)}. \tag{5.3.5}$$

Differentiating both sides of the above equation with respect to x yields

$$\frac{d\theta(x)}{dx} = \beta\theta(x)(1-\theta(x)).$$

Since $\theta(x)(1-\theta(x))$ is always positive, therefore, $\theta(x)$ is a strictly increasing function when $\beta > 0$; $\theta(x)$ is a constant when $\beta = 0$; $\theta(x)$ is a strictly decreasing function when $\beta < 0$. Especially when $\beta > 0$, we can consider $\theta(x)$ as a distribution function which characterizes the probability distribution of success and is called a **Logistic distribution**. Generally, the distribution function of the Logistic distribution is

$$F(x) = \frac{\exp(x-u)/c}{1+\exp(x-u)/c}, \tag{5.3.6}$$

where u and c are parameters, and denote the mean and the variances, respectively. (5.3.6) is called the **standard Logistic distribution** when $u = 0$ and $c = 1$.

In sum, it is very important to discuss the properties of the parameter β. As far as hypothesis-testing theory is concerned, we can consider the problem

$$H_0 : \beta = 0 \quad \text{vs.} \quad H_1 : \beta \neq 0. \tag{5.3.7}$$

or

$$H_0 : \beta \leq 0 \quad \text{vs.} \quad H_1 : \beta > 0. \tag{5.3.8}$$

Usually there are two methods to treat this kind of problem, *i.e.* **empirical Logistic method** and **maximum likelihood method**.

5.3.1.1 *Empirical Logistic method*

From the discussion in Section 1.4, the MLE of θ is $\hat{\theta} = y/n$. Substituting this estimator into the left side of (5.3.4) yields

$$\ln\left(\frac{\hat{\theta}}{1-\hat{\theta}}\right) = \ln\left(\frac{y}{n-y}\right).$$

Since the above model becomes meaningless for $y = 0$ or $y = n$, the right side of the equation is modified as follows:

$$\ln\left(\frac{y+\frac{1}{2}}{n-y+\frac{1}{2}}\right). \tag{5.3.9}$$

Recall the discussion in Section 2.4, we have used a similar modification, and the computation results show that this kind of modification is effective. Next we give a general proof.

Theorem 5.3.1. *Let y_n be a $Bi(n,\theta)$ random variable and*

$$z_n(a) = \ln\left(\frac{y_n+a}{n-y_n+a}\right).$$

Then $z_n(a) - \ln[\theta/(1-\theta)]$ is minimized at $a = 1/2$.

Proof. Let $h_n = \sqrt{n}(y_n/n - \theta)$. From the discussion in Section 1.4, $E_\theta H_n = 0$ and $E_\theta H_n^2 = \theta(1-\theta)$. Since $y_n = n\theta + \sqrt{n}h_n$, we have

$$\begin{aligned} z_n(a) &= \ln(y_n + a) - \ln(n - y_n + a) \\ &= \ln(n\theta + \sqrt{n}h_n + a) - \ln(n(1-\theta) - \sqrt{n}h_n + a). \end{aligned}$$

Making the Taylor series expansion of $\ln(1+x)$ yields

$$\begin{aligned} z_n(a) - \ln\left[\frac{\theta}{1-\theta}\right] &= \ln\left[1 + \frac{h_n}{\sqrt{n}\theta} + \frac{a}{n\theta}\right] - \ln\left[1 - \frac{h_n}{\sqrt{n}(1-\theta)} + \frac{a}{n(1-\theta)}\right] \\ &= \frac{h_n}{\sqrt{n}\theta(1-\theta)} + \frac{(1-2\theta)a}{n\theta(1-\theta)} - \frac{(1-2\theta)h_n^2}{2n\theta^2(1-\theta)^2} + o\left(\frac{1}{n}\right). \end{aligned}$$

Taking the expectation on both sides of the above equation, we have

$$E_\theta z_n(a) - \ln\left(\frac{\theta}{1-\theta}\right) = \frac{(1-2\theta)}{n\theta(1-\theta)}\left(a - \frac{1}{2}\right) + o\left(\frac{1}{n}\right).$$

Therefore, the above bias attains its minimum when $a = 1/2$. □

Usually (5.3.9) is called as the **empirical Logistic transformation**. Now suppose for model (5.3.4) we obtain k pairs of samples $(x_1, y_1), \cdots, (x_k, y_k)$, where y_i has a binomial distribution $Bi(n_i, \theta_i)$ with $\theta_i = \theta(x_i)$. By the empirical Logistic transformation, let

$$z_i = \ln\left(y_i + \frac{1}{2}\right) - \ln\left(n_i - y_i + \frac{1}{2}\right). \tag{5.3.10}$$

Thus, we can obtain the data model corresponding to model (5.3.4)

$$z_i = \alpha + \beta x_i + \varepsilon_i, \quad i = 1, \cdots, k,$$

where the ε_i's are mutually independent, satisfying $E\varepsilon_i=0$ and $V\varepsilon_i = \sigma^2, i = 1, \cdots, k$. By (5.1.17), we can obtain the LSEs of α and β

$$\begin{aligned} \hat{\alpha} &= \bar{z} - \hat{\beta}\bar{x}, \\ \hat{\beta} &= S_{xy}/S_{xx}, \end{aligned} \tag{5.3.11}$$

respectively, where $\bar{z} = \sum_{i=1}^{k} z_i/k$, $\bar{x} = \sum_{i=1}^{k} x_i/k$, $S_{xz} = \sum_{i=1}^{k}(x_i - \bar{x})(z_i - \bar{z})$, and $S_{xx} = \sum_{i=1}^{k}(x_i - \bar{x})^2$. Furthermore, a $1-\alpha$ confidence interval for β is

$$\left(\hat{\beta} - t_{\frac{\alpha}{2}}(k-2)\sqrt{\frac{\text{RSS}}{(k-2)S_{xx}}},\ \hat{\beta} + t_{\frac{\alpha}{2}}(k-2)\sqrt{\frac{\text{RSS}}{(k-2)S_{xx}}}\right), \tag{5.3.12}$$

where RSS$=\sum_{i=1}^{k}(z_i - \hat{z}_i)^2$ and $\hat{z}_i = \hat{\alpha} + \hat{\beta}x_i$. Based on the interval estimate, we can solve the testing problems given by (5.3.7) and (5.3.8).

5.3.1.2 *Maximum likelihood method*

We still assume that for the Logistic model (5.3.4), we can obtain k pairs of samples $(x_1, y_1), \cdots, (x_k, y_k)$, where y_i has a $Bi(n_i, \theta_i)$ with $\theta_i = \theta(x_i)$. Then the log-likelihood is

$$l(y;\theta) = \sum_{i=1}^{n}\left[y_i \ \ln\frac{\theta_i}{1-\theta_i} + n_i \ln(1-\theta_i)\right] + c$$

where c is a constant independent of the parameters. By (5.3.5), we can obtain

$$l(y;\alpha,\beta) = \sum_{i=1}^{n}\left\{y_i(\alpha+\beta x_i) - n_i \ln\left[1+\exp(\alpha+\beta x_i)\right]\right\}.$$

Differentiating the above equation with respect to α and β respectively, we obtain

$$\begin{aligned} \frac{\partial l(y;\alpha,\beta)}{\partial\alpha} &= \sum_{i=1}^{n}(y_i - n_i\theta_i) \\ &= \sum_{i=1}^{n}\left[y_i - \frac{n_i\exp(\alpha+\beta x_i)}{1+\exp(\alpha+\beta x_i)}\right]. \end{aligned} \tag{5.3.13}$$

$$\begin{aligned} \frac{\partial l(y;\alpha,\beta)}{\partial\beta} &= \sum_{i=1}^{n}(y_i - n_i\theta_i)x_i \\ &= \sum_{i=1}^{n}\left[y_i - \frac{n_i\exp(\alpha+\beta x_i)}{1+\exp(\alpha+\beta x_i)}\right]x_i. \end{aligned} \tag{5.3.14}$$

Let $\hat{\alpha}$ and $\hat{\beta}$ denote the MLEs of α and β, respectively. Setting Eqs. (5.3.13) and (5.3.14) equal to zero, we can obtain a solution satisfying the following equations

$$\begin{aligned} \sum_{i=1}^{n} y_i &= \sum_{i=1}^{n} n_i\theta_i, \\ \sum_{i=1}^{n} y_i x_i &= \sum_{i=1}^{n} n_i\theta_i x_i, \end{aligned} \tag{5.3.15}$$

where $\theta_i = \exp(\alpha + \beta x_i)/[1 + \exp(\alpha + \beta x_i)]$, $i = 1, \cdots, k$. We can see that (5.3.15) coincides with (4.5.17) in Chapter 4 in form. Since the above equations are nonlinear, we need to use some iterative methods to solve them. In order to explore the convergence of the methods, we consider the information matrix based on the MLEs. Since $\theta_i(1-\theta_i) = \exp(\alpha + \beta x_i)/[1 + \exp(\alpha + \beta x_i)]^2$, it is easy to obtain that

$$-\begin{pmatrix} \dfrac{\partial^2 l(y;\alpha,\beta)}{\partial\alpha^2} & \dfrac{\partial^2 l(y;\alpha,\beta)}{\partial\beta\partial\alpha} \\ \dfrac{\partial^2 l(y;\alpha,\beta)}{\partial\alpha\partial\beta} & \dfrac{\partial^2 l(y;\alpha,\beta)}{\partial\beta^2} \end{pmatrix} = \boldsymbol{X}'\boldsymbol{\Sigma}\boldsymbol{X}, \tag{5.3.16}$$

where

$$\boldsymbol{X}' = \begin{pmatrix} 1 & \cdots & 1 \\ x_1 & \cdots & x_k \end{pmatrix},$$

and

$$\boldsymbol{\Sigma} = \operatorname{diag}\{n_1\theta_1(1-\theta_1), \cdots, n_k\theta_k(1-\theta_k)\}.$$

The above matrix is also the Fisher information matrix of (α, β). From the discussion in Section 1.4, the covariance matrix of $(\hat{\alpha}, \hat{\beta})$ is $(\boldsymbol{X}'\hat{\boldsymbol{\Sigma}}\boldsymbol{X})^{-1}$, where $\hat{\boldsymbol{\Sigma}} = \operatorname{diag}\{n_1\hat{\theta}_1(1-\hat{\theta}_1), \cdots, n_k\hat{\theta}_k(1-\hat{\theta}_k)\}$. Since the matrix (5.3.16) is positive definite, the solution to (5.3.15) is unique, that is, the MLEs of α and β are unique. Next let us discuss how to get solution by using the Newton-Raphson method.

Newton-Raphson method. It is an iterative method. Let $\alpha^{(0)}$ and $\beta^{(0)}$ be the solution by using the empirical Logistic method. Let

$$\theta_i^{(t)} = \frac{\exp(\alpha^{(t)} + \beta^{(t)} x_i)}{1 + \exp(\alpha^{(t)} + \beta^{(t)} x_i)}. \tag{5.3.17}$$

Based on the t-th iteration, the iterative formula for the $t+1$-th iteration of the Newton-Raphson method is

$$\begin{pmatrix} \alpha^{(t+1)} \\ \beta^{(t+1)} \end{pmatrix} = \begin{pmatrix} \alpha^{(t)} \\ \beta^{(t)} \end{pmatrix} + (\boldsymbol{X}'\boldsymbol{\Sigma}^{(t)}\boldsymbol{X})^{-1}\boldsymbol{X}'(\boldsymbol{Y} - \boldsymbol{u}^{(t)}), \tag{5.3.18}$$

Table 5.3.1 Amount of food intake of Zucker rats.

Age in weeks	12	23	30	43	n_{i+}
Lean	153	170	192	152	667
Obese	210	257	281	293	1041
n_{+j}	363	427	473	445	1708

where $\boldsymbol{X}$ is given by Eq. (5.3.16),

$$\boldsymbol{\Sigma}^{(t)} = \text{diag}\{n_1\theta_1^{(t)}(1-\theta_1^{(t)}), \cdots, n_k\theta_k^{(t)}(1-\theta_k^{(t)})\}$$

$$\boldsymbol{Y} = (y_1, \cdots, y_k)',$$

$$\boldsymbol{u}^{(t)} = (u_1^{(t)}, \cdots, u_k^{(t)})'$$

and $u_i^{(t)} = n_i\theta_i^{(t)}$, $i = 1, \cdots, k$. It can be verified that the iteration sequences $\{\alpha^{(t)}\}$ and $\{\beta^{(t)}\}$ respectively converge to $\hat{\alpha}$ and $\hat{\beta}$ (Walker and Duncan, 1967). By using the computation results and the covariance matrix of $(\hat{\alpha}\,\hat{\beta})$, we can obtain a $1-\alpha$ confidence interval for β

$$(\hat{\beta} - Z_{\frac{\alpha}{2}}\sqrt{b^2}\ ,\ \hat{\beta} + Z_{\frac{\alpha}{2}}\sqrt{b^2}\ ), \tag{5.3.19}$$

where b^2 is the variance of $\hat{\beta}$, that is, it is the second main diagonal element of $(\boldsymbol{X}'\hat{\Sigma}\boldsymbol{X})^{-1}$.

Example 5.3.1. (Application of the Logistic model in a $2 \times k$ contingency table) Continue to consider the data given in Example 4.1.1. We want to analyze whether there is an essential difference in the food intake behaviors between the lean and obese Zucker rats or not, and then decide whether obesity is caused by some recessive gene. Average amounts consumed by four obese and four lean Zucker rats, respectively, during the light period at four different ages are listed in Table 5.3.1.

Shi (1991) has ever used the LR test under some order restrictions to analyze this problem, and rejected the null hypothesis that there is no essential difference at level 2.5%. Now we will use the Logistic model to analyze this problem again.

Let n denote the total amount consumed by the types of Zucker rats, y the amount consumed by the lean Zucker rats, θ the food intake proportion, and x the age. We use the Logistic regression model to characterize the relationship between θ and x:

$$\text{logit}(\theta) = \alpha + \beta x.$$

Obviously, when $\beta = 0$, the food intake proportion does not change with the change of age, thus there is no essential difference in the food intake behaviors between the two types of Zucker rats. Now we need to consider the problem of testing

$$H_0 : \beta = 0 \quad \text{vs.} \quad H_1 : \beta \neq 0. \tag{5.3.20}$$

Table 5.3.1 lists the results at four different ages, so $k = 4$. The choice of the age variable x is very difficult (sometimes the x's are called scores). The following two methods of choice of scores are popular in applications: natural scores, *i.e.* $x_i = i$, $i = 1, \cdots, k$; dose scores, *i.e.* $x_1 = 12$, $x_2 = 23$, $x_3 = 30$, and $x_4 = 43$. Here, we adopt the second one. Based on the data in Table 5.3.1, we have

n_i	363	427	473	445
y_i	153	170	192	152
x_i	12	23	30	43

Empirical Logistic method. By the empirical Logistic transformation and Eq. (5.3.10), we have

$$z = (-0.3158,\ -0.4123,\ -0.3800,\ -0.6547)'.$$

From Eq. (5.3.11) we can get the following linear model

$$\begin{aligned} z &= \hat{\alpha} + \hat{\beta} x \\ &= -0.1617 - 0.0103x. \end{aligned}$$

Furthermore, using Eq. (5.3.12) yields a 95% confidence interval for β as $(-0.0251, 0.0044)$. Since it does not include 0, we cannot reject the null hypothesis $H_0 : \beta = 0$ in Eq. (5.3.20), that is, we can consider that there is no essential difference in the food intake behaviors between the two types of Zucker rats.

Maximum likelihood method. Starting with the initial values $\alpha^{(0)} = -0.1617, \beta^{(0)} = -0.0103$, using the iterative formula (5.3.18), and checking the convergence of iteration process using the criterion $|\beta^{(t)} - \beta^{(t+1)}| \leq 10^{-4}$. Finally, the above iteration process stops at the second iteration, and the corresponding linear equation is

$$z = -0.1569 - 0.0104x.$$

The Fisher information matrix based on the MLEs is

$$(\boldsymbol{X}'\boldsymbol{\Sigma}^{(t)}\boldsymbol{X})^{-1} = \begin{pmatrix} 0.018076 & -0.000567 \\ -0.000567 & 0.000021 \end{pmatrix}.$$

Using (5.3.19) yields a 95% confidence interval for β as $(-0.0193, -0.0015)$. Since it does not include 0, therefore, reject H_0 at significance level $\alpha = 0.05$, that is, we can consider that there is no essential difference of the amount of food intake between the two kinds of Zucker rats, so we infer that obesity is caused by some recessive gene. This result coincides with with that of Shi (1991).

Note that the conclusions based on the above two methods are different. As a result, it reminds us that we need to select a reasonable statistical method when making statistical inference. For this example, since the maximum likelihood method is more elaborate, its corresponding conclusion is more credible.

5.3.2 *Multivariable Case*

In essence, multivariable case coincides with the one-variable case. Recall model (5.3.3), for convenience, suppose that the logistic regression model has the following form

$$\ln\left(\frac{\theta}{1-\theta}\right) = \boldsymbol{X}'\boldsymbol{\beta}, \tag{5.3.21}$$

where both $\boldsymbol{X}$ and $\boldsymbol{\beta}$ are p-dimensional vectors, and $\boldsymbol{\beta}$ is an unknown parameter. Similar to (5.3.5), we can obtain

$$\theta(\boldsymbol{X}) = \frac{\exp\{\boldsymbol{X}'\boldsymbol{\beta}\}}{1+\exp\{\boldsymbol{X}'\boldsymbol{\beta}\}}. \tag{5.3.22}$$

Let $(\boldsymbol{x}_1', y_1), \cdots, (\boldsymbol{x}_k', y_k)$ be a set of samples, where $y_i \sim Bi(n_i, \theta_i)$ and $\theta_i = \theta(\boldsymbol{x}_i), i = 1, \cdots, k$. Then the likelihood function is

$$l(\boldsymbol{y}; \boldsymbol{\beta}) = \sum_{i=1}^{k} \{y_i \boldsymbol{x}_i'\boldsymbol{\beta} - n_i \ln(1 + \exp \boldsymbol{x}_i'\boldsymbol{\beta})\}.$$

Let $\boldsymbol{X}$ be a $p \times k$ matrix with column vectors $x_i, i = 1, \cdots, k$. Corresponding to (5.3.15), the likelihood equation is

$$\boldsymbol{X}'\boldsymbol{y} = \boldsymbol{X}'\hat{\boldsymbol{\mu}}, \tag{5.3.23}$$

where $\hat{\boldsymbol{\mu}} = (\hat{\mu}_1, \cdots, \hat{\mu}_k)'$, $\hat{\mu}_i = n_i\hat{\theta}_i$ and $\hat{\theta}_i = \exp \boldsymbol{x}_i'\hat{\boldsymbol{\beta}}/(1 + \exp \boldsymbol{x}_i'\hat{\boldsymbol{\beta}})$, the Fisher information matrix of the parameters can be expressed as

$$-\frac{\partial^2 l(\boldsymbol{y}; \boldsymbol{\beta})}{\partial \boldsymbol{\beta} \partial \boldsymbol{\beta}} = \boldsymbol{X}'\Sigma\boldsymbol{X},$$

where $\boldsymbol{\Sigma} = \mathrm{diag}\{n_1\theta_1(1-\theta_1), \cdots, n_k\theta_k(1-\theta_k)\}$. So, similarly, we can use the Newton-Raphson method to get the wanted solution. Now the confidence interval for β_i is

$$\left(\hat{\beta}_i - Z_{\alpha/2}\sqrt{b_i^2},\ \hat{\beta}_i + Z_{\alpha/2}\sqrt{b_i^2}\right), \tag{5.3.24}$$

where b_i^2 is the variance of $\hat{\beta}_i$, namely, the i-th main diagonal element of $(\boldsymbol{X}'\hat{\Sigma}\boldsymbol{X})^{-1}$.

5.4 Time Series: Trend Term Models

In many practical problems, data is collected by some pre-set time period. If these data are arranged according to the time index, we can obtain a data sequence. For example, for a given time t, we obtain data x_t, $t = 1, \cdots, T$, then we can arrange them as follows:

$$x_1, \cdots, x_T. \tag{5.4.1}$$

The above serial data over an ordered time sequence is called **time series** data, or **time series**, denoted by $\{x_t\}$. Refer to the data given in Table 5.1.1, obviously it forms a time series with $T = 25$. It is imaginable that, in many fields, such as economics, finance, medicine, biology and so on, large amounts of data can form time series. In order to study in a more general framework, sometimes we need to expand the data in Eq. (5.4.1) to the following form

$$\cdots, x_{-1}, x_0, x_1, \cdots, x_T, x_{T+1}, \cdots \tag{5.4.2}$$

where $x_1, \cdots, x_T$ are observations.

Obviously, any data x_t at time t is a random variable. However, the time series data is different from the data we have studied before in the following sense, it cannot be observed repeatedly at a given time. Consequently, we cannot make an efficient inference unless we assume that there exist some relationships among the data. This assumption is usually justified since the data series is ordered. A most intuitive idea is to assume that the data is a function of time t, that is,

$$x_t = f(t; \theta) + \varepsilon_t, \tag{5.4.3}$$

where θ is an unknown parameter. Suppose that the error terms ε_t's are i.i.d. satisfying $E\varepsilon_t = 0$ and $V\varepsilon_t = \sigma^2$. This error term series $\{\varepsilon_t\}$ is called

white noise. Since $EX_t = f(t;\theta)$, this mean is called as **trend term**. In many situations, the trend term can be expressed as a polynomial of time t,

$$f(t;\boldsymbol{\theta}) = \beta_0 + \beta_1 t + \cdots + \beta_p t^p,$$

where $\boldsymbol{\theta} = (\beta_0, \beta_1, \cdots, \beta_p)'$. Of course other alternative functions, such as piecewise functions, can be also used. We will discuss this problem in detail based on the data given in Table 5.1.1. But before to do this, we need to construct a criterion to evaluate whether a trend term model is suitable or not. Based on discussion in previous sections, we can think intuitively that a better trend term model should satisfy the requirement that the **residual sum of squares (RSS) should be as small as possible, and the error terms are i.i.d.** Obviously, **the former depends on model selection, and the later depends on hypothesis test.**

AIC method. Recall the discussion in Section 4.5, when comparing the maximum likelihood of a variety of candidate models, we should consider the number of free parameters in the models. We still need to pay attention to this point in our comparison study of the RSS of a variety of candidate trend term models. Let $\hat{\theta}$ be an LSE of θ, then the residual error at time t is

$$\hat{\varepsilon}_t = x_t - f(t;\hat{\theta}), \tag{5.4.4}$$

where $t = 1, \cdots, T$. Then the residual sum of squares is

$$\text{RSS}(f) = \sum_{t=1}^{T} \hat{\varepsilon}_t^2 = \sum_{t=1}^{T} (x_t - f(t;\hat{\theta}))^2. \tag{5.4.5}$$

Let p denote the number of free parameters in the trend term $f(t;\hat{\theta})$. Let

$$\hat{\sigma}^2(f) = \text{RSS}(f)/(T-p). \tag{5.4.6}$$

From Theorem 5.1.1, $\hat{\sigma}^2(f)$ is an unbiased estimator for σ^2. If model (5.4.3) holds, then ε_t is approximately distributed as a normal distribution with mean 0 and variance $\hat{\sigma}^2(f)$. As the parameters are θ and σ^2, the number of free parameters is $p+1$. From the discussion in Section 4.5, the AIC for model (5.4.3) is approximately given by

$$\text{AIC}(f) = T \ln \hat{\sigma}^2(f) + 2(p+1). \tag{5.4.7}$$

If f_1 and f_2 are two trend terms, we can consider that f_1 is better than f_2 when $\text{AIC}(f_1) < \text{AIC}(f_2)$.

Stationary series and white noise testing problem. Let $\{x_t\}$ be the time series given by (5.4.2). As we have discussed before, since we cannot observe it repeatedly at each given time, we must assume that there exist some relationships among the data in order to make statistical analysis. The requirement of stationarity of the time series is an intuitive and reasonable assumption. Let the covariance between x_r and x_s be

$$\gamma(r,s) = CV(X_r, X_s) = E(X_r - EX_r)(X_s - EX_s). \tag{5.4.8}$$

$\{x_t\}$ is called a **stationary time series**, if for any times t, r, and s, we have

$$EX_t = \mu; \tag{5.4.9}$$

$$\gamma(r,s) = \gamma(r+t, s+t). \tag{5.4.10}$$

The above stationarity is often termed as **covariance stationarity** to stress that the covariance function depends only on the time lag. Let $r-s = h$, then

$$\gamma(r,s) = \gamma(h,0) = \gamma(h). \tag{5.4.11}$$

Since in this case, the variance of the time series is constant over time, say $\gamma(0)$, then the correlation coefficient corresponding to the lag h is

$$\rho(h) = \frac{CV(X_{t+h}, X_t)}{\sqrt{V(X_{t+h})V(X_t)}} = \frac{\gamma(h)}{\gamma(0)}. \tag{5.4.12}$$

Let $x_1, \cdots, x_T$ be the observations of the time series $\{x_t\}$, then the estimators corresponding to the above parameters are respectively

$$\hat{\mu} = \bar{x} = \frac{1}{T}(x_1 + \cdots + x_T);$$

$$\hat{\gamma}(h) = \frac{1}{T}\sum_{t=1}^{T-h}(x_t - \bar{x})(x_{t+h} - \bar{x});$$

$$\hat{\rho}(h) = \hat{\gamma}(h)/\hat{\gamma}(0). \tag{5.4.13}$$

If $\{x_t\}$ is a stationary time series, then $\hat{\mu}$ is an unbiased estimator of μ, $\hat{\gamma}(h)$ is an asymptotic unbiased estimator of $\gamma(h)$.

Obviously, if $\{x_t\}$ is a white noise, then it is a stationary time series and

$$\mu = 0; \quad \gamma(h) = 0, \text{ when } h \geq 1.$$

Therefore, under the assumption that $\{x_t\}$ is a stationary time series, we need to consider the second equality above if we want to test whether $\{x_t\}$ is a white noise or not. The first equality is not essential, since if $\mu \neq 0$, then we can consider $y_t = x_t - \mu$. The second equality implies that $\rho(h) = 0$

when $h \geq 1$. Since $\rho(\cdot)$ is the correlation coefficient corresponding to time lag, we can think that the short-time effects are greater than the long-term ones, *i.e.*

$$\rho(h) \geq \rho(h+1),$$

Thus for a small k, say $k = 2$ or 3, consider the problem of testing

$$H_0: \quad \rho(1) = \cdots = \rho(k) = 0, \tag{5.4.14}$$

the testing problem (5.4.14) is called a **white noise test problem**. To construct a test statistic, we give the following theorem at first.

Theorem 5.4.1. *For given $k \geq 1$, let $Z_h = \sqrt{T}\hat{\rho}(h)$, where $h = 1, \cdots, k$. Let $\boldsymbol{Z} = (Z_1, \cdots, Z_k)'$. If $\{X_t\}$ is a white noise series, $\boldsymbol{Z}$ converges in distribution to a k-dimensional standard normal distribution when $T \to \infty$.*

Proof. Let $\tilde{\gamma}(h) = \sum_{t=1}^{T} X_t X_{t+h}/T$ and $\tilde{\rho}(h) = \tilde{\gamma}(h)/\tilde{\gamma}(0)$. Firstly, we prove that when $T \to \infty$,

$$\sqrt{T}(\tilde{\gamma}(1), \cdots, \tilde{\gamma}(k))'/\sigma^2 \xrightarrow{d} N(\boldsymbol{0}, \boldsymbol{I}_k). \tag{5.4.15}$$

Let $Y_{th} = X_t X_{t+h}$ and $\boldsymbol{Y}_t = (Y_{t1}, \cdots, Y_{tk})'$, then by the definition of normal distribution, (5.4.15) holds if and only if for any k-dimensional vector $\boldsymbol{\lambda}$,

$$\sqrt{T}^{-1} \sum_{t=1}^{T} \boldsymbol{\lambda}' \boldsymbol{Y}_t/\sigma^2 \xrightarrow{d} N(\boldsymbol{0}, \boldsymbol{\lambda}'\boldsymbol{\lambda}). \tag{5.4.16}$$

From the given conditions, it is easy to verify that $EX_t X_{t+k} = 0$ and $V(X_t X_{t+k}) = \sigma^4$. Thus $\{X_t X_{t+h}\}$ also forms a white noise series. Since, when $h \neq h'$, we have

$$CV(X_t X_{t+h}, X_t X_{t+h'}) = 0,$$

then $\{\boldsymbol{\lambda}'\boldsymbol{Y}_t\}$ also forms a white noise series, where $E\boldsymbol{\lambda}'\boldsymbol{Y}_t = 0$ and $V(\boldsymbol{\lambda}'\boldsymbol{Y}_t) = \boldsymbol{\lambda}'V(\boldsymbol{Y}_t)\boldsymbol{\lambda} = \sigma^4\boldsymbol{\lambda}'\boldsymbol{\lambda}$. By the Central Limiting Theorem, we can get (5.4.16), consequently (5.4.15). Again by the Law of Large Numbers, $\tilde{\gamma}(0)$ converges in probability to σ^2 when $T \to \infty$. Therefore, we have

$$\sqrt{T}(\tilde{\rho}(1), \cdots, \tilde{\rho}(k))' = \sqrt{T}(\tilde{\gamma}(1), \cdots, \tilde{\gamma}(k))'/\tilde{\gamma}(0) \xrightarrow{d} N(\boldsymbol{0}, \boldsymbol{I}_k). \tag{5.4.17}$$

From the given conditions, $\bar{X}$ converges in probability to 0 when $T \to \infty$, thus, for $0 \leq h \leq k$, we have

$$\sqrt{T}(\tilde{\gamma}(h) - \hat{\gamma}(h)) \xrightarrow{d} 0.$$

This completes the proof of the theorem. □

Based on the above theorem, and refer to Example 2.5.3, w can construct some test statistics for the problem of testing (5.4.14), such as a χ^2 statistic, *i.e.*

$$\chi^2 = T \sum_{h=1}^{k} \hat{\rho}^2(h), \tag{5.4.18}$$

where $\hat{\rho}(h)$ is given by (5.4.13). By Theorem 5.4.1 and Example 1.1.8, the limiting distribution of the test statistic given by (5.4.18) is a χ^2 distribution with k degrees of freedom, and we will reject H_0 when the corresponding sample value is large. Ljung and Box (1978) gave a modified test statistic as follows

$$\text{LB} = T(T+2) \sum_{h=1}^{k} \frac{\hat{\rho}^2(h)}{T-h},$$

whose limiting distribution is still a χ^2 distribution with k degrees of freedom.

Example 5.4.1. (Analyzing the trend of China's per capita GDP) Now we apply the methods given above to analyze the per capita GDP data given in Table 5.1.1. Let x_1 denote the data of the year 1978 year, x_2 the data of the year 1979 and so on. So $T = 25$.

Exploring the trend term. First, we select three models: linear, third-order polynomial, and three-parameter Logistic. By the Least Squares method, we can obtain the following estimating equations and the corresponding residual sums of squares respectively

$$\begin{aligned}
&f_1(t;\hat{\theta}) = -3147.11 + 568.18t, && \text{RSS}(f_1) = 7.6126 \times 10^7;\\
&f_2(t;\hat{\theta}) = 1299.48 - 478.12t + 47.12t^2 - 0.30t^3, && \text{RSS}(f_2) = 8.104 \times 10^6;\\
&f_3(t;\hat{\theta}) = 0.02 + \frac{1-0.02}{1+\exp(-0.35t+6.62)}, && \text{RSS}(f_3) = 1.9012 \times 10^6.
\end{aligned}$$

Figures 5.4.1-5.4.3 respectively show the three kinds of trend terms and the corresponding fitting errors.

Based on Figs. 5.4.1-5.4.3 in conjunction with comparing the residual sums of squares, we can see that f_3 is better than both f_1 and f_2 obviously. So we can think that it is more reasonable to use the three-parameter Logistic model to characterize the growth trend of China's per capita GDP.

Exploring piecewise trend. In fact, we can use some piecewise functions to study the trend term. Based on the above figures, we briefly review China's economic development in recent years, which can be divided into

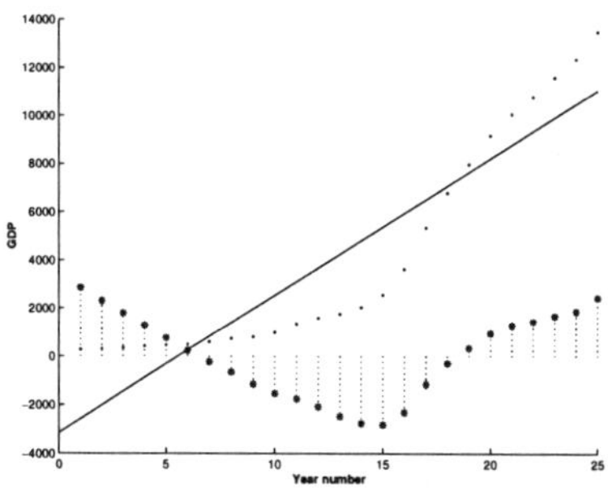

Fig. 5.4.1 Linear model and the fitting errors

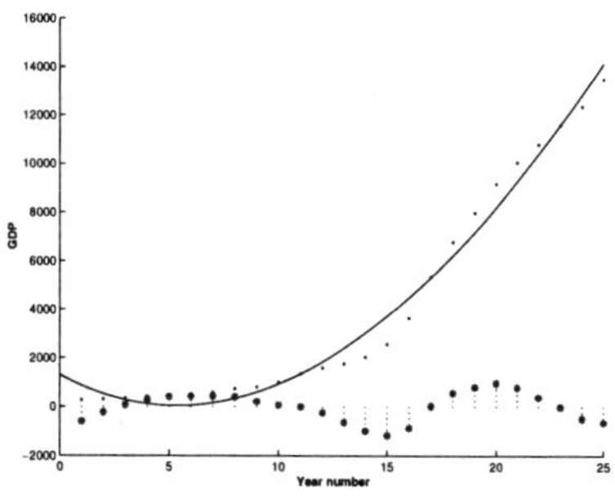

Fig. 5.4.2 Third-order polynomial model and the fitting errors

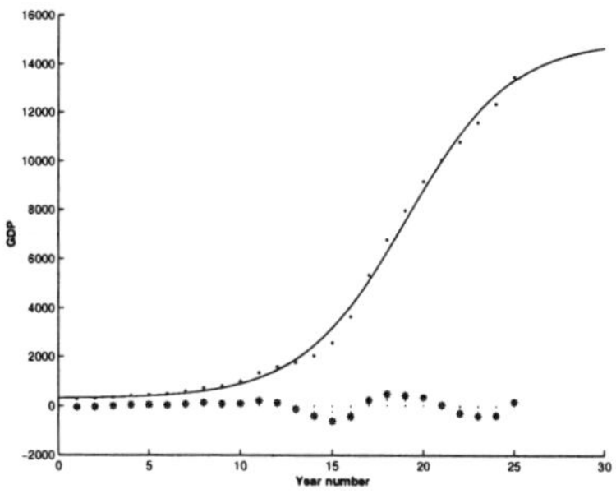

Fig. 5.4.3 Three-parameter Logistic model and the fitting errors

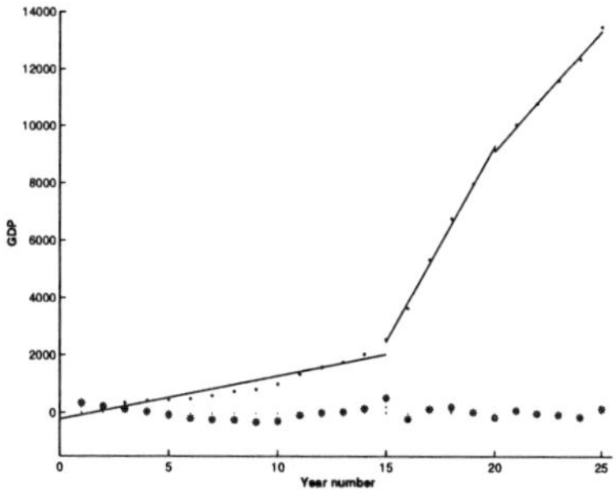

Fig. 5.4.4 Piecewise trend and the fitting errors

three stages: before the year 1992, the economic growth rate increased steadily. Deng Xiaoping's "Southern Tour Speech" in the year 1992 greatly promoted China's economic reform and development, and China had been achieving remarkably high economic growth. In the year 1997, a financial and economic crisis erupted in Asia, which threatened economic and financial stability in the region. China's government changed some economic policies accordingly. Based on the above review, using the least squares method, we can obtain the following piecewise linear model:

$$f_4(t;\hat{\theta}) = \begin{cases} -222.33 + 150.43t, & t = 1, \cdots, 15; \\ -18119.80 + 1379.70t, & t = 16, \cdots, 20; \\ -7717.50 + 842.70t, & t = 21, \cdots, 25, \end{cases}$$

where the three ranges of t correspond to the 1978-1992, 1993-1997, and 1998-2002 periods, respectively. So the residual sum of squares is $\text{RSS}(f_4) = 1.016 \times 10^6$. The corresponding trend is shown in Fig. 5.4.4.

From Fig. 5.4.4, we can observe that it is also reasonable to use a piecewise-linear model to characterize the growth trend of China's per capita GDP. Since the difference between the residual sums of squares is not significant, we can use the AIC criterion to make a comparison. It is easy to obtain that $\hat{\sigma}^2(f_3) = 76048$ and $\hat{\sigma}^2(f_4) = 40640$. By Eq. (5.4.7) we can get

$$\begin{aligned}\text{AIC}(f_3) &= 25\ln\hat{\sigma}^2(f_3) + 2\times 4 = 288.98,\\ \text{AIC}(f_4) &= 25\ln\hat{\sigma}^2(f_4) + 2\times 7 = 279.31.\end{aligned}$$

Since $\text{AIC}(f_4) < \text{AIC}(f_3)$, so by the AIC criterion, we can think that to use f_4 as a trend term is better than to use f_3. But as a predictor model, f_4 is inferior to f_3, since the piecewise function essentially destroys the integration trend.

White noise test. Corresponding to the trend term $f(t;\theta)$, let

$$\hat{\varepsilon}_t(f) = x_t - f(t;\hat{\theta}).$$

We hope that the time series $\{\varepsilon_t(f)\}$ can form a white noise, *i.e.* an i.i.d. random series with mean zero. Note that the asterisk in Figs. 5.4.1-5.4.4 corresponds to the point with residual error $\hat{\varepsilon}_t(f)$. It is easy to see that both $\{\hat{\varepsilon}_t(f_1)\}$ and $\{\hat{\varepsilon}_t(f_2)\}$ show too regular patterns to be white noises. However, for $\{\hat{\varepsilon}_t(f_3)\}$ and $\{\hat{\varepsilon}_t(f_4)\}$, it is necessary to make a careful analysis before decision-making.

Consider the two cases of $k = 2$ and 3, or equivalently,

$$\begin{aligned}H_{01} &: \rho(1) = \rho(2) = 0,\\ H_{02} &: \rho(1) = \rho(2) = \rho(3) = 0.\end{aligned} \tag{5.4.19}$$

We use the χ^2 statistic given by (5.4.18). When $k = 2$, the critical values are $\chi^2(0.05) = 5.99$ and $\chi^2(0.01) = 9.21$; when $k = 3$, the critical values are $\chi^2(0.05) = 7.81$ and $\chi^2(0.01) = 11.34$. As far as f_3 and f_4 are concerned, when $k = 2$, the observed values of the two corresponding statistics are $\chi^2(f_3) = 10.58$ and $\chi^2(f_4) = 4.77$; when $k = 3$, the corresponding observed values are $\chi^2(f_3) = 16.02$ and $\chi^2(f_4) = 5.88$. We can see that the test statistic given by using f_3 as the trend term reject H_0 significantly. Hence we cannot regard $\{\hat{\varepsilon}_t(f_3)\}$ as a white noise. However, when f_4 is used as the trend term, the corresponding statistic cannot reject H_0, more specifically, we have $p = 0.09$ and 0.12 for $k = 2$ and 3 respectively. Therefore, we can think that $\{\hat{\varepsilon}_t(f_4)\}$ forms a white noise time series. In summary, we can say that it is more reasonable to use a piecewise function to characterize China's per capita GDP during the 1978-2002 period. Finally, we would like to point out that it is also reasonable to choose the years 1992 and 1998 as the cut-points although we have chosen the years 1993 and 1998 in the previous discussion.

5.5 Time Series: Autoregressive Models

In this section, we will continue to discuss the testing problem of time series. Suppose that the data is given by (5.4.2). In many practical problems, data points in a time series not only depends on time, but also are correlated with each other. For example, the GDP of this year must be related with that of last year, or even the year before that. For this type of problem, we construct the following mathematical model.

Let $\{x_t\}$ be a time series, consider the relation between the data points x_t at time t and the ones at times $t-1, \cdots, t-p$, which can written as

$$x_t = f(x_{t-1}, \cdots, x_{t-p}) + \varepsilon_t, \tag{5.5.1}$$

where ε_t is the error term, which is usually assumed to be a white noise series; p, called **lag**, is a time index that indicates the degree of correlation. First let us consider a simple but important case.

5.5.1 *Random Walk and Brownian Motion*

In Economy and Finance, under the assumption of independence, the series of price fluctuations is just like a random walk of a drunk (Pearson, 1905), so it is called a **random walk model**, that is, the time series $\{x_t\}$ satisfies

$$x_t = x_{t-1} + \varepsilon_t, \; t = 1, \cdots, T, \tag{5.5.2}$$

where $\{\varepsilon_t\}$ is a white noise. It is the simplest case of (5.5.1). If we let $x_0 = 0$, for $t \geq 1$, it is easy to get

$$x_t = \varepsilon_1 + \cdots + \varepsilon_t. \tag{5.5.3}$$

Based on the properties of white noise, we can obtain the variance of x_t as

$$VX_t = t\sigma^2.$$

Since it is an increasing function of time, so the random walk model is nonstationary.

Next let us discuss an important property of the random walk model. For two times $t < s$, we obtain

$$x_s - x_t = \varepsilon_{t+1} + \cdots + \varepsilon_s, \tag{5.5.4}$$

and $V(X_s - X_t) = (s-t)\sigma^2$. For other two times $r < q$, if $q < t$ or $s < r$, then we can get the following results

$$\begin{cases} (X_q - X_r) \perp\!\!\!\perp (X_s - X_t), \\ X_s - X_t \sim N(0, (s-t)\sigma^2), \\ X_q - X_r \sim N(0, (q-r)\sigma^2). \end{cases} \tag{5.5.5}$$

Next let us discuss some limiting properties of random walk model which is related to Brownian motion.

Let $[a, b]$ be a time interval. If for any time $t \in [a, b]$, there exists a corresponding random variable x_t, then $\{x_t\}$ is a **continuous time process** on $[a, b]$, whose simplest case is so-called **Brownian motion**, or **Wiener process**. A continuous time process $W(\cdot)$ is called as a **standard Brownian motion** or a **standard Wiener process**, if for any time $t \in [0, 1]$, $W(t)$ satisfies

(1) $W(0) = 0$ a.s.;

(2) For any time partition $0 \leq t_1 < t_2 < \cdots < t_k \leq 1$,

$$W(t_2) - W(t_1), \cdots, W(t_k) - W(t_{k-1})$$

are mutually independent, and $W(t_{i+1}) - W(t_i)$ has a normal distribution $N(0, t_{i+1} - t_i)$;

(3) $W(t)$ is continuous in t a.s.

It can be seen that the above conditions are similar to those given in (5.5.5). By Condition (2), we can obtain that $W(t) \sim N(0, t)$ for any $t \in (0, 1]$ and $Z(t) = \sigma W(t) \sim N(0, t\sigma^2)$.

Next let us discuss the relationship between the random walk model and the Brownian motion. From (5.5.3) and the Central Limit Theorem, we can obtain

$$\sqrt{T}\bar{X}_T = \frac{1}{\sqrt{T}} \sum_{i=1}^{T} \varepsilon_i \xrightarrow{L} N(0, \sigma^2).$$

For $r \in (0, 1]$, we use $[rT]$ to denote the largest integer not exceeding rT. Let

$$\bar{X}_T(r) = \frac{1}{\sqrt{T}} \sum_{i=1}^{[rT]} \varepsilon_i. \tag{5.5.6}$$

It is a step function in r. By

$$\frac{1}{\sqrt{[rT]}} \sum_{i=1}^{[rT]} \varepsilon_i \xrightarrow{L} N(0, \sigma^2)$$

and $\sqrt{[rT]}/\sqrt{T} \longrightarrow \sqrt{r}$, we can obtain that

$$\sqrt{T}\bar{X}_T(r) \xrightarrow{L} N(0, r\sigma^2).$$

For any partition of the interval $[0, 1]$, say, $0 \leq r_1 < r_2 < \cdots < r_k \leq 1$. By a discussion similar to that of (5.5.5), we obtain that

$$\sqrt{T}[\bar{X}_T(r_2) - \bar{X}_T(r_1)], \ \cdots, \ \sqrt{T}[\bar{X}_T(r_k) - \bar{X}_T(r_{k-1})]$$

are mutually independent, and $\sqrt{T}[\bar{X}_T(r_{i+1}) - \bar{X}_T(r_i)] \xrightarrow{L} N(0, (r_{i+1} - r_i)\sigma^2)$. So we have the following theorem.

Theorem 5.5.1. *Suppose that $\{x_t\}$ is a random walk model, and $\bar{x}_T(r)$ is defined as in* (5.5.6) *with $r \in [0,1]$. Then when $T \longrightarrow \infty$,*

$$\frac{\sqrt{T}}{\sigma}\bar{X}_T(\cdot) \xrightarrow{L} W(\cdot). \tag{5.5.7}$$

Sometimes the above theorem is called the **Functional Central Limit Theorem**.

5.5.2 *Autoregressive Model*

If the function given by (5.5.1) is linear, then the relation among variables can be expressed as

$$x_t = \alpha + \beta_1 x_{t-1} + \cdots + \beta_p x_{t-p} + \varepsilon_t, \tag{5.5.8}$$

where $\{\varepsilon_t\}$ is a white noise series, and both α and β_i are unknown parameters. Because this model is very similar to the regression model, so (5.5.8) is called an **autoregressive model**, or AR(p) model.

5.5.2.1 *AR*(1) *model*

First we discuss the simplest case, *i.e.* AR(1) model. Then model (5.5.8) reduces to

$$x_t = \alpha + \beta x_{t-1} + \varepsilon_t. \tag{5.5.9}$$

Obviously when $\alpha = 0$ and $\beta = 1$, then it is a random walk model which is nonstationary. Next we derive conditions that the parameters α and β must satisfy in order to ensure that AR(1) is a stationary time series. Let

$$EX_t = EX_{t-1} = \mu.$$

Taking the expectation on both sides of (5.5.9), we can easily obtain that

$$\mu = \alpha/(1-\beta), \tag{5.5.10}$$

which implies that $\beta \neq 1$. This result also shows that the random walk model is nonstationary from another point of view. Next we consider the variance, following the notations in the previous section, let

$$E(X_t - \mu)^2 = E(X_{t-1} - \mu)^2 = \gamma_0.$$

By $(X_t - \mu) = \beta(X_{t-1} - \mu) + \varepsilon_t$, we can obtain

$$\gamma_0 = \beta^2\gamma_0 + 2\beta E(X_{t-1} - \mu)\varepsilon_t + \sigma^2. \tag{5.5.11}$$

From (5.5.9), we have

$$X_{t-1} - \mu = \varepsilon_{t-1} + \beta\varepsilon_{t-2} + \beta^2\varepsilon_{t-3} + \cdots .$$

Since $\{\varepsilon_t\}$ is a white noise series, the cross-term in (5.5.11) is 0. Then (5.5.11) becomes

$$\gamma_0 = \sigma^2/(1 - \beta^2). \tag{5.5.12}$$

In order to ensure that the AR(1) model is a stationary time series, the parameter β must satisfy

$$\mid \beta \mid < 1.$$

Thus, we get a very important testing problem of time series on the autoregressive model

$$H_0: \ \beta = 1 \quad \text{vs.} \quad H_1: \ \beta < 1, \tag{5.5.13}$$

that is, to test whether a time series is a random walk model and the alternative hypothesis that the time series is stationary. Considering the form of (5.5.12), this test is sometimes called a **unit root test**. This testing problem has been widely studied in modern econometrics research. First we will discuss some parameter estimation problems related to this testing problem.

5.5.2.2 *MLEs of parameters*

The outstanding feature of the autoregressive model is that the value of variable at time t depends on the values before t, so we can construct a conditional likelihood function.

From Eqs. (5.5.10) and (5.5.12), we have

$$X_1 \sim N\left(\frac{\alpha}{1-\beta}, \frac{\sigma^2}{1-\beta^2}\right). \tag{5.5.14}$$

By (5.5.9), the conditional distribution of X_2 given that $X_1 = x_1$ is

$$X_2 \mid X_1 = x_1 \sim N(\alpha + \beta x_1, \sigma^2). \tag{5.5.15}$$

By using the expression as (1.2.19) in Section 1.2, the joint density function of x_2 and x_1 can be written as

$$f_\theta(x_2, x_1) = f_\theta(x_2 \mid x_1) f_\theta(x_1),$$

the two density functions in the right side of the above equation are known, where $\theta = (\alpha, \beta, \sigma^2)$. Similarly,

$$\begin{aligned} f_\theta(x_3, x_2, x_1) &= f_\theta(x_3 \mid x_2, x_1) f_\theta(x_2, x_1) \\ &= f_\theta(x_3 \mid x_2) f_\theta(x_2 \mid x_1) f_\theta(x_1). \end{aligned}$$

The above second equality follows from the independence of X_3 and X_1 given X_2. Accordingly, we obtain the log-likelihood function

$$l(x;\theta) = \ln f_\theta(x_1) + \sum_{t=2}^{T} \ln f_\theta(x_t \mid x_{t-1}). \tag{5.5.16}$$

By Eqs. (5.5.14) and (5.5.15), we can obtain the analytic expression of the log-likelihood function. For the convenience of computation, when T is large, we can assume that x_1 is a given initial value, thus the conditional log-likelihood function formed in this way has only the second term in the right-hand side of (5.5.16), *i.e.*

$$\begin{aligned} l(x|x_1;\theta) &= \sum_{t=2}^{T} \ln f_\theta(x_t \mid x_{t-1}) \\ &= c - \frac{T-1}{2} \ln \sigma^2 - \sum_{t=2}^{T} \frac{(x_t - \alpha - \beta x_{t-1})^2}{2\sigma^2}. \end{aligned}$$

It is easy to verify that maximizing the above equation is equivalent to minimizing

$$\sum_{t=2}^{T} (x_t - \alpha - \beta x_{t-1})^2.$$

From the discussion in the previous sections (cf. (5.1.40)), we can obtain the LSEs (or the conditional likelihood estimators) of α and β are respectively

$$\begin{aligned} \hat{\alpha} &= \bar{y} - \hat{\beta}\bar{x}, \\ \hat{\beta} &= s_{xy}/s_{xx}, \end{aligned} \tag{5.5.17}$$

where

$$\begin{aligned} \bar{x} &= \sum_{t=2}^{T} x_{t-1}/(T-1), \quad & \bar{y} &= \sum_{t=2}^{T} x_t/(T-1), \\ s_{xx} &= \sum_{t=2}^{T} (x_{t-1} - \bar{x})^2, \quad & s_{xy} &= \sum_{t=2}^{T} (x_{t-1} - \bar{x})(y_t - \bar{y}). \end{aligned}$$

Then the conditional MLE of σ^2 is

$$\hat{\sigma}^2 = \frac{1}{T-1} \sum_{t=2}^{T} (x_t - \hat{\alpha} - \hat{\beta} x_{t-1})^2. \tag{5.5.18}$$

Based on the above estimators, we can discuss the testing problem now.

5.5.2.3 *Unit root test*

Now we discuss the unit root test of AR(1) model, *i.e.* the testing problem (5.5.13) for model (5.5.9). We can see that the parameter α in model (5.5.9) is a nuisance parameter, so we can use the centering method in Section 5.1 to simplify discussion. Then (5.5.9) can be written as

$$x_t = \beta x_{t-1} + \varepsilon_t. \tag{5.5.19}$$

We still want to test the hypothesis (5.5.13), *i.e.*

$$H_0: \ \beta = 1 \quad \text{vs.} \quad H_1: \ \beta < 1. \tag{5.5.20}$$

By (5.5.17), we obtain that the LSE of β is

$$\hat{\beta} = \sum_{t=2}^{T} x_t x_{t-1} \Big/ \sum_{t=2}^{T} x_{t-1}^2. \tag{5.5.21}$$

For the problem of testing (5.5.20), we construct the following test statistic

$$R = T(\hat{\beta} - 1), \tag{5.5.22}$$

which is called a unit root test statistic. We have the following theorem.

Theorem 5.5.2. *When H_0 is true, we have*

$$R \xrightarrow{L} \frac{[W(1)]^2 - 1}{2\int_0^1 [W(s)]^2 ds}, \tag{5.5.23}$$

where $W(\cdot)$ is a Brownian motion on $[0, 1]$.

Proof. It is easy to see that, when H_0 is true, (5.5.19) is a random walk model. From Theorem 5.5.1,

$$\sqrt{T}\bar{X}_T(\cdot) \xrightarrow{L} \sigma W(\cdot),$$

where $\bar{x}_T(\cdot)$ is defined by (5.5.6). By (5.5.21) and the hypothesis H_0, we can obtain that

$$\begin{aligned} R &= T(\hat{\beta} - 1) \\ &= \tfrac{1}{T}\sum_{t=2}^{T} \varepsilon_t x_{t-1} \Big/ \left(\tfrac{1}{T^2}\sum_{t=2}^{T} x_{t-1}^2\right). \end{aligned} \tag{5.5.24}$$

Next we compute the limiting distributions of the numerator and the denominator in the right side of (5.5.24) respectively.

For the numerator, we have

$$\begin{aligned}\frac{1}{T}\sum_{t=2}^{T}\varepsilon_t x_{t-1} &= \frac{1}{2T}\sum_{t=2}^{T}(x_t^2 - x_{t-1}^2 - \varepsilon_t^2)\\ &= \frac{1}{2T}(x_T^2 - x_1^2) - \frac{1}{2T}\sum_{t=2}^{T}\varepsilon_t^2\\ &= \frac{T}{2}\left\{[\bar{x}_T(1)]^2 - \left[\bar{x}_T\left(\frac{1}{T}\right)\right]^2\right\} - \frac{1}{2T}\sum_{t=2}^{T}\varepsilon_t^2\\ &\xrightarrow{L} \frac{\sigma^2}{2}[W(1) - 1].\end{aligned}$$

For the denominator, by Eqs. (5.5.3) and (5.5.6), we have

$$\frac{1}{T^2}\sum_{t=2}^{T}x_{t-1}^2 = \sum_{t=2}^{T}\left[\bar{x}_T\left(\frac{t-1}{T}\right)\right]^2 = \sum_{h=0}^{T-1}\left[\bar{x}_T\left(\frac{h}{T}\right)\right]^2.$$

Recall again $\bar{x}_T(r)$ defined by (5.5.6), when $h/T \le r < (h+1)/T$, $\bar{x}_T(r) = \bar{x}_T(h/T)$ is a constant, so

$$\begin{aligned}\frac{1}{T^2}\sum_{t=2}^{T}x_{t-1}^2 &= \sum_{h=0}^{T-1}\frac{1}{T}\left[\sqrt{T}\bar{x}_T\left(\frac{h}{T}\right)\right]^2\\ &= \sum_{h=0}^{T-1}\int_{\frac{h}{T}}^{\frac{h+1}{T}}\left[\sqrt{T}\bar{x}_T(r)\right]^2 dr\\ &= \int_0^1 \left[\sqrt{T}\bar{x}_T(r)\right]^2 dr\\ &\xrightarrow{L} \sigma^2\int_0^1 [W(s)]^2 ds.\end{aligned}$$

This completes the proof of the theorem. □

For discussion on unit root test, see Fuller (1976), Dickey and Fuller (1979), and Evans and Savin (1981), and so on. For more detailed discussion, see Hamilton (1994). Sometimes the test statistic R given by (5.5.22) is called a **Dickey-Fuller test**, or **DF test**. Dickey and Fuller (1981) gave some critical values for the R test statistic by using the Monte Carlo method, see Table 5.5.1.

In economic research, we usually use the DF t-test and call it as τ **test** (Gujarati, 1995). For model (5.5.19) the unit root test statistic is

$$\tau = \frac{\hat{\beta} - 1}{s}, \tag{5.5.25}$$

Table 5.5.1 Simulated asymptotic critical values for the R statistic for unit root tests.

Sample size	Significance level							
T	0.01	0.025	0.05	0.10	0.90	0.95	0.975	0.99
25	−11.9	−9.3	−7.3	−5.3	1.01	1.40	1.79	2.28
50	−12.9	−9.9	−7.7	−5.5	0.97	1.35	1.70	2.16
100	−13.3	−10.2	−7.9	−5.6	0.95	1.31	1.65	2.09
250	−13.6	−10.3	−8.0	−5.7	0.93	1.28	1.62	2.04
500	−13.7	−10.4	−8.0	−5.7	0.93	1.28	1.61	2.04
∞	−13.8	−10.5	−8.1	−5.7	0.93	1.28	1.60	2.03

where $s^2 = \hat{\sigma}^2 / \sum_{t=2}^{T} x_{t-1}^2$ and $\hat{\sigma}^2$ is given by Eq. (5.5.18). Similar to Eq. (5.5.24), the τ statistic can be written as

$$\tau = \frac{1}{T}\sum_{t=2}^{T} \varepsilon_t x_{t-1} \Big/ \left[\frac{1}{T^2}\sum_{t=2}^{T} x_{t-1}^2 \cdot \hat{\sigma}^2\right]^{\frac{1}{2}}.$$

Since $\hat{\sigma}^2$ converges to σ^2 in probability, similar to the proof of Theorem 5.5.2, we can obtain

$$\tau \xrightarrow{L} \frac{[W(1)]^2 - 1}{2\left\{\int_0^1 [W(s)]^2 ds\right\}^{\frac{1}{2}}}.$$

Example 5.5.1. (Analyzing China's GDP time series) Table 5.5.2 records China's GDP data during the 1978-2002 period, the data are cited from China Statistical Yearbook. In the table, the nominal GDP measures the value of all the goods and services produced expressed in current prices, *i.e.* the price of GDP in that year. On the other hand, removing the effects of price changes from the nominal GDP results in what is known as the real GDP, thus making it possible to compare the GDP from one year to the next. For this reason, the real GDP is also called as comparable GDP. In practical analysis, logarithm transformation is usually taken for the raw data to stabilize the variance.

(1) Analyzing the trend term. Let $y_1, \cdots, y_{25}$ denote the log-real GDP data during the 1978-2002 period, then $T = 25$. Figure 5.5.1 shows the growth trend of the log-real GDP data.

We can see that the trend is very obvious, so we need to eliminate the effect of trend at first, then we will discuss the stationarity of the series. Construct the following linear model

$$y_t = \alpha + \beta t + \mu_t, \ t = 1, \cdots, T.$$

Table 5.5.2 China's GDP during the 1978-2002 period.

Year	Nominal GDP	Log-nominal GDP	Real GDP	Log-real GDP
1978	3624.10	8.20	3202.40	8.07
1979	4038.20	8.30	3445.78	8.14
1980	4517.80	8.42	3714.55	8.22
1981	4862.40	8.49	3907.71	8.27
1982	5294.70	8.57	4263.31	8.36
1983	5934.50	8.69	4728.01	8.46
1984	7171.00	8.88	5446.67	8.60
1985	8964.40	9.10	6181.97	8.73
1986	10202.20	9.23	6725.99	8.81
1987	11962.50	9.39	7506.20	8.92
1988	14928.30	9.61	8354.40	9.03
1989	16909.20	9.74	8696.93	9.07
1990	18547.90	9.83	9027.41	9.11
1991	21617.80	9.98	9857.94	9.20
1992	26638.10	10.19	11257.76	9.33
1993	34634.40	10.45	12777.56	9.46
1994	46759.40	10.75	14387.54	9.57
1995	58478.10	10.98	15898.23	9.67
1996	67884.60	11.13	17424.46	9.77
1997	74462.60	11.22	18957.81	9.85
1998	78345.20	11.27	20436.52	9.93
1999	82067.50	11.32	21887.51	9.99
2000	89468.10	11.40	23638.51	10.07
2001	97314.80	11.49	25411.40	10.14
2002	105172.30	11.56	27520.54	10.22

Using the method introduced in Section 5.4, we obtain the LSEs of α and β as follows

$$\hat{\alpha} = 7.95 \text{ and } \hat{\beta} = 0.09.$$

Let $\hat{\mu}_t = y_t - \hat{\alpha} - \hat{\beta}t,\ t = 1, \cdots, T$, then $\{\hat{\mu}_t\}$ is a time series that has been eliminated the effect of trend, its change trend is shown in Fig. 5.5.2.

Next we discuss the stationarity of $\{\hat{\mu}_t\}$. Suppose

$$\hat{\mu}_t = \rho\hat{\mu}_{t-1} + \varepsilon_t,\ t = 1, \cdots, T, \tag{5.5.26}$$

where $\{\varepsilon_t\}$ is a white noise series. Consider the problem of testing

$$H_0 : \rho = 1 \quad \text{vs.} \quad H_1 : \rho < 1$$

By Eq. (5.5.21), we obtain $\hat{\rho} = 0.76$. So by Eq. (5.5.22), we obtain the DF statistic $R = 25 \times (0.76 - 1) = -6$. From Table 5.5.1 the DF statistic cannot reject the assumption of random walk at the significance level 0.05. We can calculate this probability further and obtain that

$$P\{R \leq -6\} = 0.074.$$

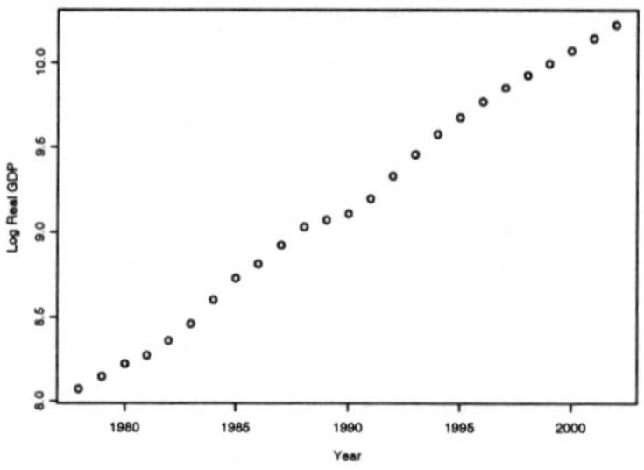

Fig. 5.5.1 Scatter diagram of the log-real GDP data during the 1978-2002 period

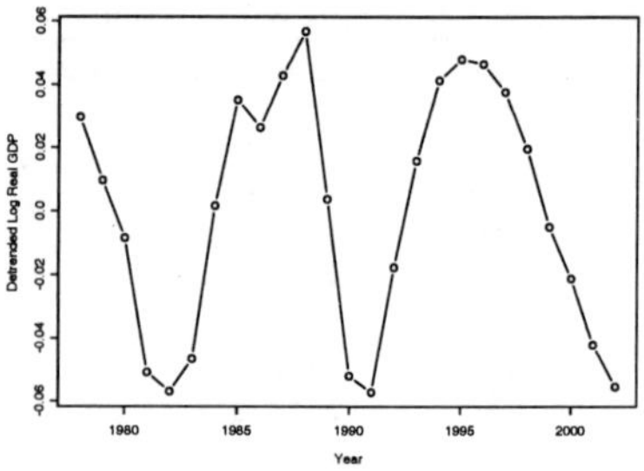

Fig. 5.5.2 Line diagram of the detrended log-real GDP data during the 1978-2002 period

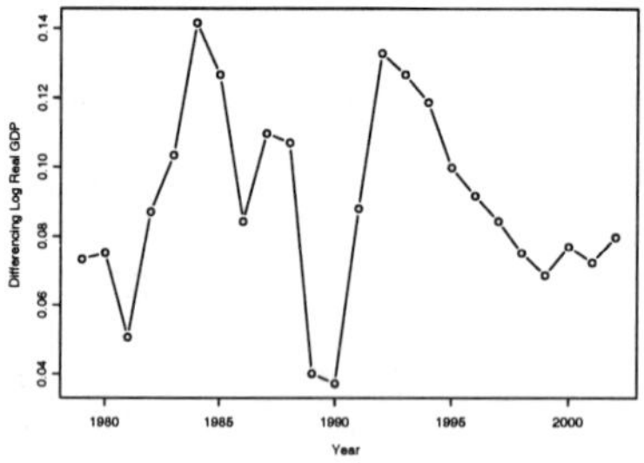

Fig. 5.5.3 Line diagram of the differencing log-real GDP data during the 1978-2002 period

This relatively larger probability shows that the detrended GDP time series data during the 1978-2002 period is nonstationary, and the fluctuation is relatively large.

(2) Difference stationarity. When a time series is nonstationary, we usually consider the difference stationarity, which is a common method in economic time series. Testing the stationarity of the first and second difference are equivalent to testing the stationarity of economic growth and its acceleration, respectively. Let

$$\triangle y_t = y_t - y_{t-1}$$

where $t = 2, \cdots, 25$, we can obtain the change regularity of difference, see Fig. 5.5.3.

Comparing with Fig. 5.5.2, we can see that the change of $\triangle y_t$ shows no obvious regularity. Next we test the difference stationarity, recall the AR(1) model defined by (5.5.9):

$$\triangle y_t = \alpha + \beta \triangle y_{t-1} + \varepsilon_t$$

Consider the problem of testing

$$H_0 : \beta = 1 \quad \text{vs.} \quad H_1 : \beta < 1 \tag{5.5.27}$$

After centralization, the estimator of β is

$$\hat{\beta} = \frac{\sum\limits_{t=3}^{25} (\triangle y_t - \alpha)(\triangle y_{t-1} - \alpha)}{\sum\limits_{t=3}^{25} (\triangle y_{t-1} - \alpha)^2}$$

$$= 0.53$$

where α denotes the mean of the difference, namely, $\alpha = \sum\limits_{t=2}^{24} \triangle y_t / 24$. By Eq. (5.5.22) the DF statistic is $R = 24 \times (0.53 - 1) = -11.24$, so the p-value is

$$P\{R \leq -11.24\} = 0.01.$$

At the significance level 0.01 we reject the null hypothesis given in (5.5.27), in other words, we can consider that the time series formed by China's GDP difference during the 1978-2002 period is stationary, that is, China's economic growth trend is stationary in recent years.

Based on the above discussion, we can see that two aspects of this research deserve special attention: one is the trend term, the other is the differences.

5.6 Granger Causality Tests

5.6.1 *Granger Causality*

Granger causality is a technique for determining whether one time series is useful in forecasting another. For two series x_t and y_t, y_t is said to Granger-cause x_t if x_t can be better predicted using the histories of both y_t and x_t than it can using the history of x_t alone. Otherwise, if

$$E(x_t | x_{t-1}, y_{t-1}, x_{t-2}, y_{t-2}, \cdots) = E(x_t | x_{t-1}, x_{t-2}, \cdots),$$

then y_t is said to to fail to Granger-cause x_t. As an example, if x_t and y_t are generated by the following model:

$$\begin{bmatrix} x_t \\ y_t \end{bmatrix} = \begin{bmatrix} a_{11}x_{t-1} + \cdots + a_{1p}x_{t-p} + b_{11}y_{t-1} + \cdots + b_{1p}y_{t-p} + \epsilon_{1t} \\ a_{21}x_{t-1} + \cdots + a_{2p}x_{t-p} + b_{21}y_{t-1} + \cdots + b_{2p}y_{t-p} + \epsilon_{2t} \end{bmatrix} \tag{5.6.1}$$

where $(\epsilon_{1t}, \epsilon_{2t})'$ are white noises with mean vector $\mathbf{0}$ and covariance matrix

$$\mathbf{\Sigma} = \begin{bmatrix} \sigma_{11} & \sigma_{12} \\ \sigma_{21} & \sigma_{22} \end{bmatrix},$$

then y_t does not Granger-cause x_t if and only if

$$b_{11} = b_{12} = \cdots = b_{1p} = 0. \tag{5.6.2}$$

We shall initialize Eq. (5.6.1) at $t = -p+1, \cdots, 0$. Furthermore, since the initial values $\{x_{-p+1}, \cdots, x_0, y_{-p+1}, \cdots, y_0\}$ do not affect asymptotics, we can let them be any random variables. Let $\boldsymbol{z}_t^{(1)} = (x_{t-1}, \cdots, x_{t-p})'$, $\boldsymbol{z}_t^{(2)} = (y_{t-1}, \cdots, y_{t-p})'$, $\boldsymbol{Z}^{(1)} = (\boldsymbol{z}_1^{(1)\prime}, \cdots, \boldsymbol{z}_T^{(1)\prime})'$, $\boldsymbol{Z}^{(2)} = (\boldsymbol{z}_1^{(2)\prime}, \cdots, \boldsymbol{z}_T^{(2)\prime})'$, $\boldsymbol{X} = (x_1, \cdots, x_T)'$, $\boldsymbol{Y} = (y_1, \cdots, y_T)'$ and

$$\boldsymbol{Z} = \begin{bmatrix} \boldsymbol{Z}^{(1)} & \boldsymbol{Z}^{(2)} \end{bmatrix},$$

then for $\boldsymbol{a}^{(1)} = (a_{11}, \cdots, a_{1p})'$, $\boldsymbol{b}^{(1)} = (b_{11}, \cdots, b_{1p})'$, $\boldsymbol{a}^{(2)} = (a_{21}, \cdots, a_{2p})'$ and $\boldsymbol{b}^{(2)} = (b_{21}, \cdots, b_{2p})'$, the ordinary least square (OLS) estimator is given by

$$\begin{bmatrix} \hat{\boldsymbol{a}}^{(1)\prime} & \hat{\boldsymbol{b}}^{(1)\prime} \\ \hat{\boldsymbol{a}}^{(2)\prime} & \hat{\boldsymbol{b}}^{(2)\prime} \end{bmatrix} = \begin{bmatrix} \boldsymbol{X}' \\ \boldsymbol{Y}' \end{bmatrix} \boldsymbol{Z}(\boldsymbol{Z}'\boldsymbol{Z})^{-1},$$

and

$$\hat{\mathbf{\Sigma}} = \frac{1}{T}\left\{\begin{bmatrix} \boldsymbol{X}' \\ \boldsymbol{Y}' \end{bmatrix} - \begin{bmatrix} \hat{\boldsymbol{a}}^{(1)\prime} & \hat{\boldsymbol{b}}^{(1)\prime} \\ \hat{\boldsymbol{a}}^{(2)\prime} & \hat{\boldsymbol{b}}^{(2)\prime} \end{bmatrix} Z'\right\}\left\{\begin{bmatrix} X' \\ Y' \end{bmatrix} - \begin{bmatrix} \hat{\boldsymbol{a}}^{(1)\prime} & \hat{\boldsymbol{b}}^{(1)\prime} \\ \hat{\boldsymbol{a}}^{(2)\prime} & \hat{\boldsymbol{b}}^{(2)\prime} \end{bmatrix} Z'\right\}'.$$

For two stationary series x_t and y_t, let

$$\mathbf{\Gamma} = \begin{bmatrix} \mathbf{\Gamma}_{11} & \mathbf{\Gamma}_{12} \\ \mathbf{\Gamma}_{21} & \mathbf{\Gamma}_{22} \end{bmatrix},$$

where $\mathbf{\Gamma}_{11} = \mathrm{Var}(\boldsymbol{z}_1^{(1)})$, $\mathbf{\Gamma}_{12} = \mathrm{Cov}(\boldsymbol{z}_1^{(1)}, \boldsymbol{z}_1^{(2)})$, $\mathbf{\Gamma}_{22} = \mathrm{Var}(\boldsymbol{z}_1^{(2)})$ and $\mathbf{\Gamma}_{21} = \mathbf{\Gamma}_{12}'$. And it can be estimated by

$$\hat{\mathbf{\Gamma}} = \frac{1}{T}\begin{bmatrix} \sum_{t=1}^{T} \boldsymbol{z}_t^{(1)}\boldsymbol{z}_t^{(1)\prime} & \sum_{t=1}^{T} \boldsymbol{z}_t^{(1)}\boldsymbol{z}_t^{(2)\prime} \\ \sum_{t=1}^{T} \boldsymbol{z}_t^{(2)}\boldsymbol{z}_t^{(1)\prime} & \sum_{t=1}^{T} \boldsymbol{z}_t^{(2)}\boldsymbol{z}_t^{(2)\prime} \end{bmatrix}.$$

Theorem 5.6.1. *When x_t and y_t generated by Eq. (5.6.1) are stationary with $E(|\epsilon_{1t}|^{2+\delta}) < \infty$ and $E(|\epsilon_{2t}|^{2+\delta}) < \infty$ for some $\delta > 0$,*

$$\sqrt{T}\begin{bmatrix} \hat{\boldsymbol{a}}^{(1)} - \boldsymbol{a}^{(1)} \\ \hat{\boldsymbol{b}}^{(1)} - \boldsymbol{b}^{(1)} \\ \hat{\boldsymbol{a}}^{(2)} - \boldsymbol{a}^{(2)} \\ \hat{\boldsymbol{b}}^{(2)} - \boldsymbol{b}^{(2)} \end{bmatrix} \xrightarrow{d} N(\boldsymbol{0}, \boldsymbol{\Phi})$$

where $\boldsymbol{\Phi}^{-1} = \boldsymbol{\Sigma} \bigotimes \boldsymbol{\Gamma}^{-1}$, i.e., the Kronecker product of $\boldsymbol{\Sigma}$ and $\boldsymbol{\Gamma}^{-1}$.

Proof. See Proposition 11.2 given by Hamilton (1994, p. 301). □

For the hypothesis

$$H_0 : y_t \text{ does not Granger-cause } x_t,$$

the Wald test statistic is

$$W = \hat{\boldsymbol{b}}^{(1)\prime}[\hat{\boldsymbol{\sigma}}_{11}^{-1}(\boldsymbol{Z}^{(2)\prime}\boldsymbol{R}\boldsymbol{Z}^{(2)})]\hat{\boldsymbol{b}}^{(1)}, \tag{5.6.3}$$

where $\boldsymbol{R} = \boldsymbol{I} - \boldsymbol{Z}^{(1)}(\boldsymbol{Z}^{(1)\prime}\boldsymbol{Z}^{(1)})^{-1}\boldsymbol{Z}^{(1)\prime}$.

Theorem 5.6.2. *If x_t and y_t generated by Eq. (5.6.1) are stationary with $E(|\epsilon_{1t}|^{2+\delta}) < \infty$ and $E(|\epsilon_{2t}|^{2+\delta}) < \infty$ for some $\delta > 0$, the Wald test statistic W given in Eq. (5.6.3) is asymptotically chi-square distributed with p degree of freedom under $\boldsymbol{b}^{(1)} = 0$.*

Proof. If x_t and y_t generated by Eq. (5.6.1) are stationary with $E(|\epsilon_{1t}|^{2+\delta}) < \infty$ and $E(|\epsilon_{2t}|^{2+\delta}) < \infty$ for some $\delta > 0$, then we have

$$\hat{\sigma}_{11} \xrightarrow{P} \sigma_{11},$$

and

$$\hat{\boldsymbol{\Gamma}}_{22} - \hat{\boldsymbol{\Gamma}}_{21}\hat{\boldsymbol{\Gamma}}_{11}^{-1}\hat{\boldsymbol{\Gamma}}_{12} \xrightarrow{P} \boldsymbol{\Gamma}_{22} - \boldsymbol{\Gamma}_{21}\boldsymbol{\Gamma}_{11}^{-1}\boldsymbol{\Gamma}_{12}.$$

By Theorem 5.6.1, the conclusion follows. □

When x_t and y_t are not stationary, the test given in Eq. (5.6.3) are more complex. It has been studied by Dickey and Fuller (1979), Phillips and Durlauf (1986), Phillips and Perron (1988), Park and Phillips (1988, 1989), Sim *et al.* (1990). Toda and Phillips (1993) demonstrated that the Wald test has an asymptotic distribution in terms of nonlinear functions of χ^2 variates. Later Toda and Yamamoto (1995) also discussed some related testing problems for nonstationary time series.

5.6.2 Integration of Order d

x_t is said to be integrated of order d if $(1-B)^d x_t$ is stationary and $(1-B)^{d-1}x_t$ is nonstationary, denoted by $x_t \sim I(d)$, where B is the lag operator.

Let $\triangle$ denote the familiar difference operator, *i.e.* $\triangle x_t = (1-B)x_t$. For $x_t \sim I(1)$ and $y_t \sim I(1)$ in Eq. (5.6.1), we have

$$\begin{aligned} x_t &= a_{11}^*\triangle x_{t-1} + \cdots + a_{1p}^*\triangle x_{t-p} + a_{1p+1}^* x_{t-p-1} \\ &\quad + b_{11}^*\triangle y_{t-1} + \cdots + b_{1p}^*\triangle y_{t-p} + b_{1p+1}^* y_{t-p-1} + \epsilon_{1t}, \\ y_t &= a_{21}^*\triangle x_{t-1} + \cdots + a_{2p}^*\triangle x_{t-p} + a_{2p+1}^* x_{t-p-1} \\ &\quad + b_{21}^*\triangle y_{t-1} + \cdots + b_{2p}^*\triangle y_{t-p} + b_{2p+1}^* y_{t-p-1} + \epsilon_{2t}, \end{aligned}$$

where $a_{ij}^* = -\sum_{l=1}^{j} a_{il}$, $b_{ij}^* = -\sum_{l=1}^{j} b_{il}$, $a_{i,p+1}^* = \sum_{j=1}^{p} a_{ij}$, and $b_{i,p+1}^* = \sum_{j=1}^{p} b_{ij}$ for $j = 1, \cdots, p$ and $i = 1, 2$.

If $\boldsymbol{a}^{(i*)} = (a_{i1}^*, \cdots, a_{ip}^*)'$ and $\boldsymbol{b}^{(i*)} = (b_{i1}^*, \cdots, b_{ip}^*)'$, the relationships between $\boldsymbol{a}^{(i*)}$ and $\boldsymbol{a}^{(i)}$, $\boldsymbol{b}^{(i*)}$ and $\boldsymbol{b}^{(i)}$ are one-to-one for $i = 1, 2$. Thus if we get the asymptotic distribution for estimators of $\boldsymbol{a}^{(1*)}$, $\boldsymbol{b}^{(1*)}$, $\boldsymbol{a}^{(2*)}$ and $\boldsymbol{b}^{(2*)}$, we can also obtain that of $\boldsymbol{a}^{(1)}$, $\boldsymbol{b}^{(1)}$, $\boldsymbol{a}^{(2)}$ and $\boldsymbol{b}^{(2)}$. By OLS regression, we have

$$\begin{bmatrix} \hat{\boldsymbol{a}}^{(1*)} - \boldsymbol{a}^{(1*)} \\ \hat{\boldsymbol{b}}^{(1*)} - \boldsymbol{b}^{(1*)} \end{bmatrix} = \boldsymbol{D}[\triangle \boldsymbol{Z}'\boldsymbol{\epsilon}^{(1)} - \triangle \boldsymbol{Z}'\boldsymbol{Z}^{(3)}(\boldsymbol{Z}^{(3)\prime}\boldsymbol{Z}^{(3)})^{-1}\boldsymbol{Z}^{(3)\prime}\boldsymbol{\epsilon}^{(1)}],$$

where

$$\boldsymbol{D} = \left(\triangle \boldsymbol{Z}'\triangle \boldsymbol{Z} - \triangle \boldsymbol{Z}'\boldsymbol{Z}^{(3)}(\boldsymbol{Z}^{(3)\prime}\boldsymbol{Z}^{(3)})^{-1}\boldsymbol{Z}^{(3)\prime}\triangle \boldsymbol{Z}\right)^{-1},$$

$$\boldsymbol{Z}^{(3)} = \begin{bmatrix} x_{-p} & y_{-p} \\ \vdots & \vdots \\ x_{T-p-1} & y_{T-p-1} \end{bmatrix},$$

and $\boldsymbol{\epsilon}^{(1)} = (\epsilon_{11}, \cdots, \epsilon_{1T})'$.

Lemma 5.6.1. *If x_t and y_t generated by Eq. (5.6.1) are $I(1)$ with $E(|\epsilon_{1t}|^{2+\delta}) < \infty$ and $E(|\epsilon_{2t}|^{2+\delta}) < \infty$ for some $\delta > 0$, we have*

$$\triangle \boldsymbol{Z}'\boldsymbol{Z}^{(3)}(\boldsymbol{Z}^{(3)\prime}\boldsymbol{Z}^{(3)})^{-1}\boldsymbol{Z}^{(3)\prime}\boldsymbol{\epsilon}^{(1)}/\sqrt{T} \xrightarrow{P} \boldsymbol{0},$$

and

$$\frac{\boldsymbol{D}^{-1}}{T} \xrightarrow{P} \boldsymbol{\Lambda},$$

where

$$\boldsymbol{\Lambda} = Var\begin{pmatrix} \triangle \boldsymbol{z}_1^{(1)} \\ \triangle \boldsymbol{z}_1^{(2)} \end{pmatrix}.$$

Proof. By the Representation Theorem given by Engle and Granger (1987) and Johansen (1992), we have

$$\frac{\boldsymbol{Z}^{(3)\prime}\boldsymbol{Z}^{(3)}}{T^2} = O_p(1), \quad \frac{\triangle \boldsymbol{Z}'\boldsymbol{Z}^{(3)}}{T} = O_p(1), \quad \frac{\boldsymbol{Z}^{(3)\prime}\boldsymbol{\epsilon}^{(1)}}{T} = O_p(1),$$

and

$$\begin{aligned}\frac{\triangle \boldsymbol{Z}'\boldsymbol{Z}^{(3)}(\boldsymbol{Z}^{(3)\prime}\boldsymbol{Z}^{(3)})^{-1}\boldsymbol{Z}^{(3)\prime}\boldsymbol{\epsilon}^{(1)}}{\sqrt{T}} &= \frac{1}{\sqrt{T}}\frac{\triangle \boldsymbol{Z}'\boldsymbol{Z}^{(3)}}{T}\left(\frac{\boldsymbol{Z}^{(3)\prime}\boldsymbol{Z}^{(3)}}{T^2}\right)^{-1}\frac{\boldsymbol{Z}^{(3)\prime}\boldsymbol{\epsilon}^{(1)}}{T}\\ &= o_p(1).\end{aligned}$$

Furthermore, we have

$$\frac{\boldsymbol{D}^{-1}}{T} = \frac{\triangle \boldsymbol{Z}'\triangle \boldsymbol{Z}}{T} - \frac{\triangle \boldsymbol{Z}'\boldsymbol{Z}^{(3)}(\boldsymbol{Z}^{(3)\prime}\boldsymbol{Z}^{(3)})^{-1}Z^{(3)\prime}\triangle \boldsymbol{Z}}{T} = \frac{\triangle \boldsymbol{Z}'\triangle \boldsymbol{Z}}{T} + o_p(1),$$

which converges to the variance of $(\triangle \boldsymbol{z}_1^{(1)\prime}, \triangle \boldsymbol{z}_1^{(2)\prime})'$ by Eq. (5.6.1). □

Theorem 5.6.3. *If x_t and y_t generated by Eq. (5.6.1) are $I(1)$ with $E(|\epsilon_{1t}|^{2+\delta}) < \infty$ and $E(|\epsilon_{2t}|^{2+\delta}) < \infty$ for some $\delta > 0$, then we have*

$$\sqrt{T}\begin{bmatrix}\hat{\boldsymbol{a}}^{(1*)} - \boldsymbol{a}^{(1*)}\\ \hat{\boldsymbol{b}}^{(1*)} - \boldsymbol{b}^{(1*)}\end{bmatrix} \xrightarrow{d} N(\boldsymbol{0}, \sigma_{11}\boldsymbol{\Lambda}^{-1}).$$

Proof. Since $\triangle \boldsymbol{Z}'\boldsymbol{\epsilon}^{(1)}$ is stationary, we have by Theorem 5.6.1

$$\frac{\triangle \boldsymbol{Z}'\boldsymbol{\epsilon}^{(1)}}{\sqrt{T}} \xrightarrow{d} N(\boldsymbol{0}, \sigma_{11}\boldsymbol{\Lambda}),$$

and thus we have

$$\sqrt{T}\begin{bmatrix}\hat{\boldsymbol{a}}^{(1*)} - \boldsymbol{a}^{(1*)}\\ \hat{\boldsymbol{b}}^{(1*)} - \boldsymbol{b}^{(1*)}\end{bmatrix} \xrightarrow{d} N(\boldsymbol{0}, \sigma_{11}\boldsymbol{\Lambda}^{-1})$$

by Lemma 5.6.1. □

For the test of Granger-causality between y_t and x_t, the Wald test statistic based on $\hat{b}^{(1*)\prime}$, which is asymptotically chi-square distributed with p degrees of freedom, is given by

$$W = \hat{\boldsymbol{b}}^{(1*)\prime}[\hat{\sigma}_{11}^{-1}(\triangle \boldsymbol{Z}^{(2)\prime}\boldsymbol{Q}^*\triangle \boldsymbol{Z}^{(2)})]\hat{\boldsymbol{b}}^{(1*)\prime}, \tag{5.6.4}$$

where

$$\boldsymbol{Q}^* = \boldsymbol{Q}_I - \boldsymbol{Q}_I\triangle \boldsymbol{Z}^{(1)}(\boldsymbol{Z}^{(1)\prime}\boldsymbol{Q}_I\triangle \boldsymbol{Z}^{(1)})^{-1}\triangle \boldsymbol{Z}^{(1)\prime}\boldsymbol{Q}_I,$$

and

$$\boldsymbol{Q}_I = \boldsymbol{I} - \boldsymbol{Z}^{(3)}(\boldsymbol{Z}^{(3)\prime}\boldsymbol{Z}^{(3)})^{-1}\boldsymbol{Z}^{(3)\prime}.$$

By simple matrix computation, we have

$$\hat{\boldsymbol{b}}^{(1*)'}[\hat{\sigma}_{11}^{-1}(\triangle \boldsymbol{Z}^{(2)'}\boldsymbol{Q}^{*}\triangle \boldsymbol{Z}^{(2)})]\hat{\boldsymbol{b}}^{(1*)'} = \hat{\boldsymbol{b}}^{(1)'}[\hat{\sigma}_{11}^{-1}(\boldsymbol{Z}^{(2)'}\boldsymbol{Q}\boldsymbol{Z}^{(2)})]\hat{\boldsymbol{b}}^{(1)'}, \quad (5.6.5)$$

where $\hat{\boldsymbol{b}}^{(1)} = (\hat{b}_{12}, \cdots, \hat{b}_{1p})'$ is the OLS estimator $\boldsymbol{b}^{(1)} = (b_{11}, \cdots, b_{1p})'$ in the following equation:

$$\begin{bmatrix} x_t \\ y_t \end{bmatrix} = \begin{bmatrix} a_{11}x_{t-1} + \cdots + a_{1p+1}x_{t-p-1} + b_{11}y_{t-1} + \cdots + b_{1p+1}y_{t-p-1} + \epsilon_{1t} \\ a_{21}x_{t-1} + \cdots + a_{2p+1}x_{t-p-1} + b_{21}y_{t-1} + \cdots + b_{2p+1}x_{t-p-1} + \epsilon_{2t} \end{bmatrix},$$

$$\boldsymbol{Q} = \boldsymbol{Q}_I - \boldsymbol{Q}_I\boldsymbol{Z}^{(1)}(\boldsymbol{Z}^{(1)'}\boldsymbol{Q}_I\boldsymbol{Z}^{(1)})^{-1}\boldsymbol{Z}^{(1)'}\boldsymbol{Q}_I,$$

and

$$\boldsymbol{Q}_I = \boldsymbol{I} - \boldsymbol{Z}^{(3)}(\boldsymbol{Z}^{(3)'}\boldsymbol{Z}^{(3)})^{-1}\boldsymbol{Z}^{(3)'}.$$

Thus if x_t and y_t in Eq. (5.6.1) are integrated of order 1, we can estimate related parameters of $b^{(1)}$ with one more lag by OLS to test the Granger causality betweeny_t and x_t, and then construct the Wald test of the right-hand side of Eq. (5.6.5), which is asymptotically chi-square distributed with p degrees of freedom. For x_t and y_t integrated of order $d > 1$, one can refer to general results given by Toda and Yamamoto (1995).

5.6.3 *Real Data Analysis*

We use the above-mentioned methods to study the relationship between China economic development and USA trade, and the relationship between USA economic development and China trade. The data (the price in 2000 as constant price) are given in Table 5.6.1, where Openness = (Import + Export)/GDP. The line diagrams about the log per capita GDP and trade of China and USA respectively are plotted in Figs. 5.6.1-5.6.4.

Let X and XO denote the log per capita GDP of China and the log product of its per capita and openness respectively, and Y and YO denote the log per capita GDP of USA and the log product of its per capita and openness respectively. By the augmented Dickey-Fuller test (Dickey and Fuller, 1979) and Phillips-Perron test (Phillips and Perron, 1988), all economic variables above are $I(1)$ for significant level 1%.

For variables X and YO, by the BIC, they satisfy the following vector autoregression (VAR) models:

$$X_t = a_{11}X_{t-1} + a_{12}X_{t-2} + b_{11}YO_{t-1} + b_{12}YO_{t-2} + \epsilon_{1t},$$

and

$$YO_t = a_{21}X_{t-1} + a_{22}X_{t-2} + b_{21}YO_{t-1} + b_{22}YO_{t-2} + \epsilon_{2t}.$$

Table 5.6.1 Per capita GDP and openness of USA and China during the 1978-2004 period (US$).

Year	Per capita GDP of USA	Openness of USA	Per capita GDP of China	Openness of China
1978	10054.21	17.53	323.97	10.30
1979	11128.03	18.97	388.62	11.74
1980	11990.45	20.75	452.28	13.13
1981	13313.43	20.06	510.41	15.73
1982	13704.03	18.16	586.16	15.29
1983	14736.47	17.26	648.01	15.14
1984	16221.44	18.13	770.93	18.36
1985	17228.98	17.17	876.87	25.14
1986	18034.63	17.48	966.25	27.48
1987	18963.55	18.57	1075.81	28.28
1988	20222.18	19.72	1176.58	27.94
1989	21514.28	20.12	1227.00	26.95
1990	22530.02	20.54	1411.05	29.90
1991	23029.18	20.53	1603.14	33.01
1992	24086.88	20.74	1812.04	35.46
1993	25031.18	20.85	1996.41	33.31
1994	26311.62	21.88	2278.06	41.63
1995	27233.77	23.37	2534.95	39.06
1996	28484.40	23.62	2809.36	38.15
1997	29956.03	24.39	3122.93	39.37
1998	31235.27	23.83	3343.42	36.63
1999	32766.51	24.34	3591.47	37.76
2000	34364.50	26.34	4001.82	43.90
2001	35107.52	24.14	4389.44	42.82
2002	35944.96	23.39	4847.36	47.38
2003	37313.33	23.74	5321.27	54.28
2004	39535.28	25.44	5771.69	47.37

Similarly, for variables Y and XO, they satisfy the following VAR models:

$$Y_t = c_{11}Y_{t-1} + d_{11}XO_{t-1} + \eta_{1t},$$

and

$$XO_t = c_{21}Y_{t-1} + d_{21}XO_{t-1} + \eta_{2t}.$$

We are interested in whether or not YO_t Granger-causes X_t, and XO_t Granger-causes Y_t respectively, that is,

$$H_0: \quad b_{12} = b_{22} = 0,$$

and

$$H_0': \quad d_{12} = 0$$

respectively. By the Wald test give in Eq. (5.6.5), the p-values are 0.308 and 0.725 under H_0 and H_0' respectively. So there are no Granger causalities.

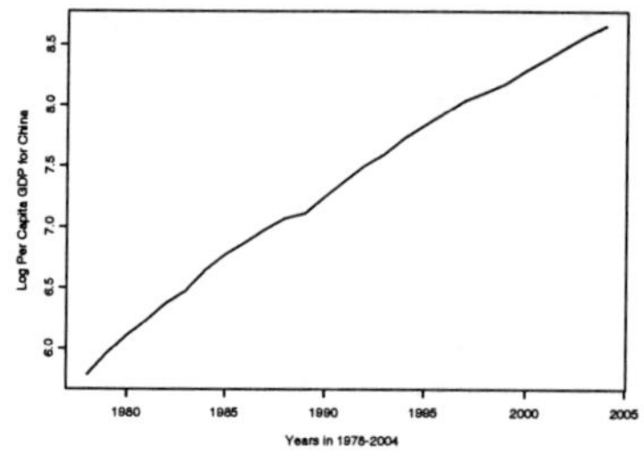

Fig. 5.6.1 Log per capita GDP for China

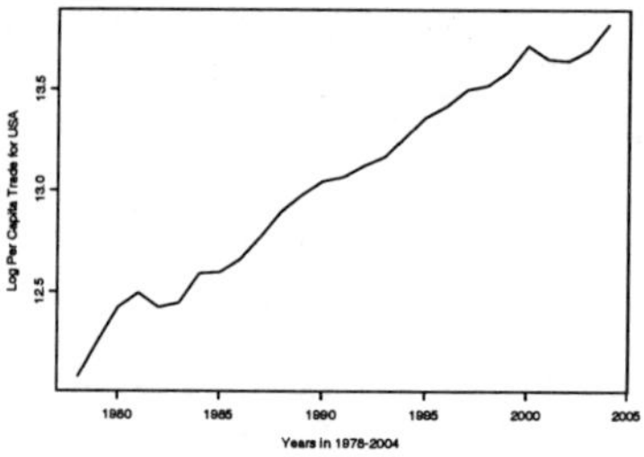

Fig. 5.6.2 Log per capita trade for USA

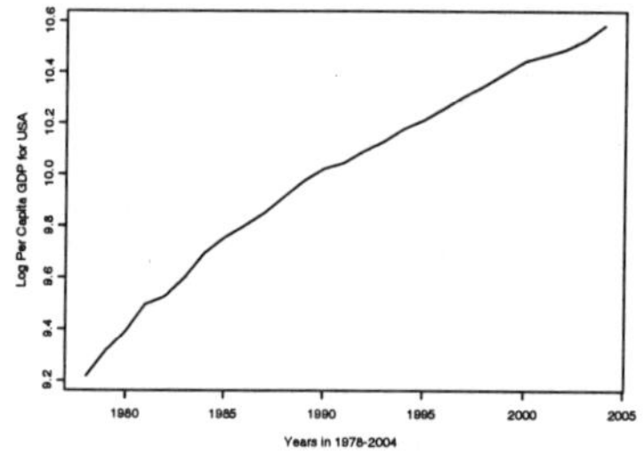

Fig. 5.6.3 Log per capita GDP for USA

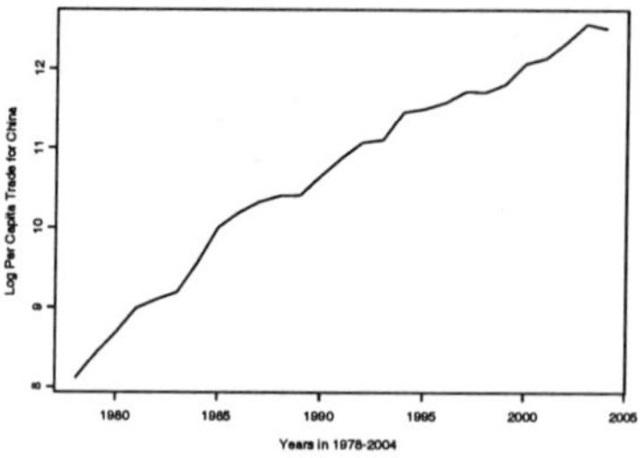

Fig. 5.6.4 Log per capita trade for China

5.7 Problems

5.1. According to the data of China's per capita income and expenditure during the 1978-2002 period Table 5.1.1.

(a) Determine the least-squares estimates of each of the following two linear regression lines: the per capita income of rural and urban households to the per capita living expenditure respectively;

(b) Sketch each estimated line on the appropriate scatter diagram;

(c) Using your sketch and the Keynes' Psychological Law, make an empirical analysis of the relationship between the per capita income of the rural (or urban) households and the per capita living expenditure.

5.2. Using the data given in Table 5.1.2 to test the hypothesis (5.1.12).

5.3. For the data model

$$Y = X\beta + \epsilon,$$

where Y is an n-dimensional vector, X is an $n \times p$ known matrix, β is a p-dimensional parameter vector, and ϵ is an n-dimensional random vector. $E\epsilon = 0$ and $CV(\epsilon, \epsilon) = \Sigma$ where Σ is a known $n \times n$ positive definite matrix. Figure out the LSE of β, and the MLE of β when $\epsilon \sim N(\mathbf{0}, \Sigma)$.

5.4. Prove Eqs. (5.2.7) and (5.2.8).

5.5. Let X be p-dimensional random variable. Prove that X has a p-dimensional normal distribution if and only if $a'X$ has a univariate normal distribution for any $a \in R^p$.

5.6. Consider a sociological study about the relationship between intelligence and delinquency. A delinquency index (ranging from 0 to 50) was formulated to account for both the severity and the frequency of crimes committed, while intelligence was measured by IQ. The following table displays the delinquency index and IQ of a sample of 18 convicted minors (Kleinbaum *et al.*, 1998).

IQ (X)	Delinqency index (Y)	IQ (X)	Delinquency index (Y)
110	26.20	92	22.10
89	33.00	116	18.60
102	17.50	85	35.50
98	25.25	73	38.00
110	20.30	90	30.00
98	31.90	104	19.70
122	21.10	82	41.10
119	22.70	134	39.60
120	10.70	114	25.15

(a) Plot the scatter diagram for the paired data;

(b) Given that $\hat{\beta}_0 = 52.273$ and $\hat{\beta}_1 = -0.249$, draw the estimated regression line on the scatter diagram;

(c) How do you account for the fact that $\hat{Y} = 52.273$ when IQ$= 0$, even though the delinquency index goes no higher than 50?

(d) Suppose

$$S_{Y|X} = \sqrt{\frac{1}{n-2}\sum_{i=1}^{n}(Y_i - \hat{Y}_i)^2} = 7.704,$$

and

$$S_{XX} = \sqrt{\frac{1}{n-1}\sum_{i=1}^{n}(X_i - \bar{X})^2} = 16.192,$$

Construct a 95% confidence interval for the slope β_1;

(e) Interpret this confidence interval with regard to testing the null hypothesis of zero slope at significance level $\alpha = 0.05$;

(f) Notice that the convicted minor with IQ= 134 and the delinquency index 39.60 appears to be quite out of place in the data. Decide whether this outlier has any effect on your estimate by omitting it;

(g) When the outlier is removed, for given $\alpha = 0.05$, $n = 17$,

$$S_{Y|X} = \sqrt{\frac{1}{n-2}\sum_{i=1}^{n}(Y_i - \hat{Y}_i)^2} = 4.933,$$

and

$$S_{XX} = \sqrt{\frac{1}{n-1}\sum_{i=1}^{n}(X_i - \bar{X})^2} = 14.693.$$

Test $H_0 : \beta_1 = 0$;

(h) For these data, would you conclude that the delinquency index decreases as IQ increases?

5.7. The following table gives the data of height (HGT), age (AGE), and weight (WGT) of a random sample of twelve nutritionally deficient children (Kleinbaum, *et al.*, 1998):

Number	1	2	3	4	5	6	7	8	9	10	11	12
WGT (Y)	29	32	24	30	25	26	35	26	25	23	35	31
HGT (X_1)	145	150	125	158	130	127	140	122	107	107	155	145
AGE (X_2)	8	10	6	11	8	7	10	9	10	6	12	9

where Y = WGT(kg) is a response variable, X_1 = HGT(cm) and X_2 = AGE(year) are two predictor variables.

(a) Figure out the least-squares estimated equation for the following predictive model:

$$Y = \beta_0 + \beta_1 X_1 + \beta_2 X_2 + \beta_3 X_2^2 + \epsilon.$$

(b) For the given level $\alpha = 0.01$, test the hypothesis

$$H_0 : \beta_1 = \beta_2 = \beta_3 = 0.$$

5.8. **(Driver age and highway sign-reading distance)** In a study on the legibility and visibility of highway signs, a New South Wales research firm determined the maximum distance at which each 9 drivers could read a newly designed sign. The 9 participant ranged in age from 18 to 66 years old. The New South Wales government funded that study to improve highway safety for older drivers and wanted to examine the relationship between age and sign legibility distance.

Age (x)	18	24	30	36	42	48	54	60	66
Distance (y)	51.2	47.7	42.3	46.1	31.8	43.6	37.9	32.1	30.3

(a) Plot the data and describe the pattern of the plot.
(b) Estimate the coefficients b and a of the least square regression line $y = a + bx$. [Hint: $S_{xx} = 2160, S_{yy} = 395.1, S_{xy} = -850.2$.]
(c) Calculate the standard errors corresponding to the estimates of b and a.
(d) Test the hypothesis (H) that $b = 0$.
(e) Do you think that sign-reading distance is significantly correlated to age?

5.9. Let $Y_1, \cdots, Y_n$ be independent and identically distributed with model function $f(y;\theta)$, where $\theta \in \Theta \subseteq \mathbf{R}^d$, and let θ_0 denote the true parameter value. Derive the asymptotic distribution of the maximum likelihood estimator $\hat{\theta}_n$.
[You may assume that the usual regularity conditions hold. In particular, you may assume a Taylor series expansion for the score function $U(\theta)$, of the form

$$0 = U(\hat{\theta}_n) = U(\theta_0) - j(\theta_0)(\hat{\theta}_n - \theta_0) + o_p(n^{\frac{1}{2}})$$

as $n \to \infty$, where $j(\theta)$ is the observed information matrix at θ.]
Describe how this asymptotic result is related to the Wald test of $H_0 : \theta = \theta_0$ against $H_1 : \theta \neq \theta_0$. Now suppose that $\theta = (\psi, \lambda)$, where only ψ is of interest. Describe the Wald test of $H_0 : \theta = \theta_0$ vs. $H_1 : \theta \neq \theta_0$.

5.10. Let $Y_1, \cdots, Y_n$ be independent and identically distributed with inverse Gaussian density

$$f(y;\psi,\lambda) = \left(\frac{\psi}{2\pi y^3}\right)^{1/2} \exp\left\{-\frac{\psi}{2\lambda^2 y}(y-\lambda)^2\right\}, \quad y > 0, \psi > 0, \lambda > 0.$$

Show that the maximum likelihood estimator of ψ is

$$\hat{\psi} = \left\{\frac{1}{n}\sum_{i=1}^{n}\left(\frac{1}{Y_i} - \frac{1}{\bar{Y}}\right)\right\}^{-1}$$

where $\bar{Y} = (Y_1 + \cdots + Y_n)/n$.
Using the fact that $E_{\psi,\lambda}(Y_1) = \lambda$, show further that the Wald statistics for testing $H_0 : \psi = \psi_0$ against $H_1 : \psi \neq \psi_0$ coincide in the two cases where λ is known and where λ is unknown.

5.11. Consider the linear regression

$$Y = X\beta + \varepsilon,$$

where Y is an n-dimensional observation vector, X is an $n \times p$ matrix of rank p, and ε is an n-dimensional vector with components $\varepsilon_1, \cdots, \varepsilon_n$. Here, $\varepsilon_1, \cdots, \varepsilon_n$ are normally and independently distributed, each with mean zero and variance σ^2; we write this as $\varepsilon \sim N_n(0, \sigma^2 I_n)$.

(a) Define $Q(\beta) = (Y - X\beta)^T(Y - X\beta)$. Find an expression for $\hat{\beta}$, the least squares estimator for β and state without proof the joint distribution of $\hat{\beta}$ and $Q(\hat{\beta})$.
(b) Define $\hat{\varepsilon} = Y - X\hat{\beta}$. Find the distribution of $\hat{\varepsilon}$.
(c) Suppose $\beta^T = (\beta_1^T : \beta_2^T)$. How would you test $H_0 : \beta_2 = 0$?

5.12. Suppose that (B_t) is a standard one-dimensional Brownian motion with $B_0 = 0$. Let $x > 0$ and $T_x = \inf\{t > 0;\ B_t = x\}$. Use the reflection principle to show that

$$P(T_x \leq t) = \frac{\sqrt{2}}{\sqrt{\pi t}} \int_x^\infty e^{-\frac{u^2}{2t}}\, du$$

By making the substitution $y = x^2 t/u^2$, show that the probability density function of T_x is

$$\frac{x e^{x^2/2y}}{\sqrt{2\pi} y^{3/2}} \quad (0 < y < \infty).$$

Now suppose that B_t and W_t are independent standard one-dimensional Brownian motions with $B_0 = 0$ and $W_0 = 1$. Let $H = \inf\{t : B_t = W_t\}$ be the first time B_t hits W_t. Find the probability density function of H.

5.13. Let $Y_1, \cdots, Y_n$ be independent, identically distributed $N(\mu, \sigma^2)$, where the parameter of interest is μ, with σ^2 a nuisance parameter.
Consider testing $H_0 : \mu = \mu_0$. Show that the likelihood ratio statistic for testing H_0 may be expressed as

$$W = n \ln\{1 + t^2/(n-1)\},$$

where t is the usual Student's t statistic, and derive appropriate forms of the score and Wald statistics for testing H_0.
Explain in detail why the notion of Bartlett correction leads to consideration of the statistic

$$W' = W/\{1 + 3/(2n)\}.$$

5.14. Suppose that $Y_1, \cdots, Y_n$ are independent binomial observations, with

$$Y_i \sim Bi(t_i, \pi_i) \text{ and } \ln(\pi_i/(1-\pi_i)) = \beta^T x_i, \text{ for } 1 \le i \le n,$$

where $t_1, \cdots, t_n$ and $x_1, \cdots, x_n$ are given. Discuss carefully the estimation of β. Your solution should include

(a) the method of checking the fit of the above logistic model; and
(b) the method for finding an approximate 95% confidence interval for β_2, the second component of the rector β.

5.15. Suppose (r_{ij}) are independent observations, with

$$r_{ij} \sim Bi(n_{ij}, p_{ij}), \quad 1 \le i, j \le 2,$$

where $n_{11}, n_{12}, n_{21}, n_{22}$ are given totals. Consider the model

$$w : \text{Logit}(p_{ij}) = \mu + \alpha_i + \beta_j, \quad 1 \le i, j \le 2$$

where $\alpha_1 = \beta_1 = 0$.
Write down the log-likelihood under w, and discuss carefully how $\hat{\alpha}_2, \hat{\beta}_2$ and their corresponding standard errors may be derived. [Do not attempt to find analytical expression for $\hat{\alpha}_2, \hat{\beta}_2$ and their standard errors.]

5.16. Let $Y_1, \cdots, Y_n$ be independent variables, such that $\boldsymbol{Y} = \boldsymbol{X\beta} + \boldsymbol{\epsilon}$, where $\boldsymbol{X}$ is a given $n \times p$ matrix, of rank p, $\boldsymbol{\beta}$ is an unknown vector of dimension p, and $\epsilon_1, \cdots, \epsilon_n$ are independent normal variables, each with mean 0 and unknown variance σ^2.

(a) Derive an expression for $\hat{\boldsymbol{\beta}}$, the least squares estimator of $\boldsymbol{\beta}$, and derive the distribution of $\hat{\boldsymbol{\beta}}$.
(b) How would you estimate σ^2?
(c) In fitting the model

$$Y_i = \mu + \alpha x_i + \beta z_i + \gamma t_i + \epsilon_i, \quad 1 \le i \le n,$$

where $(x_i), (z_i), (t_i)$ are given vectors, and $\epsilon_1, \cdots, \epsilon_n$ has the distribution given above, explain carefully how you would test

$$H_0 : \beta = \gamma = 0.$$

5.17. (a) Suppose we have independent observations $y_1, \cdots, y_n$, and Y_i has density function of the form

$$f(y_i|\theta_i, \phi) = \exp\left\{\frac{y_i\theta_i - b(\theta_i)}{\phi} + c(y_i, \phi)\right\}.$$

Suppose further that

$$E(Y_i) = \mu_i, \quad g(\mu_i) = \beta^T x_i,$$

where $g(\cdot)$ is a known link function, and (x_i) is a set of known covariances. Show that

$$b'(\theta_i) = \mu_i, \quad \phi b''(\theta_i) = V(Y_i),$$

and hence find equations for the maximum likelihood estimator $\hat{\beta}$. Briefly discuss the method of solution of these equations.

(b) By writing the log-likelihood $L(\beta, \phi)$ as a function of the two unknown parameters β, ϕ, show that for large n, the estimates $\hat{\beta}$ and $\hat{\phi}$ are approximately independent.

5.18. Describe the Wald, score and likelihood ratio tests for hypotheses concerning a multidimensional parameter θ. Explain briefly how they can be used to construct confidence regions for θ of approximate $(1-\alpha)$-level coverage.

5.19. Let $Y_0, Y_1, \cdots, Y_n$ be a sequence of random variables such that Y_0 has a Poisson distribution with mean θ and for $i \geq 1$, conditional on $Y_0, \cdots, Y_{i-1}$, the random variable Y_i has a Poisson distribution with mean θY_{i-1}. The parameter θ satisfies $0 < \theta \leq 1$. Find the log-likelihood for θ, and show that the maximum likelihood estimator, $\hat{\theta} = \hat{\theta}(Y_0, Y_1, \cdots, Y_n)$, may be expressed as $\hat{\theta} = \min(\tilde{\theta}, 1)$, where $\tilde{\theta} = \tilde{\theta}(Y_0, Y_1, \cdots, Y_n)$ is a function which should be specified.

For $\theta \in (0,1)$, compute the Fisher information $I(\theta)$, and show that

$$I(\theta) \leq \frac{1}{\theta(1-\theta)}$$

for all n.

Deduce that the Wald statistic for testing $H_0 : \theta = \theta_0$ vs. $H_1 : \theta \neq \theta_0$, where $0 < \theta_0 < 1$, does not have an asymptotic chi-squared distribution under the null hypothesis.

5.20. Let $Y_1, \cdots, Y_n$ be independent variables, such that

$$\boldsymbol{Y} = \boldsymbol{\mu}\mathbf{1} + \boldsymbol{X\beta} + \boldsymbol{\epsilon},$$

where $\boldsymbol{X}$ is a given $n \times p$ matrix of rank p, $\boldsymbol{\beta}$ is an unknown vector of dimension p, $\boldsymbol{\mu}$ is an unknown constant, and $\mathbf{1}$ is the n-dimensional vector with every element 1. Assume that $\boldsymbol{X}'\mathbf{1} = \mathbf{0}$, and that $\boldsymbol{\epsilon} \sim N(\mathbf{0}, \sigma^2 \boldsymbol{I})$, where σ^2 is unknown.

(a) Derive an expression for $\hat{\boldsymbol{\beta}}$, the least squares estimator of $\boldsymbol{\beta}$, and derive its distribution.
(b) How would you test $H_0 : \boldsymbol{\beta} = \mathbf{0}$?
(c) How would you check the assumption $\boldsymbol{\epsilon} \sim N(\mathbf{0}, \sigma^2 \boldsymbol{I})$?

Bibliography

Abelson, R. P. and Tukey, J. W. (1963). Efficient utilization of non-numerical information in quantitative analysis: general theory and the case of the simple order, *Ann. Math. Statist.* **34**, pp. 1347–1369.

Agresti, A. (2002). *Categorical Data Analysis, 2nd Edition* (Wiley, New York).

Akkerboom, J. C. (1990). *Testing Problems with Linear or Angular Inequality Constraints* (Springer-Verlag, New York).

Anderson, S. (1984). *The organization of phonology* (Academic press, New York).

Arbuthnott, J. (1710). An argument for divine providence, taken from the constant regularity observed in the births of both sexes, *Phil. Trans.* **27**, pp. 186–190.

Ash, R. B. (1972). *Real Analysis and Probability* (Academic press, New York).

Bahadur, R. R. (1954). Sufficiency and statistical decision function, *Ann. Math. Statist.* **25**, pp. 423–462.

Bar-Lev, S. and Enis, P. (1988). On the classical choice of variance stabilizing transformations and an application for a poisson variate, *Biometrika* **75**, pp. 803–804.

Bar-Lev, S. and Enis, P. (1990). On the construction of classes of variance stabilizing transformations, *Statist. Prob. Let.* **10**, pp. 95–100.

Barlow, R. E., Bartholomew, D. J., Bremner, J. M. and Brunk, H. D. (1972). *Statistical Inference under Order Restrictions: The theory and application of isotonic regression* (Wiley, New York).

Bartholomew, D. J. (1959a). A test of homogeneity for ordered alternatives i, *Biometrika* **46**, pp. 36–48.

Bartholomew, D. J. (1959b). A test of homogeneity for ordered alternatives ii, *Biometrika* **46**, pp. 328–335.

Barton, D. E. and Mallows, C. L. (1961). The randomization bases of the problem of the amalgamation of weighted means, *J. R. Statist. Soc. B* **23**, pp. 303–305.

Basu, D. (1958). On statistics independent of sufficient statistics, *Sankhya* **A20**, pp. 223–226.

Basu, D. (1959). The family of ancillary statistics, *Sankhya* **A21**, pp. 247–256.

Benjamini, Y. and Hochberg, Y. (1995). Controlling the false discovery rate: A practical and powerful approach to multiple testing, *J. R. Stat. Soc. B* **57**, pp. 289–300.

Benjamini, Y. and Yekutieli, D. (2001). The control of false discovery rate under dependency, *Ann. Stat.* **29**, pp. 1165–1188.

Bickel, D. (2004). Degrees of differential gene expression: detecting biologically significant expression differences and estimating their magnitudes, *Bioinformatics* **20**, pp. 682–688.

Bishop, Y. M. M., Fienberg, S. E. and Holland, P. W. (1975). *Discrete Multivariate Analysis* (MIT Press, Cambridge).

Breslow, N. E. and Day, N. E. (1980). *Statistical Methods in Cancer Research II: The Design and Analysis in Case-Control Studies* (INRC, Lyon).

Burnham, K. P. and Anderson, D. R. (1998). *Model Selection and Inference: A practical Information Theoretic Approach* (Springer-Verlag, New York).

Carroll, R. J., Ruppert, D. and Stefanski, L. A. (1995). *Nonlinear Measurement Error Models* (Chapman & Hall, New York).

Casella, G. and Berger, R. L. (2002). *Statistical inference, 2nd Edition* (Duxbury Press, New York).

Chuang, C.-S. and Chan, N. H. (2002). Empirical likelihood for autoregressive models, with applications to unstable time series, *Statist. Sinica* **12**, pp. 387–407.

Chung, K. L. (1974). *A course in Probability Theory* (Academic Press, New York).

Cramér, H. (1946). *Mathematical methods of Statistics* (Princeton University Press, New Jersey: Princeton).

Dickey, D. A. and Fuller, W. A. (1979). Distribution of the estimators for autoregressive time series with a unit root, *J. Amer. Statist. Assoc* **74**, pp. 427–431.

Dickey, D. A. and Fuller, W. A. (1981). Likelihood ratio statistics for autoregressive time series with a unit root, *Econometrica* **49**, pp. 1057–1072.

Engle, R. F. and Granger, C. W. J. (1987). Cointegration and error correction: Representation, estimation and testing, *Econometrica* **55**, pp. 251–271.

Evans, G. B. and Savin, N. E. (1981). Testing for unit roots: I, *Econometrica* **49**, pp. 753–779.

Feller, W. (1968). *An Introduction to Probability Theory and Its Applications, Volume I* (Wiley, New York).

Fellingham, S. A. and Stoker, D. (1964). An approximation for the exact distribution of the wilcoxon test for symmetry, *J. Amer. Statist. Assoc.* **59**, pp. 899–905.

Fix, E. and Hodges, J. L. J. (1955). Significance probabilities of the wilcoxon test, *Ann. Math. Statist.* **26**, pp. 301–312.

Fuller, W. A. (1976). *Introduction to Statistical Time Series* (Wiley, New York).

Fuller, W. A. (1987). *Measurement Error Models* (Wiley, New York).

Galton, F. (1886). Regression towards mediocrity in hereditary stature, *Journal of the Anthropological Institute* **15**, pp. 1351–1362.

Gleser, L. and Hwang, J. T. (1987). The nonexistence of $100(1-\alpha)\%$ confidence set of finite expected diameter in errors-in-variables and related models, *Ann. Statist.* **15**, pp. 1351–1362.

Gujarati, D. N. (1995). *Basic Econometrics, 4th Edition*, 4th edn. (Mcgraw-hill, New York).

Halmos, P. R. (1957). *Measure Theory* (Springer, New York).

Halmos, P. R. and Savage, L. J. (1949). Application of the radom-nikodym theorem to sufficient statistics, *Ann. Math. Statist.* **20**, pp. 225–241.

Hamilton, J. D. (1994). *Time Series Analysis* (Princeton University Press, New Jersey: Princeton).

Hayter, A. J. (1984). A proof of the conjecture that the tukey-kramer multiple comparisons procedure is conservative, *Ann. Statist.* **12**, pp. 61–75.

Henshall, J. M. and Goddard, M. E. (1999). Multiple trait mapping of quantitative trait loci after selective genotyping using logistic regression, *Genetics* **151**, pp. 885–894.

Hettmansperger, T. P. (1984). *Statistical Inference Based on Ranks* (Wiley, New York).

Hochberg, Y. (1988). A sharper bonferroni procedure for multiple tests of significance, *Biometrika* **75**, pp. 800–803.

Hodges, J. L. J. and Lehmann, E. L. (1962). Rank methods for combinations of independent experiments in analysis of variance, *Ann. Math. Statist.* **33**, pp. 482–497.

Holm, S. (1979). A simple sequential rejective multiple test procedure, *Scand. J. Stat.* **6**, pp. 65–70.

Hsu, J. C. (1996). *Multiple Comparisons: Theory and Methods* (Chapman & Hall, London).

Hsu, P. L. (1938). Notes on hotelling's generalized t, *Ann. Math. Statist.* **9**, pp. 231–243.

Johansen, S. (1992). A representation of vector autoregressive processes integrated of order 2, *Econometric Theory* **8**, pp. 188–202.

Kendall, M. G. and Stuart, A. (1979). *The advanced theory of statistics, 4nd edition* (Griffin, London).

Keynes, J. M. (1936). *The General Theory of Employment, Interest and Money* (Macmillan Cambridge University Press, London).

Kitamura, Y. (2001). Asymptotic optimality of empirical likelihood for testing moment restrictions, *Econometrica* **69**, pp. 1661–1672.

Kleinbaum, D. G., Kupper, L. L., Muller, K. E. and Nizam, A. (1998). *Applied Regression Analysis and Other Multivariate Methods*, 3rd edn. (Brooks/Cole Publishing Company, Pacific Grove, California).

Koehn, U. and Thomas, D. L. (1975). On statistics independent of a sufficient statistic: Basu's lemma, *The American Statistician* **29**, pp. 40–42.

Kramer, C. Y. (1956). Extension of multiple range tests to group means with unequal numbers of replications, *Biometrics* **12**, pp. 309–310.

Kuhn, H. W. and Tucker, A. E. (eds.) (1956). *Linear inequalities and related systems, Annals of Mathematics Studies, No.38* (Princeton Univ. Press, Princeton).

Lee, C. I. C. (1981). The quadratic loss of isotonic regression under normality, *Ann. Statist.* **9**, pp. 686–688.

Lehmann, E. L. (1959). *Testing Statistical Hypotheses* (Wiley, New York).

Lehmann, E. L. (1963). Nonparametric confidence intervals for a shift parameter, *Ann. Math. Statist.* **34**, pp. 1507–1512.

Lehmann, E. L. (1980). An interpretation of completeness and basu's theorem, *J. Amer. Statist. Assoc.* **76**, pp. 335–340.

Lehmann, E. L. (1986). *Testing Statistical Hypotheses*, 2nd edn. (Wiley, New York).

Levinson, D. F., Holmans, P., Straub, R. E., Owen, M. J., Wildenauer, D., Gejman, P. V., Pulver, A. E., Laurent, C., Kendler, K. S., Walsh, D., Norton, N., Williams, N. M., Schwab, S. G., Lerer, B., Mowry, B. J., Sanders, A. R., Antonarakis, S. E., Blouin, J. L., DeLeuze, J. F. and Mallet, J. (2000). Multicenter linkage study of schizophrenia candidate regions on chromosomes 5q, 6q, 10p, and 13q: schizophrenia linkage collaborative group iii, *Am. J. Hum. Genet.* **67**, pp. 652–663.

Ljung, G. M. and Box, G. P. E. (1978). On a measure of lack of fit in time series models, *Biometrika* **65**, pp. 297–303.

Loève, M. (1963). *Probability Theory*, 3rd edn. (Van Nostrand, Princeton).

Maatta, J. M. and Casella, G. (1987). Conditional properties of interval estimators of the normal variance, *Ann. Statist.* **15**, pp. 1372–1388.

Madansky, A. (1962). More on length of confidence intervals, *J. Amer. Statist. Assoc.* **57**, pp. 586–589.

Mann, H. B. and Whitney, D. R. (1947). On a test of whether one of two random variables is stochastically larger than the other, *Ann. Math. Statist.* **18**, pp. 50–60.

Mantel, N. (1963). Chi-square tests with one degree of freedom: extensions of mantel-haenszel procedure, *J. Amer. Statist. Assoc.* **58**, pp. 690–700.

Miller, R. G. (1981). *Simultaneous Statistical Inference* (Springer-Verlag, Berlin).

Narum, S. R. (2006). Beyond bonferroni: Less conservative analyses for conservation genetics, *Conservation Genetics* **7**, pp. 783–787.

Noether, G. E. (1955). On a theorem of pitman, *Ann. Math. Statist.* **26**, pp. 64–68.

Owen, A. (1988). Empirical likelihood ratio confidence intervals, *Biometrika* **75**, pp. 237–249.

Owen, A. (1990). Empirical likelihood confidence regions, *Ann. Statist.* **18**, pp. 90–120.

Owen, A. (1991). Empirical likelihood for linear models, *Ann. Statist.* **19**, pp. 1725–1747.

Park, J. Y. and Phillips, P. C. B. (1988). Statistical inference in regressions with integrated processes:part i, *Econometric Theory* **4**, pp. 468–497.

Park, J. Y. and Phillips, P. C. B. (1989). Statistical inference in regressions with integrated processes:part ii, *Econometric Theory* **5**, pp. 95–132.

Pearson, K. (1905). The problem of the random walk, *Nature* **72**, pp. 72.1865: 294 and 72.1867: 342, with Rayleigh's answer in 72.1866: 318.

Pearson, K. and Lee, A. (1903). On the laws of inheritance in man: I. inheritance of physical characters, *Biometrika* **2**, pp. 357–462.

Phillips, P. C. B. and Durlauf, S. N. (1986). Multiple time regression with integrated processes, *Review of Economic Studies* **53**, pp. 473–495.

Phillips, P. C. B. and Perron, P. (1988). Testing for a unite root in time series regression, *Biometrika* **73**, pp. 335–346.

Pratt, J. W. (1961). Length of confidence intervals, *J. Amer. Statist. Assoc.* **56**, pp. 549–567.

Rao, C. R. (1973). *Linear Statistical Inference and Its Applications*, 2nd edn. (Wiley, New York).

Reid, C. (1982). *Neyman - From Life* (Springer-Verlag, New York).

Reiner, A., Yekutieli, D. and Benjamini, Y. (2003). Identifying differentially expressed genes using false discovery rate controlling procedures, *Bioinformatics* **19**, pp. 368–375.

Robertson, T., Wright, F. T. and Dykstra, R. L. (1988). *Order Restricted Statistical Inference* (Wiley, New York).

Sakamoto, Y., Ishiguro, M. and Kitagawa, G. (1999). *Akaike Information Criterion Statistics* (KTK Scientific Publishers, Tokyo).

Schaafsma, W. (1968). A comparison of the most stringent and the most stringent somewhere most powerful test for certain problems with restricted alternative, *Ann. Math. Statist.* **39**, pp. 531–546.

Schaafsma, W. and Smid, L. J. (1966). Most stringent somewhere most powerful test against alternatives restricted by a number of liner inequalities, *Ann. Math. Statist.* **37**, pp. 1161–1172.

Scheffé, H. (1953). A method for judging all contrasts in the analysis of variance, *Biometrika* **40**, pp. 87–104.

Serfling, R. J. (1980). *Approximation Theorems of Mathematical Statistics* (Wiley, New York).

Shapiro, A. (1988). Toward a unified theory of inequality constrained testing in multivariate analysis, *Internat. Statist. Rev.* **56**, pp. 49–62.

Shi, N.-Z. (1987). Testing a normal mean vector against the alternative determined by a convex cone, *Mem. Fac. Sc. Kyushu Univ.* **A41**, pp. 133–145.

Shi, N.-Z. (1988a). A test of homogeneity for umbrella alternatives, *Commun. Statist. A* **17**, pp. 657–670.

Shi, N.-Z. (1988b). Rank test statistics for umbrella alternatives, *Commun. Statist. A* **17**, pp. 2059–2073.

Shi, N.-Z. (1988c). Testing the hull hypothesis that a normal mean vector lies in the positive orthant, *Mem. Fac. Sc. Kyushu Univ.* **A42**, pp. 109–122.

Shi, N.-Z. (1991). A test of homogeneity of odds ratios against order restrictions, *J. Amer. Statist. Assoc.* **86**, pp. 154–158.

Shi, N.-Z. (2008). A conjecture on maximum likelihood estimation, *IMS Bulletin* **37(4)**, p. 4.

Shi, N.-Z., Kudo, A., Fukagawa, M. and Sakata, T. (1988). Testing equality of proportions against trend and applications in study of aging effects, *Bulletin of the Biometric Society of Japan* **9**, pp. 61–72.

Silvapulle, M. J. and Sen, P. K. (2004). *Constrained Statistical Inference: Inequality, Order, and Shape Restrictions* (Wiley, New York).

Sim, C. A., Stock, J. H. and Watson, M. W. (1990). Inference in linear time series models with some unit roots, *Econometrica* **58**, pp. 113–144.

Simes, R. J. (1986). An improved bonferroni procedure for multiple tests of significance, *Biometrika* **73(3)**, pp. 751–754.

Simpson, D. G. and Margolin, B. H. (1986). Recursive nonparametric testing for dose-response relationship subject to downturns at high doses, *Biometrika* **73**, pp. 589–596.

Simpson, E. H. (1951). The interpretation of interaction in contingency tables, *J. Roy. Statist. Soc.* **B13**, pp. 238–241.

Snijders, T. A. B. (1979). *Asymptotic Optimality Theory for Testing Problems With Restricted Alternatives*, Ph.D. thesis, Mathematisch Centrum, Amsterdam.

Storey, J. D. (2002). A direct approach to false discovery rates, *J. R. Stat. Soc. B* **64**, pp. 479–498.

Stuart, A., Ord, J. K. and Arnold, S. (1999). *Advanced Theory of Statistics, Volume 2A: Classical Inference and the linear Model*, 6th edn. (Oxford University Press, London).

Sun, H. J. (1988). A fortran subroutine for computing normal orthant probabilities of dimensions up to nine, *Commun. Statist.* **B17**, pp. 1097–1111.

Toda, H. Y. and Phillips, P. C. B. (1993). Vector autoregressions and causality, *Econometrica* **61**, pp. 1367–1393.

Toda, H. Y. and Yamamoto, T. (1995). Statistical inference in vector autoregressions with possibly integrated processes, *Journal of Econometrics* **66**, pp. 225–250.

Tukey, J. W. (1953). The problem of multiple comparisons, Unpublished manuscripts.

Tukey, J. W. (1960). A survey of sampling from contaminated distributions, in I. Olkin (ed.), *Contributions to Probability and Statistics* (Standford University Press, Stanford, California).

Tukey, J. W. (1994). The problem of multiple comparisons, in H. I. Braun (ed.), *The Collected Works of John W. Tukey*, Vol. VIII (Chapman & Hall, London).

Walker, S. H. and Duncan, D. B. (1967). Estimation of the probability of an event as a function of several independent variables, *Biometrika* **54**, pp. 167–179.

Wang, Q. H. and Rao, J. N. K. (2002a). Empirical likelihood-based inference in linear errors-in-covariables models with validation data, *Biometrika* **89**, pp. 345–358.

Wang, Q. H. and Rao, J. N. K. (2002b). Empirical likelihood-based inference under imputation for missing response data, *Ann. Statist.* **30**, pp. 896–924.

Widder, D. V. (1946). *The Laplace Transform* (Princeton University Press, Princeton).

Wilcoxon, F. (1945). Individual comparisons by ranking methods, *Biometrics* **1**, pp. 80–83.

Wilks, S. S. (1938). The large-sample distribution of the likelihood ratio for testing composite hypotheses, *Ann. Math. Statist.* **9**, pp. 60–62.

Wright, T. (1992). Lagrange's identity reveals correlation coefficient and straight line connection, *Amer. Statist.* **46**, pp. 106–107.

Zucker, L. and Zucker, T. F. (1961). Fatty, a new mutation in the rat, *J. Heredity* **52**, pp. 275–278.

Index